THE NATIONAL ACADEMIES
Advisers to the Nation on Science, Engineering, and Medicine

The **National Academy of Sciences** is a private, nonprofit, self-perpetuating society of distinguished scholars engaged in scientific and engineering research, dedicated to the furtherance of science and technology and to their use for the general welfare. Upon the authority of the charter granted to it by the Congress in 1863, the Academy has a mandate that requires it to advise the federal government on scientific and technical matters. Dr. Ralph J. Cicerone is president of the National Academy of Sciences.

The **National Academy of Engineering** was established in 1964, under the charter of the National Academy of Sciences, as a parallel organization of outstanding engineers. It is autonomous in its administration and in the selection of its members, sharing with the National Academy of Sciences the responsibility for advising the federal government. The National Academy of Engineering also sponsors engineering programs aimed at meeting national needs, encourages education and research, and recognizes the superior achievements of engineers. Dr. Charles M. Vest is president of the National Academy of Engineering.

The **Institute of Medicine** was established in 1970 by the National Academy of Sciences to secure the services of eminent members of appropriate professions in the examination of policy matters pertaining to the health of the public. The Institute acts under the responsibility given to the National Academy of Sciences by its congressional charter to be an adviser to the federal government and, upon its own initiative, to identify issues of medical care, research, and education. Dr. Harvey V. Fineberg is president of the Institute of Medicine.

The **National Research Council** was organized by the National Academy of Sciences in 1916 to associate the broad community of science and technology with the Academy's purposes of furthering knowledge and advising the federal government. Functioning in accordance with general policies determined by the Academy, the Council has become the principal operating agency of both the National Academy of Sciences and the National Academy of Engineering in providing services to the government, the public, and the scientific and engineering communities. The Council is administered jointly by both Academies and the Institute of Medicine. Dr. Ralph J. Cicerone and Dr. Charles M. Vest are chair and vice chair, respectively, of the National Research Council.

www.national-academies.org

Acknowledgments

The Committee on Population was established in 1983 by the National Research Council (NRC), under the charter of the National Academy of Sciences, to bring population sciences to bear on issues affecting public policy. A dozen years ago, with sponsorship from the U.S. National Institute on Aging (NIA), the Committee embarked on a series of projects relating to the emerging field of biodemography. In 1997, the Committee on Population published *Between Zeus and the Salmon: The Biodemography of Longevity*, edited by K.W. Wachter and C.E. Finch. This pioneering volume brought together demographers, evolutionary theorists, genetic epidemiologists, anthropologists, and biologists, describing implications of their disciplines for understanding and foreseeing the trajectory of human longevity. With support from the U.S. National Institute on Child Health and Human Development, the Committee went on to explore biodemographic aspects of fertility and family formation in the 2003 volume *Offspring: Human Fertility Behavior in Biodemographic Perspective*, edited by K.W. Wachter and R.A. Bulatao.

The call in the 1997 volume for more interdisciplinary work contributed to demand for collecting biological data in the context of large, population-based social and demographic surveys. Advances in biodemography would require data with better linkages between social and biological domains. Techniques under development made the collection of biological measurements and samples in nonclinical settings more feasible. With renewed support from the NIA, the Committee on Population held workshops that led in 2001 to the volume, *Cells and Surveys: Should*

Biological Measures Be Included in Social Science Research?, edited by C.E. Finch, J.W. Vaupel, and K. Kinsella.

The volume is a sequel to *Cells and Surveys*. It takes stock of the rapid advances made in the field since 2001. The volume is based on a workshop that was held at the National Research Council's Keck Center in Washington, D.C., in June 2006. In the forefront is the question, what has been learned so far from the inclusion of biological indicators in social surveys? What changes in perspective are emerging from the interdisciplinary communication associated with the enterprise? What biological and genetic data promise to be most useful? How can better models integrate biological information with social, behavioral, and demographic information?

The chapters of this volume were enriched by free-flowing discussion and debate at the workshop. In response to suggestions, several additional chapters were added after the workshop. We owe a debt of gratitude to the individuals who gave of their time to evaluate and strengthen the contributions, providing authors with candid comments to assist them with revisions. The independent review also seeks to ensure that the volume meets the institutional standards of the National Research Council for objectivity, balance, faithfulness to evidence, and responsiveness to the original charge. The review comments and draft manuscript remain confidential to protect the integrity of the process.

We thank the following individuals for their participation in the review: Dan G. Blazer, Psychiatry, Duke University Medical Center; Floyd E. Bloom, Department of Molecular and Integrative Neuroscience (emeritus), The Scripps Research Institute; James R. Carey, Department of Entomology, University of California, Davis; Kaare Christensen, Institute of Public Health, University of Southern Denmark; Christopher L. Coe, Department of Psychology, University of Wisconsin; Caleb E. Finch, Davis School of Gerontology, University of Southern California; Vicki A. Freedman, Department of Health Systems and Policy, University of Medicine and Dentistry of New Jersey School of Public Health; Guang Guo, Department of Sociology, University of North Carolina; Judith R. Kidd, Department of Genetics, Yale University School of Medicine; Chris Kuzawa, Department of Anthropology, Laboratory for Human Biology Research, Northwestern University; Margie E. Lachman, Psychology Department, Brandeis University; Partha P. Majumder, Human Genetics Unit, Indian Statistical Institute, Kolkata, India; Carlos F. Mendes de Leon, Rush Institute for Healthy Aging and the Department of Preventive Medicine, Rush-Presbyterian-St. Luke's Medical Center, Chicago; Robert Millikan, School of Public Health, University of North Carolina; Kathleen A. O'Connor, Department of Anthropology and Center for Studies in Demography and Ecology, University of Washington, Seattle;

Jose M. Ordovas, Nutrition and Genomics Laboratory, Tufts University; Alberto Palloni, Department of Sociology, Northwestern University; Germán Rodríguez, Office of Population Research, Princeton University; Luis Rosero-Bixby, Centro Americano de Población, University of Costa Rica; Michael L. Rutter, Institute of Psychiatry, Social, Genetic and Development, Psychiatry Centre, London; Carol D. Ryff, Institute on Aging, Department of Psychology, University of Wisconsin; Nicolas J. Schork, Research and Scripps Genomic Medicine and Department of Molecular and Experimental Medicine, The Scripps Research Institute; Christopher L. Seplaki, Center on Aging and Health and Department of Population, Family, and Reproductive Health, The Johns Hopkins Bloomberg School of Public Health; Mikhail S. Shchepinov, Office of the President, Retrotope, Inc., Oxford, U.K.; Jean Chen Shih, Department of Molecular Pharmacology and Toxicology, Pharmaceutical Science Center, University of Southern California; Burton H. Singer, Office of Population Research, Princeton University; MaryFran Sowers, Department of Epidemiology, Center for Integrated Approaches to Complex Diseases, University of Michigan; Duncan Thomas, Department of Economics, Duke University; Kenneth W. Wachter, Department of Demography, University of California, Berkeley; and Keith E. Whitfield, Department of Psychology and Neuroscience, Duke University.

The Committee on Population expresses its warm appreciation to Jim Vaupel, who chaired the planning meetings and workshop and took charge as editor of the volume. Special thanks are also due to the members of the Steering Committee who advised and assisted the chair: Kaare Christensen, Susan Hankinson, Teresa Seeman, Kenneth Wachter, and Kenneth Weiss. Their intellectual contributions can be found throughout this volume. Jennifer Harris at the NIA also provided valuable input and guidance. Committee on Population member Eileen Crimmins, of the Davis School of Gerontology, oversaw the review process.

Particular thanks go to Maxine Weinstein who served as consultant on the project and took responsibility for the broad range of practical and intellectual tasks that have gone into shaping it and bringing it to completion. From identifying and recruiting participants to putting the final touches on the work, her efforts have been indispensable.

Funding from the NIA has made this volume possible. Richard Suzman, director of the NIA for Behavioral and Social Research, has long been a lively supporter of NRC endeavors, relying on the NRC to assemble appropriate scholars and craft reliable, influential reports. His vision has been crucial in launching and developing the field of biodemography. John Haaga, Georgeanne Patmios, and Erica Spotts at the NIA have encouraged and guided us in our biodemographic emphases.

Thanks are also due to the staff of the NRC. Anthony Mann coordinated the logistics and travel arrangements for the meetings and prepared the final manuscript. Christine McShane edited the manuscript. Kirsten Sampson-Snyder coordinated the review of the volume. Development and execution of the project occurred under the guidance of the director of the Committee on Population, Barney Cohen.

Kenneth W. Wachter
Chair, Committee on Population

Barney Cohen
Director, Committee on Population

Contents

xi

PART II: THE POTENTIAL AND PITFALLS OF GENETIC INFORMATION

PART III: NEW WAYS OF COLLECTING, APPLYING, AND THINKING ABOUT DATA

Introduction

James W. Vaupel, Kenneth W. Wachter, and *Maxine Weinstein*

New kids on the block! This volume spreads the word. Population-based sample surveys that combine demographic, social, and behavioral data with biological indicators have arrived in town. The subtitle of the precursor to this volume, *Cells and Surveys*, was the question "Should biological measures be included in social science research?" In practice, that question seems to be already answered: yes. Social surveys that include the collection of biological data have proliferated since that volume's publication in 2001. Are these new studies going to be our friends?

Speculation gives way to assessment, as publications emerge from the dozen or more large-scale social surveys that have moved into collecting biological measurements and materials ranging from grip strength to DNA. Unlike biomedical surveys and clinical studies, always rich in biological indicators, these social surveys probe demographic characteristics, economic and health-related behaviors, resources, constraints, and life-course transitions in depth. Consequently, they hold out promise for shining a spotlight on interactions between biological factors and "environment" in its personalized complexity. They hold out promise for distinguishing causal pathways that turn out only statistically significant in experiments and trials from causal pathways that have real "oomph" in the squishy, messy, diverse, adapting world of whole populations.

But along with a chorus of greeting are glances of doubt. These new arrivals on the social science scene bring out feelings of caution among many, especially around analysis of genetic indicators in the study of complex social traits. As practical challenges to the inclusion of biological indicators are mastered and as technology rushes ahead, questions of meaning call for ongoing debate.

Like the earlier one, this volume was sponsored by the Behavioral and Social Research Division of the National Institute on Aging and put together under the auspices of the Committee on Population of the National Research Council. *Cells and Surveys* was influential, but not because the book caused social scientists to add biological measures to their surveys. Doing that was an idea whose time had come. Rather, *Cells and Surveys* was influential because it provided an authoritative overview of the kinds of biological information that could be collected and how and what the problems and pitfalls were.

In this volume the focus shifts to what has been and can be learned. The authors of most of the chapters have firsthand experience with incorporating biomarkers into social science research and have followed the rapid development of the endeavor. They have thought long and hard about the critical issues. Although there are still many open questions, the time seems ripe for taking stock, not for advocacy but for aspiration to mature judiciousness and nuanced wisdom.

The book emerges from a series of discussions and interchanges that started in 2005 and ended in mid-2007. A workshop, held in June 2006, enabled many of the authors to exchange information and perspectives. Participants in the workshop considered (and the contributions to the book reflect) a few fundamental questions: What has been learned from what has already been done? What is the place of genetic information in social research? What new concepts and methods are being developed or need to be developed?

Biodemography (National Research Council, 1997; Carey and Vaupel, 2005) and its related disciplines have expanded rapidly; one book cannot begin to provide a comprehensive summary of its findings and challenges. We had to pick and choose. Our focus in this volume is on the inclusion of biomarkers in social science surveys: what does (or should) one include and how does (or can) one use the data. The volume grew as we proceeded; in particular, we found ourselves asking more questions about the collection and analysis of genetic information. We also made a conscious choice not to solicit manuscripts on some subjects: notably absent are chapters that specifically focus on ethics. This omission is not because we felt the concerns were unimportant, but because the enormity of the subject made it unfeasible to include. However, both explicitly and implicitly, almost every chapter discusses concerns relating to privacy, informed consent, and the appropriate treatment of study participants and their information.

Hunting for genes that determine health or behavior is not the preoccupation of the authors of these chapters. This may surprise readers who follow the news about science but who have not kept up with social science research that includes biological indicators. Most of the research being carried out in conjunction with such surveys does not even involve DNA: the biological indicators pertain to grip strength, pulmonary functioning, clinical measurements of various substances in blood, saliva, or urine, blood pressure, heart rate variability, weight and height, perceived age, and various other physiological and anthropometric measures of risk factors, exposures, and health outcomes.

It is sometimes argued that scientists should wait for some more advanced technology before trying to discover genes that influence some trait. The darling of the moment is genome-wide association studies, an approach that is powerful and rapidly becoming less expensive. This argument, however, is largely irrelevant for social science research that incorporates biological markers. Some social science surveys have collected DNA, but the intent has not been to discover genes that determine health or behavior. The genetic information that is collected generally pertains to genes that have already been discovered to have (or reportedly have) important effects. Social scientists can verify that the effect is indeed important: this is what Harald Göring cogently argues is a more appropriate task for social surveys than gene discovery (Chapter 11), and this is what Kaare Christensen and colleagues report as one of the many uses of data from surveys of elderly Danish twins, nonagenarians, and centenarians (Chapter 1). Social scientists can also use the genetic information to better estimate environmental and behavioral effects on health: George Davey Smith and Shah Ebrahim describe an ingenious method to exploit genetic variants to make causal inferences about environmental variables (Chapter 16). Other chapters describe how knowledge about a person's genotype can be used to study how genes and nongenetic variables interact to produce behavioral and health outcomes. Accounting for genetic variation among individuals permits more accurate estimation of how environmental and behavioral factors influence health and longevity. As summarized by David Abrams (quoted by Davey Smith and Ebrahim), "The more we learn about genes, the more we see how important environment and lifestyle really are." These critically important areas of research should not be delayed because better methods of gene discovery are being developed.

A recurring question throughout the volume pertains to the utility of DNA markers for survey-based social and behavioral analysis. Skeptics predominate over enthusiasts. It seems appropriate, however, to include here one version of an enthusiast's perspective. The key phrase in this discussion is "complex traits." Many contributors explain why it is a mistake to look for genes or for simple genetic determinants of complex traits.

Cautiously put—and most contributors put it cautiously—this argument is a counterbalance to the extravagance of the press and of scientists seeking celebrity billing. *The* gene for financial success, for tennis stardom, for obesity, for inanity, or for extroversion is not a target for reputable research. There is a continuing need for social science to test and bury false claims. Vehemently put, however, this kind of argument about complex traits can amount to a radical critique.

Suppose the decisions and outcomes now coming onto the explanatory horizon of social science are driven by large assemblages of interacting factors, each with a hunk of its own variation. Then even the substantial sample sizes of existing surveys will be too small to pin down the proliferating combinations. Multiway tables with too many multiways have too many cells to estimate. Nonlinear functional relationships with too many arguments are not identifiable with modest batteries of questions. This argument sees the enterprise of survey research coming up against its limits.

Two modes of complexity need to be kept quite separate. One is complexity in the basic units that act as determinants. Single-nucleotide polymorphisms or single genes or epistatic pairs or triples or regulators or promoters may none of them be the right units to be studying. But there may still be some sort of complex units, gene networks, regulatory feedback systems—something else, perhaps as yet undiscovered, for which relevant causal variation can be specified by a modest number of bits of information. Low-dimensional structure may be hiding in high-dimensional systems. This kind of complexity makes the problems of social scientists hard but soluble. Progress in genetics should translate into progress in behavioral analysis.

The other mode of complexity is the one implicated in the radical critique: many interlinked factors, each with variation of its own, each only sometimes a rate-limiting factor, each only sometimes a catalyst or inhibitor in feedback loops. This kind of complexity is akin to the complexity of ordinary daily life. Relevant variation irreducibly requires large numbers of bits of information to specify. If such complexity is the rule in the behaviorally relevant structure of the genome, prospects for using genetic indicators from social surveys would seem daunting.

There are, however, countervailing reasons for encouragement. A complex structure with enough interacting but partly independently varying elements can often be modeled by a random system. In such a system, laws of large numbers often operate, and many detailed interactions cancel out. Physicists successfully model complex particle systems with random matrices whose spectra obey simple, discoverable laws. Genetic influences that figure importantly in trends or in population differentials may prove to be accessible to research, without the necessity of fully untangling the underlying complex causal structure.

We go to the beach and ask, "Why does this wave surge up far enough to wash away our sandcastle?" Hopeless. But we could ask, and answer, the other question, "Why do waves surge up far enough to wash away our sandcastle?" Pursuing knowledge of causal processes and pathways is important, but usable knowledge depends ultimately on finding something simple in the picture. The goal is not a secular theology of predestination. We are not aspiring to understand the full determinants of complex traits or tell out the person-specific reasons for each person's path to troubled or successful aging. Survey research has to be a gamble on simplicity, no less so as we start to try to take advantage of our first samples of DNA.

The kind of complexity that might defeat social science research is also a kind of complexity that might have defeated natural selection. The genomic information that mattered for behavior and reproductive success over evolutionary time should not have been so contingent that selective pressure could not impinge on it. Environments matter, and it seems plausible that some good part of existing functional genetic variation persists because variation has been adaptive in variable environments. Despite the complexity of complex traits, what natural selection manages to notice might be out there for us to find. Why hesitate to look?

So it is that some social scientists hope that genes might be discovered that influence behavior—and that they can play a role in their discovery. Parts of Chapter 15 by Daniel Benjamin and colleagues point in this direction, albeit with cautious recognition of problems and difficulties. In particular, the authors hypothesize that genes with known effects on neurotransmission pathways or on memory (and some other cognitive functions and abilities) might influence labor force participation, wealth, and other economic behaviors and outcomes. The four chapters in Part II provide detailed explanations of how problematic and questionable such a quest is. The reader can make a judgment about whether Benjamin et al. are brave or reckless. The more general point is clear: most social science research using biological information does not pertain to genes, and the research that does include genes largely focuses on using this information to better assess nongenetic influences.

WHAT HAVE WE LEARNED SO FAR?

Part I, roughly "What we have learned from what we have already done," includes chapters on six different research programs, a chapter that reviews other relevant surveys with an emphasis on biological indicators, and a chapter that more generally reviews the theory and practice of biomarkers in social science research on health and aging. The first three chapters cover large, well-established studies of health and aging that have long included biological information. Kaare Christensen and colleagues describe the main research directions and findings of the Danish studies of elderly twins, nonagenarians, and centenarians. Michael Marmot and Andrew Steptoe provide a review of two important English surveys, Whitehall II and ELSA (the English Longitudinal Study of Ageing). Jack Chang and colleagues describe and discuss the Taiwan biomarker project. These three chapters provide a wealth of details and insights based on deep experience.

The directors of the Health and Retirement Study, one of the largest and most widely used surveys of aging, were considering the addition of biomarkers at the time of *Cells and Surveys* (Weinstein and Willis, 2001). The study is cautiously adding the collection of a few markers. The reasons for caution and lessons from recent experience are discussed by David Weir.

Robert Wallace calls social scientists' attention to the Women's Health Initiative, a very large, intricate set of clinical trials. He describes the difficulties in managing such an endeavor: the complexity of the nested studies and the extent of the collaborative networks constitute a Petri dish for nurturing administrative complications. Wallace emphasizes the potential value of the Women's Health Initiative for social science research, a potential that has not yet been realized. Duncan Thomas and Elizabeth Frankenberg demonstrate that it is indeed possible for social scientists to use clinical trials to learn about behaviors and social outcomes. The experiment involved adults in Indonesia who were randomly assigned to two groups, one receiving iron supplements and the other a placebo. The health of those receiving the iron supplements improved. Thomas and Frankenberg studied how this affected workers' productivity and time allocations.

Not every significant study is or could be represented in this volume—a testament to how widespread the collection of biomarkers has become. Among others, important studies that do not have their own chapters include MIDUS (National Survey of Midlife Development in the United States), the Women's Health and Aging Study, SHARE (Study of Health and Retirement in Europe), and SAHR (Stress, Aging, and Health in Russia). The chapters by Jennifer Harris, Tara Gruenewald, and Teresa Seeman and by Douglas Ewbank provide overviews of some of these other studies and their connections with the studies covered in the earlier chapters. Harris and colleagues critically review biomarker research from community and population-based studies, with an emphasis on physiological parameters and genes. Ewbank reviews theory and practice, with an emphasis on studies of social behaviors and environments. He concentrates on studies of the effects of chronic social and psychological stress.

Taken together, the eight chapters on what we have learned so far include discussion of findings that have been made possible by the combination of biological and social data. They also include candid commentaries on the logistical problems that the researchers encountered. These hurdles include consent processes; challenges in the field; difficulties in storing, archiving, and reusing specimens; and management of collaborations across institutions. Ewbank reminds us that it is probably too early to draw generalizations from the studies. A corollary to this caution is that however great the temptation, it is also too early to restrict ourselves to "cookie cutter" surveys that simply replicate the biomarker collections of previous studies.

Still, the broad conclusion is clear: the studies to date provide new ways of understanding the interactions and joint effects of behavior, the social environment, and physiology on health. Of particular interest for future understanding of these effects is the growing number of studies that are being done outside the United States. For example, as noted by Marmot and Steptoe, the collection of biological and social data in both England and the United States has contributed to understanding of the higher rates of morbidity in the United States. The chapters by Harris et al. and Chang et al. note differences between results from the MacArthur study in the United States and the Taiwan biomarker project; these variations have the potential to open a window onto macro-level influences on the links between social experience and health. As additional studies are added to the "arsenal"—nascent and ongoing studies, for example, in Japan, Korea, mainland China, Mexico, and Russia—there will be opportunity to explore these factors.

WHAT ABOUT GENETICS IN SOCIAL SCIENCE RESEARCH?

Traditionally, social scientists have studied people's behavioral, psychological and demographic characteristics, how these characteristics interact, and how they are affected by the social environment. More recently, social scientists have begun to collect and analyze data on individuals' health and on physiological and morphological factors that affect health. In the past few years, it has become possible not only to study phenotypes—observable traits of individuals—but also to study genotypes, the genetic makeup of individuals. This ability presents a new set of scientific opportunities, practical difficulties, and ethical challenges. Part II offers an introduction to these issues. We expect that this primer will be of particular interest and value to social scientists.

The chapter by Mary Jane West-Eberhard provides a "reader's guide for how to relate genes to phenotypic traits . . . to better interpret research results and public discussions. . . ." She explains why it is deeply misleading to claim that there is a gene for a complex trait, such as a gene for obesity or laughter. She stresses the role of the environment in shaping the action of genes and the enormous intricacy of the genetic architecture of complex traits, such as longevity or intelligence.

George Vogler and Gerald McClearn provide a complementary perspective on the limited opportunities, subtle complexities, and numerous pitfalls in analyzing genetic markers in social science research. They dismiss the outmoded nature versus nurture distinction and outline various types of correlations and interactions among genetic and environmental factors.

Harald Göring covers many of the same issues of polygenes and complicated gene-environment interactions and correlations. He does so, however, from a different perspective: he focuses on explaining why large-scale social surveys are not well suited to the discovery of genes that influence complex traits related to behavior or health. Social surveys can, however, sometimes play an important role—not in gene discovery but in the validation of purported discoveries. Many of the published discoveries are false positives. Many do not accurately estimate effect sizes. And many are based on study of special populations that are not representative of the general population. Large social surveys can provide additional information to address these deficiencies.

The fourth chapter on genetics in social science research, by Kenneth Weiss, also emphasizes that many genes may influence a trait and that those genes interact with each other and with the environment in extremely complex ways. Furthermore, "genes typically harbor tens to hundreds or more alleles," genetic variants that differ from individual to individual. Somatic mutations (that is, changes in genes over the course of life) and "stochasticities of countless sorts" further complicate analysis. Weiss also raises some uncomfortable and far-reaching ethical concerns. His questions are particularly important for demography, economics, and other domains of social science that have strong influences on policy formulation. The hazards of entering into an arena with a history of political and social abuse should not be ignored.

Understanding the profound complexities of analyzing genes in social science research is enhanced by the complementary perspectives of the four chapters in this section. The chapters are a tutorial for social scientists, valuable for naïve researchers intrigued by publicity about obesity or longevity "genes," and even more valuable for experienced social science researchers who have recognized the usefulness of incorporating (nongenetic) biological indicators in surveys and are now considering studying genes.

WHAT'S NEXT?

Part III explores concepts, methods, and modeling strategies that are emerging as biomarkers are increasingly collected. Although incorporation of biological indicators in social surveys has burgeoned over the past several years, the field is still young. New kinds of biomarkers and new ways of thinking about rich combinations of social and biological data are needed.

Stacy Tessler Lindau and Thomas McDade provide an overview of minimally invasive and innovative methods for collecting biological information in population-based research. Some of the strategies they discuss are being implemented. For instance, information from MRIs is being collected by a subproject headed by Richie Davidson as part of the MIDUS II study led by Carol Ryff and Burt Singer. Extensive electrocardiogram data from Holter monitoring of heart patterns over a 24-hour period are being gathered from a sample of hundreds of older individuals in Moscow as part of the study of Stress, Aging, and Health in Russia (SAHR) under the direction of Maria Shkolnikova. The use of Holter monitoring represents an innovative application of a well-established method to collect information on the dynamics of physiological variation. Most of the biological indicators used to date are based on a single measurement at a single time. Multiple measurements two or more years apart are sometimes taken in longitudinal surveys, and these data can be valuable in reducing measurement error as well as in tracking trends. Temporal patterns of variation over hours or days in an individual's physiology are often much more informative than point values. Holter monitoring and other methods for continuously assessing a person's biological activity and exposures to environmental influences will undoubtedly become a major feature of future social surveys, especially if comfortable, minimally invasive instruments can be developed to gather such data.

The salience of nutrition and its relationships with social and economic characteristics are increasingly being recognized by social scientists; the research described by Thomas and Frankenberg (Chapter 7) is an example. The potential importance of genetic influences on an individual's response to nutrition and biomarkers for nutritional intake are described by John Milner and colleagues in their chapter on nutrigenomics. Given the well-documented and extensive problems in collecting complete and reliable data on diet (see, for example, comments in the chapter by Davey Smith and Ebrahim), the development of sensitive and specific biomarkers of nutritional intake would be a major advance.

The chapter by Daniel Benjamin and colleagues describes the nascent field of genoeconomics. The field includes three kinds of contributions. "First, economics can contribute a theoretical and empirical framework for understanding how market forces and behavioral responses mediate the influence of genetic factors. Second, incorporating genetics into economic analysis can help economists identify and measure important causal pathways (which may or may not be genetic). Finally, economics can aid in analyzing the policy issues raised by genetic information." As an example, the authors present their ongoing work exploring candidate genes that may influence decision making and their associations with economic characteristics. As the authors acknowledge, it is a daunting enterprise, one that underscores the relevance of the warnings, both scientific and ethical, raised in Part II.

The social scientists whose work is reviewed in this volume are primarily interested in biological indicators in order to better understand how other kinds of factors—behavioral, social, environmental—affect health and aging. Social scientists would like to understand causal relationships. It is, however, difficult to distinguish causation from correlation, association, or reverse causation in social science research. This difficulty is documented in the chapter by George Davey Smith and Shah Ebrahim. One strategy for identifying causal links is to conduct randomized, controlled experiments or clinical trials. Social scientists have only occasionally been able to do so, but Thomas and Frankenberg describe how a clinical trial with iron supplements could be used to study how health influenced workers' productivity and time allocations in Indonesia. Davey Smith and Ebrahim describe an alternative strategy, in which quasi-randomization is produced by use of instrumental variables related to individuals' genetic variants.

How can social scientists move beyond simplistic associations to more sophisticated understanding of mechanisms and causal linkages? This overarching question is also the topic of the final two papers of the volume. John Cacioppo and colleagues consider the difficulties of interdisciplinary research that crosses biological and social levels of organization. They discuss how a scholar "can productively think about concepts, hypotheses, theories, theoretical conflicts, and theoretical tests" in multilevel investigations. They describe concepts to aid thinking about the "mapping of biological measures to social and behavioral constructs in surveys." John Hobcraft also considers general issues in the way one models and thinks about research questions. Social science research is increasingly drawing on multiple levels of biological, behavioral, social, and environmental observation; disparate sources of data; and diverse perspectives and areas of expertise. Hobcraft argues that "progress in understanding human behavior (or health) requires an integrative approach." He recommends "greater attention to pathways within the individual and their interplays with the processes and progresses whereby the individual interplays with multiple contexts over the life course," concluding that "a concentration on chains or sequences of events, greater awareness of contingent relationships . . . and elaboration of partial mid-level frameworks or mechanisms" is required.

Such an approach is difficult, and most research studies will continue to focus on fragments. Including biological indicators in social surveys will, however, encourage broader integrative thinking and produce deeper understanding of mechanisms and causal linkages. This volume documents how difficult and how promising the endeavor is—and provides practical advice and wise insights. The report concludes with an appendix that provides biographical sketches of contributors to this volume

REFERENCES

Carey, J.R., and Vaupel, J.W. (2005). Biodemography. In D.L. Poston and M. Micklin (Eds.), *Handbook of population* (pp. 625-658). New York: Kluwer Academic/Plenum.

National Research Council. (1997). *Between Zeus and the salmon: The biodemography of longevity.* Committee on Population, K.W. Wachter and C.E. Finch, Eds. Commission on Behavioral and Social Sciences and Education. Washington, DC: National Academy Press: .

National Research Council. (2001). *Cells and surveys: Should biological measures be included in social science research?* Committee on Population, C.E. Finch, J.W. Vaupel, and K. Kinsella, Eds. Commission on Behavioral and Social Sciences and Education. Washington, DC: National Academy Press.

Weinstein, M., and Willis, R.J. (2001). Stretching social surveys to include bioindicators: Possibilities for the Health and Retirement Study, experience from the Taiwan Study of the Elderly. In National Research Council, *Cells and surveys: Should biological measures be included in social science research?* (pp. 250-275). Committee on Population, C.E. Finch, J.W. Vaupel, and K. Kinsella, Eds. Commission on Behavioral and Social Sciences and Education. Washington, DC: National Academy Press.

Part I: What We've Learned So Far

1

Biological Indicators and Genetic Information in Danish Twin and Oldest-Old Surveys

Kaare Christensen, Lise Bathum, and *Lene Christiansen*

Over the last 10 years, biological indicators and genetic information have been added to ongoing nationwide longitudinal surveys among twins and elderly people in Denmark. This chapter summarizes experiences and some important results obtained by adding both traditional and new biological indicators to these surveys in order to get a better understanding of the predictors and determinants of healthy aging.

The overall strategy on biological indicators in these Danish studies was designed to achieve two goals. First, we decided to collect a few easily obtainable biological indicators from nearly all participants in the surveys to be able to make valid and statistically powerful tests of new (or old) findings from other studies—in particular, genetic association studies, which constitute a research field that suffers from many false positive findings due to multiple testing in many small-scale studies. Second, we decided to obtain more detailed biological indicators from individuals who were identified through the survey as particularly informative for a given research purpose, meaning that we used the survey as a screening instrument for follow-up ancillary studies.

The chapter starts with a short description of the challenges and biases in the search for biological and nonbiological determinants of healthy aging. An overview of the Danish surveys follows, with a focus on the limited impact the inclusion of biological indicators in the surveys had on response rates and the experience that the small amount of DNA obtained by a finger prick or a cheek brush probably is enough for hundreds (if not thousands) of genotypes. Then examples of cross-sectional and longitudinal large-scale genetic association studies in the Danish cohort are given, focusing on physical and cognitive functioning and survival. Studies using more detailed biological indicators for particularly interesting subsets of the survey participants, such as twin pairs, will also be covered. Finally, a new biological indicator, perceived age based on photographs, is described and we show how this biological indicator—which is at the other end of the biological spectrum from molecular indicators of aging—correlates to environmental exposures and predicts survival.

CHALLENGES IN THE SEARCH FOR DETERMINANTS OF HEALTHY AGING AND LONGEVITY

A large ongoing research effort is under way to identify the genetic, environmental, and behavioral determinants of extreme survival (longevity) by comparing exceptional survivors (often centenarians) with younger cohorts—that is, case-control or association studies. A characteristic that is more common in centenarians than in the younger control group is interpreted as a determinant for exceptional longevity. A weakness of this approach is that, for all unfixed characteristics, it cannot be distinguished whether an observed characteristic among exceptional long-lived persons is a cause or an adaptive change. Another reason for the limited success of centenarian studies in identifying factors (even fixed characteristics such as genotype) of importance for survival until extreme ages may also be the lack of an appropriate control group, as cohort-specific characteristics may confound the comparison between the centenarians and younger cohorts.

The interference from an observation from such cross-sectional studies (e.g., that a fixed characteristic like a gene variant is decreasing in frequency with age) is dependent on a stable population with little migration into or out of it—that is, no population stratification (Lewis and Brunner, 2004). Remote islands or other isolated populations will therefore often be very well suited for such cross-sectional age dependency studies (also because of better preserved linkage disequilibrium between markers and putative causal mutations). Immigrant countries like the United States and Australia are less suited due to greater genetic and environmental heterogeneity. The case-control studies may also suffer from publication bias because many of the reported associations are often found in subgroups (defined by age or sex) and usually without an a priori hypothesis of which polymorphism is the advantageous one. Consequently, most associations fail to be replicated in independent studies (Ioannidis, 2003). Large longitudinal studies in genetically rather homogeneous populations, such as that in Denmark, avoid many of these biases.

THE DANISH SURVEYS OF TWINS AND THE OLDEST-OLD

The Longitudinal Study of Aging Danish Twins (LSADT) and the Danish 1905 Cohort Study are longitudinal surveys aimed at understanding the determinants of healthy aging and exceptional survival. They have been conducted in parallel and are both nationwide in scope and population based.

Longitudinal Study of Aging Danish Twins

LSADT began in 1995 with the assessment of members of same-sex twin pairs born in Denmark prior to 1920 (i.e., at least 75 years old at the beginning of 1995). The surviving members of the initial cohort were followed up every two years in 1997, 1999, 2001, 2003, and 2005. Additional cohorts were added to the 1997, 1999, and 2001 assessments and subsequently followed at two-year intervals in 2003 and 2005. Twin pairs, in which both were alive and born between 1920 and 1923 (i.e., at least 73 years old at the beginning of 1997), were added in 1997; twins born between 1921 and 1928 (i.e., at least 70 years old at the beginning of 1999) were added in 1999; and twins born between 1929 and 1930 (i.e., at least 70 years old at the beginning of 2001) were added in 2001. An overview of the LSADT cohort-sequential design is given in Figure 1-1. These studies comprised a home-based two-hour multidimensional interview focusing on health issues, assessment of functional and cognitive abilities, and DNA sampling.

Biological Indicators in LSADT

Full Blood Samples on Selected Individuals in 1997

Shortly after the interview in 1997, a trained phlebotomist collected a blood sample from all same-sex twin pairs in which both members were alive and willing to participate. In all a total of 689 subjects donated a blood sample, including 290 same-sex twin pairs and 109 pairs from whom we received blood from only one twin. The samples were sent by ordinary mail from all over Denmark and each sample, within two days of receipt, was separated into cells and plasma layers by centrifugation.

DNA Samples on All Nonproxy Participants from 1999

For the 1999 survey, all participants (except those who participated by proxy—spouse, children, or caretaker) were asked to provide a DNA sample by means of a finger prick blood spot sample. The blood was stored

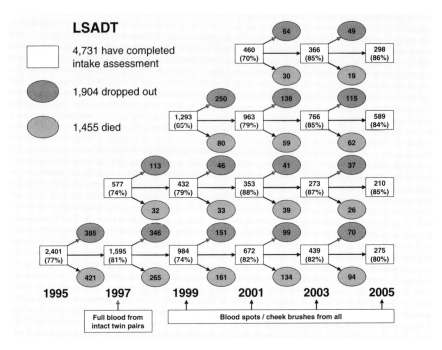

FIGURE 1-1 Overview of the Longitudinal Study of Aging Danish Twins (LSADT). Participation rates are given in percentage in the white boxes.

on filter paper, and the procedure is described in detail in Christensen (2000, pp. 56-60). The blood spot card was the preferred DNA collection method; however, the participants could provide biological material by using cheek brushes if they disliked the idea of a finger prick. A total of 90 percent of the nonproxy participants provided a biological sample, of which 91 percent were blood spots. The same procedure was repeated in all subsequent waves (2001, 2003, and 2005) in order to study changes over time and to obtain more DNA. As seen in Figure 1-1, the inclusion of blood spots/cheek brushes did not affect the participation rates in a negative way. Furthermore, more demanding measurements, such as grip strength and pulmonary functioning, were also included for the first time in 1999, still with no apparent influence on the participation rate.

Photographs of All Participants from 2001

When assessing health, physicians traditionally compare perceived and chronological age. Among adults, "looking old for your age" is often interpreted as an indicator of poor health and is used in clinical practice as a biomarker of aging. The sparse data available on the relationship between perceived age and survival indicate an inverse association (Borkan, Bachman, and Norris, 1982). Starting with the 2001 LSADT survey (and continuing in 2003 and 2005), we obtained a digital photograph of each subject in order to explore this association further.

The Danish 1905 Cohort Study

We surveyed the complete Danish 1905 cohort in 1998 when 3,600 individuals from this cohort were still alive and 2,262 participated in the survey that included an interview, physical and cognitive tests, as well as collection of biological material. Subsequently we made three in-person follow-up studies of the participating survivors in 2001, 2003, and 2005 (Figure 1-2) establishing the first large-scale centenarian study in which both environmental and genetic information is available from a noncen-

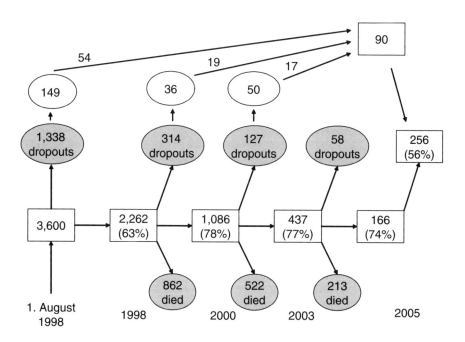

FIGURE 1-2 Overview of the Danish 1905 Cohort Study. All nonproxy participants were eligible for blood spots/cheek brushes at all waves. Participation rates are given in percentage in the white boxes.

tenarian control group from the same birth cohort. The participants were age 92-93 when they entered the study, which is only "halfway to becoming a centenarian" in terms of selection: only about 1 in 20 of the 1905 cohort made it to age 92-93, and only about 1 in 20 of these survivors at age 92-93 lived to celebrate their 100th birthday. Hence, the Danish 1905 cohort provides a powerful and unique opportunity for studying the determinants of survival in the second leg of the long trip to becoming a centenarian.

DNA Samples from All Participants in All Waves

All nonproxy participants, in all waves, were asked to provide blood spots or cheek brushes, and approximately 90 percent of the nonproxy respondents agreed. Of these, around 80 percent agreed to blood spots and 20 percent to cheek brushes.

Full Blood Samples from All Participants in 2005

To obtain comparable data to a Danish centenarian study conducted 10 years earlier, that of the 1895-1896 cohort (Andersen-Ranberg, Schroll, and Jeune, 2001), full blood samples were taken in 2005 (shortly after the interview) from all participants who consented. Also to enhance comparability to the previous Danish study, we recontacted previous nonresponders (a group comprised of both nonresponders and study dropouts) and were able to add 90 additional subjects to our study (Figure 1-2). The participation in 2005 of the former nonresponders in the 1905 cohort is likely to be influenced by the common misconception among the elderly (and their relatives) that, whereas studies of centenarians are interesting and important for science, this is not the case for studies of 93-year-olds.

How Many Genotypes Can Be Obtained by
Blood Spots and Cheek Brushes?

Dried blood spot specimens are made by carefully placing a few drops of blood, freshly drawn by finger prick with a lancet, onto specially manufactured absorbent specimen collection (filter) paper. Cheek brushes collect cells by gently rubbing the inside of the cheek (Christensen, 2000). For both DNA sampling methods, the amount of collected cells and hence the amount of isolated DNA are very variable, especially in the elderly. Using ordinary methods based on polymerase chain reactions, we have until now genotyped the entire 1905 cohort for 20 different single-nucleotide polymorphisms (SNPs, microsatellites and smaller insertions/deletions). We estimate that it is possible to genotype at least 2,000 SNPs from a fully filled blood spot card, whereas each cheek brush is estimated to provide DNA for 200 genotypes if the analyses are carefully optimized for the use of a low DNA content. However, the development of whole genome amplification techniques may perhaps in the future make it feasible to conduct large genetic epidemiological studies using small DNA sources, such as blood spots and cheek brushes. Several reports have shown that whole genome amplification methods are indeed applicable for low DNA yield samples; however, a higher rate of genotyping errors has been reported when using these low DNA yield samples, possibly due to allele dropout. The use of these techniques requires increased attention to genotyping quality control and caution when interpreting results (Silander et al., 2005).

IMPORTANT FINDINGS DUE TO INCLUSION OF BIOLOGICAL INDICATORS IN THE DANISH SURVEYS

Studies of Candidate Genes for Healthy Aging Using the Full Surveys

A very large number of candidate genes have been investigated for putative associations with human aging and longevity (Christensen, Johnson, and Vaupel, 2006). Based on the promising results from animal studies, genes from the insulin/IGF-1 pathway, stress response genes (heat and oxidative stress), and genes influencing mitochondrial functioning have been obvious candidates. So have immune system–regulating genes (e.g., interleukines), as it is a biological system with a sufficiently broad spectrum of functions and hence likely to be associated with aging and survival (Finch and Crimmins, 2004; Franceschi et al., 2000). Other candidates have been selected based on known downstream functioning and association with early or late-in-life pathology and diseases, such as metabolic genes (e.g., lipoproteins), iron metabolism–regulating genes (e.g., HFE), genes affecting the cardiovascular system (e.g., ACE), and tumor suppressor genes called p53. A distinct group of candidates are the genes for the premature aging syndromes, such as the DNA repair helicase Werner syndrome (Yu et al., 1996) in which "leaky mutations" are considered—that is, the syndrome-causing gene is investigated to test if "milder" common mutations could be associated with aging and survival generally.

One candidate gene, microsomal triglyceride transfer protein (MTP) was identified after a genome-wide linkage scan in long-lived families had provided evidence for a locus for longevity on chromosome 4 (Puca et al., 2001; Geesaman et al., 2003). This finding received considerable coverage in both the scientific and the lay communities as a promising "longevity gene." As illustrated by the examples below, one of the most important aspects of having biological material in surveys like the Danish is that they can readily provide valid and powerful statistical tests of such findings on putative determinants of healthy aging or exceptional survival.

Microsomal Triglyceride Transfer Protein

A genome-wide linkage scan in long-lived families provided evidence for a locus for longevity at chromosome 4 near microsatellite marker D4S1564 (Puca et al., 2001; Reed, Dick, Uniacke, Foroud, and Nichols, 2004), although a follow-up study in a French population did not confirm this observation (Geesaman et al., 2003). Fine mapping of the region identified MTP as the gene most probably responsible for the chromosome 4 linkage peak (Geesaman et al., 2003). The two SNPs found to account for the majority of the variation at the locus are rs2866164—an SNP in complete linkage disequilibrium with rs1800591 (-493G/T), an MTP promoter mutation—and MTP Q/H—a semiconservative mutation in exon 3 of MTP (glutamine to histidine at aminoacid 95) (Geesaman et al., 2003). The rs2866164-G allele and the haplotype composed of rs2866164-G/MTP-Q9 had a significantly lower frequency in long-lived individuals compared with a group of younger controls in the initial study.

MTP is necessary for the assembly of apolipoprotein B-containing lipoproteins in the liver and small intestine (Swift et al., 2003) and lack of the protein leads to abetalipoproteinemia, a rare genetic disorder that is characterized by an inability to produce chylomicrons and very low-density lipoproteins. MTP is a good candidate for a gene involved in human longevity. Its functions and critical position in lipoprotein have made it the basis for a new generation of lipid-lowering drugs that have reached trials in animal models (Wierzbicki, 2003). However, we genotyped the 1,905 cohort, and even after seven years of follow-up with more than 80 percent of the population deceased, we were not able to detect any association between MTP and survival (Bathum et al., 2005) (Figure 1-3). Furthermore, a large case-control study of German centenarians (Nebel et al., 2005) and an association study in Dutch nonagenarians failed to confirm any association of MTP variants with longevity. Finally, in 2006, a meta-analysis of all published studies, analyzing data from 4,915 long-lived cases older than 85 years and 3,002 younger controls, resulted in an odds ratio of 0.93 (95% CI, 0.77-1.14), thus confirming the lack of association between MTP and longevity and illustrating the necessity of replication studies in identifying genes that affect human longevity (Beekman et al., 2006).

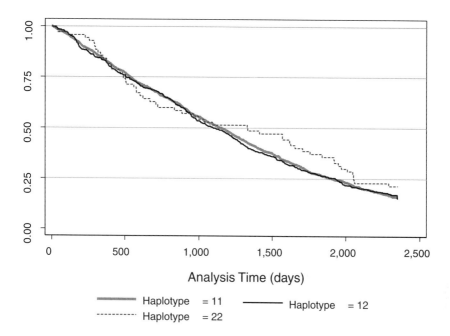

Analysis Time (days)

Haplotype = 11   Haplotype = 12
Haplotype = 22

FIGURE 1-3 Kaplan-Meier survival estimates in the 1905 Cohort Study (N = 1,651) divided into the three haplotype groups for microsomal triglyceride transfer protein (MTP). Haplotype 2 is the suggested risk haplotype composed of the rs2866164-G and Q95 allele and haplotype 1 is all other haplotypes. The analysis unit of time is days. The difference is not significant (p = 0.32).

Apolipoprotein E

Apolipoprotein E plays an important role in lipoprotein metabolism, and the protein is found in three common variants (ApoE-2, ApoE-3, and ApoE-4) that were known at the protein level before the gene was cloned. The ApoE-4 variant has been consistently associated with a moderately increased risk of both cardiovascular diseases and Alzheimer disease, while the ApoE-2 is protective (Corder et al., 1996; Panza et al., 2004; Lewis and Brunner, 2004). Not only has ApoE-4 been shown to be a risk factor for cardiovascular diseases and Alzheimer disease per se, but also several studies have found that the ApoE-4 carrier is more susceptible to some environmental exposures, making this an example of a gene-environment interaction.

For example, an increased risk of chronic brain injury after head trauma has been observed for individuals who carry the ApoE-4 gene variant, compared with non-ApoE-4 carriers (Jordan et al., 1997). ApoE-4 also seems to modulate the effect of other risk factors for cognitive decline. Individuals with ApoE-4 in combination with atherosclerosis, peripheral vascular disease, or diabetes mellitus have a substantially higher risk of cognitive decline than those without ApoE-4 (Haan, Shemanski, Jagust, Manolio, and Kuller, 1999).

Cross-sectional studies of allele frequency differences between age groups for ApoE have been remarkably consistent from study to study, in contrast to other candidate genes (Ewbank, 2004). The ApoE-4 frequency varies considerably among populations of younger adults (about 25 percent among Finns, 17-20 percent among Danes, and about 10 percent among French, Italians, and Japanese), but in all these populations the frequency among centenarians is about half the value of the younger adults (Figure 1-4). Although these changes in ApoE allele frequency with age are substantial, it is compatible with a scenario in which ApoE-2 carriers in adulthood who have an estimated average mortality risk that is only 4-12 percent lower than in ApoE-3 and ApoE-4 carriers have a risk that is only 10-14 percent higher throughout adulthood (Gerdes, Jeune, Ranberg, Nybo, and Vaupel, 2000), making ApoE a "frailty gene" rather than a "longevity gene." The longitudinal Danish 1905 cohort showed that the ApoE genotype has an association with cognitive functioning and cognitive decline (Figure 1-5) in this age group, albeit of only borderline significance. However, when defining a well-functioning group, that is, those still alive, without interview by proxy or extreme decline in the Mini Mental State Examination at follow-up, there was a significant decrease in the frequency of ApoE-4 positive subjects from intake to first and second follow-up. So ApoE genotypes have a statistically significant effect on the probability of remaining a well-functioning nonagenarian (Bathum et al., 2005). While the impact on survival is minor, the difference in survival between ApoE-4 positive and negative subjects did become statistically significant in January 2006 after more than seven years of follow-up and more than 1,200 deaths in the cohort. Preliminary analyses suggest that, at intake at age 92-93, the participants (both ApoE-4 positive and ApoE-4 negative) have mortality rates below the population average, confirming that terminally ill individuals are overrepresented among the nonparticipants and the proxy-responders. As follow-up time increases, the difference in survival becomes clear, showing that ApoE-4 positive individuals have higher mortality rates than the overall population, whereas ApoE-4 negative individuals have lower mortality rates. This highlights the need for careful analyses of the nonparticipants. In Denmark such analyses can be carried out through linkage to national health registers.

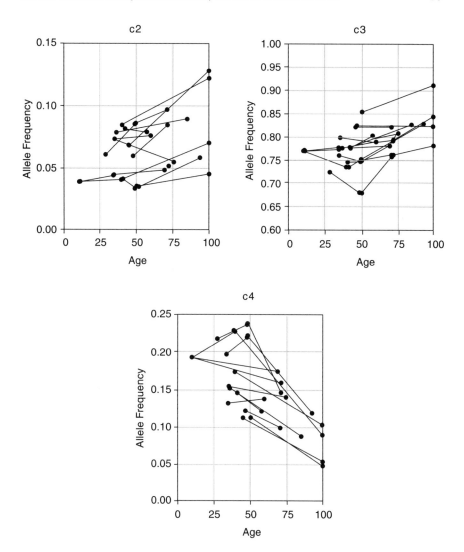

FIGURE 1-4 The frequencies of apolipoprotein E alleles vary with age. Frequencies of the three common ApoE alleles—ApoE-2, ApoE-3, and ApoE-4—are shown, taken from data in 13 published studies. Each line connects the frequencies in various age groups in a given population.
SOURCE: Reproduced from Gerdes et al. (2000).

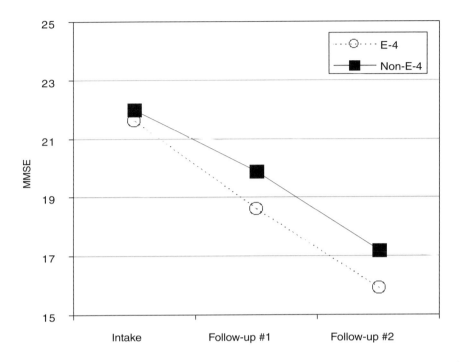

FIGURE 1-5 Decline in cognitive performance in 1,551 nonagenarians, - E-4 carriers compared to non-E-4 carriers. The genotype difference in rate of decline is borderline significant, p = 0.07.

Angiotensin I-Converting Enzyme (ACE)

The insertion/deletion polymorphism in intron 16 in the gene coding for angiotensin I-converting enzyme (ACE) is also a biologically plausible candidate for successful aging and longevity. This is because the genotype accounts for approximately half of the variance in the circulating ACE level and from the II (homozygous insertion) to the DD (homozygous deletion) genotype the presence of each D allele is associated with an additive effect on ACE activity (50 percent higher in the DD compared with the II genotype) (Rigat et al., 1990). The cleavage of angiotensin I by ACE produces the octapeptide angiotensin II, which is a potent vasoconstrictor. The association between ACE genotypes, in particular DD, and the occurrence of cardiovascular and renal diseases has therefore been thoroughly studied in the past decade. The results have been inconsistent, and, for ischemic heart disease, publication bias has been shown to be a likely explanation for the discrepancies (Keavney et al., 2000).

In recent years the ACE gene polymorphism has also been associated with the outcome of physical exercise, especially extreme endurance and performance. Presence of the II genotype has been found to be associated with improved performance (Montgomery et al., 1998; Williams et al., 2000; Gayagay et al., 1998; Myerson et al., 1999). However, the study subjects have mostly been from highly selected populations (e.g., elite athletes and military recruits) and the samples were small. The associations could not be confirmed in LSADT (Frederiksen et al., 2003), but a recent paper suggests an interaction between ACE genotype and mobility limitations among those age 70-79 who exercise (Kritchevsky et al., 2005). Again, no a priori hypothesis about existence of the interaction was found.

ACE genotype has been hypothesized to be associated with longevity since a German study found an increased frequency of the DD genotype in octogenarians (Luft, 1999), which was to some degree supported in LSADT data (Frederiksen et al., 2003). This finding could not, however, be confirmed in two large studies of centenarians and younger controls (Bladbjerg et al., 1999; Blanche, Cabanne, Sahbatou, and Thomas, 2001). ACE genotyping is currently being performed in the 1905 cohort.

Interleukin 6

Chronic low-grade inflammatory activity has been suggested to play a central role in aging processes, as it seems to be implicated in the pathology of several age-related diseases, consequently leading to increased mortality (Bruunsgaard, Pedersen, and Pedersen, 2001; Harris et al., 1999). The multifunctional cytokine interleukin 6 (IL-6) is one of the inflammatory markers central to low-grade inflammation. IL-6 is overexpressed in many of the stress conditions that are characteristic features of the aging process (Ershler and Keller, 2000; Bruunsgaard et al., 2001; Cohen, Pieper, Harris, Rao, and Currie, 1997).

It has been proposed that an increase in IL-6 is a risk factor for the development of age-related diseases, such as rheumatoid arthritis, osteoporosis, Alzheimer disease, cardiovascular disease, and type 2 diabetes (Pradhan, Manson, Rifai, Buring, and Ridker, 2001; Ershler and Keller, 2000; Licastro et al., 2000; Bonaccorso et al., 1998; Robak, Gladalska, Stepien, and Robak, 1998), as well as for functional decline (Barbieri et al., 2003; Ferrucci et al., 1999; Cohen et al., 1997) and mortality (Harris et al., 1999). However, Petersen and Pedersen (2005) suggest that IL-6 is a marker of metabolic syndrome rather than a cause and that IL-6 is actually a "myokine" released by contracting skeletal muscle fibers and disseminates the beneficial effects of exercise to other organs (Petersen and Pedersen, 2005). Twin studies have shown that interindividual variations in IL-6 expression have a substantial genetic component (de Maat et al.,

2004; de Craen et al., 2005), and these differences may be a result of base variations located in the promoter region of the IL-6 gene.

Three single-point polymorphisms (-597G/A, -572G/C, and -174G/C) and an AT-stretch polymorphism (-373(A)n(T)m) have been identified in the IL-6 promoter (Georges et al., 2001; Terry, Loukaci, and Green, 2000; Fishman et al., 1998). The potential significance of the -174G/C polymorphism for both the IL-6 level and hence disease susceptibility and mortality risk has been investigated in a large number of studies involving a variety of diseases. The results have, however, been conflicting (Nauck et al., 2002; Humphries, Luong, Ogg, Hawe, and Miller, 2001; Rauramaa et al., 2000).

Recent independent findings of a modest, but significant, increase in the frequency of interleukin 6 -174GG homozygotes both in Finland (Hume, 2005) and in our Danish surveys (Christiansen, Bathum, Andersen-Ranberg, Jeune, and Christensen, 2004) suggest that this genotype is advantageous for longevity. However, a very recent study in Italy (Ravaglia et al., 2005) was unable to confirm this.

Paraoxonase 1

Stress-response genes, such as paraoxonase, mediate protection against oxidative damage and are therefore excellent candidates for studies of successful aging and longevity. Paraoxonase 1 (PON1) is found to be associated with high-density lipoproteins (HDLs) in plasma. One physiological function of the enzyme appears to be the elimination of toxic oxidized lipids in lipoproteins, whereby it confers protection against atherosclerosis and coronary heart disease. Recent studies have demonstrated a progressive decrease of PON1 activity with age, which may be related to oxidative stress conditions developing with increased age. In general, there is a 40-fold variation in PON1 activity among individuals, which is partly explained by genetic variations in both the coding region and the promoter of the gene, although environmental influences, for example, tobacco smoking, also exert an effect (Costa, Vitalone, Cole, and Furlong, 2005; Seres, Paragh, Deschene, Fulop, and Khalil, 2004).

Investigating the influence of the two coding sequence polymorphisms, L55M and Q192R, on the efficacy of protection of low-density lipoproteins (LDL) lipid oxidation, Mackness et al. (1998) demonstrated that individuals with the PON1 192QQ/55MM genotype were most effective in protecting LDL against lipid oxidation, whereas PON1 192RR/55LL homozygotes were least effective. A meta-analysis of 43 studies involving 11,212 cases and 12,786 controls suggested a weak overall association between 192Q/R and coronary heart disease.

We explored the association between the PON1 192Q/R polymor-

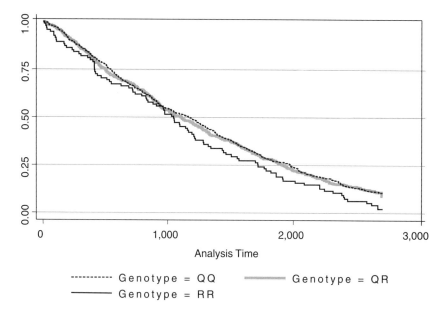

FIGURE 1-6 Kaplan-Meier survival estimates in the combined group of participants from the 1905 cohort (N = 1,265) divided into the three genotype groups for PON1 192. The analysis time unit is days (p < 0.05).

phism both in LSADT and in the 1905 cohort. Using the longitudinal study design, we first found that 192RR homozygotes had a significantly poorer survival compared with QQ homozygotes in LSADT. We extended this study by testing an independent sample of 541 individuals from the 1905 cohort and confirmed the initial findings. In both samples the effect was most pronounced in women, suggesting that PON1 192RR homozygosity is associated with increased mortality in women in the second half of life (Christiansen et al., 2004). We just completed the PON1 genotyping of the whole 1905 cohort, and the seven years of follow-up (Figure 1-6) further corroborated the initial finding. One of the biggest challenges to genetic association studies are false positive findings (Ioannidis, 2003) and one of the major strengths of including biological material in both LSADT and the Danish 1905 Cohort Study has been that independent replication can be performed.

STUDIES OF BIOLOGICAL INDICATORS FOR HEALTHY AGING USING SELECTED SUBGROUPS IN THE SURVEYS

Biological indicators from elderly twin pairs provide a number of unique research opportunities, and therefore we included a full blood sample in the LSADT 1997 wave for the "intact twin pairs" in the survey. Among the key uses of the samples were to investigate telomere length, X inactivation, and gene expression.

Telomere Length

The consistent findings of a negative correlation between telomere length and the replicative potential of cultured cells as well as a decreasing telomere length in a number of different tissues in humans with age (Cherif, Tarry, Ozanne, and Hales, 2003) have led to the suggestion that telomeres play a role in cellular aging in vivo, and ultimately even in organismal aging. A possible association between telomere length and mortality late in life in humans was indicated by Cawthon, Smith, O'Brien, Sivatchenko, and Kerber (2003), who measured telomere length on blood samples drawn about 20 years ago from 147 healthy individuals (ages 60-97). They found that, corrected for age, individuals with shorter telomeres had a poorer survival than those with longer telomeres. However, a larger independent study failed to confirm this finding and showed that telomere measurements are fluctuating over time in blood cells (Martin-Ruiz, Gussekloo, van Heemst, von Zglinicki, and Westendorp, 2005). We analyzed the 1997 LSADT samples (Bischoff et al., 2006) and also found that telomere length is not a predictor for remaining life span, once age is controlled for. The sample provided a unique opportunity to do intra-pair comparisons among twins, in which the effect of the genetic factor is controlled for (100 percent for monozygotic and 50 percent for dizygotic pairs). Also, this approach did not reveal evidence for an association between telomere length and survival among the elderly. Currently we are reexamining the blood samples using new and improved methods of measurement of telomere length to test if the lack of association could be due to methodological problems.

X Inactivation

Although females have two X chromosomes, only one of them is active in each cell; the other is inactivated in early embryonic life and stays so throughout life (a phenomenon called X inactivation). The random X inactivation makes females mosaics for two cell populations, usually with an approximate 1:1 distribution. Males have only one cell line because they receive only one X chromosome (from their mother) and one Y chromosome (from their father).

Cross-sectional studies have shown that, among younger females, it is very rare to have a skewed distribution of X inactivation, while for females over age 60, more than a third have a predominance of one of the cell lines in their blood, and among centenarian females the majority has a predominant cell line (Christensen et al., 2000). The skewing of this distribution in peripheral blood could be due to either depletion of hematopoietic stem cells followed by random differentiation or selection processes based on X-linked genetic factors.

It has been difficult to make animal experiments or human observations that could critically test the random versus selection hypotheses. The female twins who provided full blood in the LSADT survey, however, provided an excellent opportunity. On one hand, if the often observed predominance of one of the two cell lines in peripheral blood in elderly females was determined by a stochastic process with no selection, then one would expect little correlation in the X inactivation patterns between two monozygotic co-twins. A selection process based on X-linked genetic factors, on the other hand, would create a tendency for the same cell line to become predominant in two monozygotic co-twins. Our studies indicate that the peripheral blood cells from LSADT monozygotic female twins show a strong tendency for the same cell line to become predominant in two co-twins, which suggests that X-linked genetic factors influence human hematopoietic stem cell kinetics and potentially organismal survival. The fact that females have two cell lines with different potentials could be one of the reasons why women live longer than men (Christensen et al., 2000).

Gene Expression

Microarray analyses in animal and humans of gene expression and changes with age are still few, but many believe that a better understanding of gene expression will lead to insight into the mechanisms of aging. We used LSADT to identify three pairs of healthy 80+ year-old female monozygotic and dizygotic twins (six pairs in total), and even in this age group on this small sample we were able to detect a substantial genetic component to the variation in gene expression (Tan, Kruse, and Christensen, 2006).

SURVEYS AS A SCREENING INSTRUMENT TO IDENTIFY CASES FOR ANCILLARY STUDIES

The twin surveys have also been used as a screening instrument for a variety of conditions involving a two-stage study design with initial screening through the LSADT assessment and a follow-up assessment by a specialist. This has been done for movement disorders, in which the LSADT participants performed a so-called Archimedes spiral test—a drawing test to disclose individuals with movement disorders, such as essential tremor and Parkinson's disease. Individuals whose spiral test was positive were recontacted, and 119 twin pairs were interviewed and examined by a movement disorder specialist. The study revealed a nearly complete concordance for monozygotic twins for essential tremor and a substantial lower concordance rate for dizygotic twins, indicating a high heritability for essential tremor, and hence a good candidate for a phenotype to be used in linkage studies (Lorenz et al., 2004).

In Europe there is also an extensive collaboration between the major twin registers (Skytthe et al., 2003), which together have information on more than half a million twin pairs. A collaborative effort including LSADT has been used to identify particularly informative pairs, for example, dizygotic twin pairs extremely concordant or discordant for continuous traits that are especially powerful in genetic linkage studies (Risch and Zhang, 1995). Currently studies of twin pairs extremely concordant or discordant for body mass index are being conducted.

An outstanding example of the potentials of exceptional twin pairs is the cloning of the genes for Van der Woude syndrome (a syndrome with cleft lip and palate and lip pits). Family studies have for many years indicated that this monogenic disease is caused by a gene on chromosome 1, but despite large collaborative efforts, identification of the actual gene has failed. The identification of one monozygotic twin pair discordant for the disease led to the identification. The discordance could be due to either reduced penetrance or alternatively a somatic mutation after a split of a two-cell-stage zygote. A sequencing of the candidate region revealed one genetic difference between the two monozygotic twins in the gene coding for interferon-regulatory factor 6 (IRF-6). With this information in hand, the previously identified families were reinvestigated. It became clear that it was indeed mutations in this gene, IRF-6, that was causing Van der Woude syndrome (Kondo et al., 2002).

A NEW BIOLOGICAL INDICATOR OF AGING: PERCEIVED AGE

Modern medicine includes an increasing number of physiological parameters, biomarkers, and other "objective" measures through body fluids and increasingly sophisticated imaging tools. However, when assessing health, physicians in many countries traditionally still compare perceived and chronological age and record this in the patient's record. Among adults, "looking old for your age" is often interpreted as an indicator of poor health. Hence in the clinic, perceived age is used as a "biomarker of aging," yet little validation of this biomarker has been performed. The sparse data available on the relation between perceived age and survival indicate an inverse association (Christensen et al., 2004). It is not known whether looking old for your age is primarily a result of lifestyle and other environmental factors or whether genetic factors play an important role.

We used the 2001 LSADT survey to test this: 91 percent of cognitively intact participants agreed to have their picture taken (digital camera, 0.6 meter distance, neutral background, if possible). For a total of 387 same-sex twin pairs we obtained high-quality pictures of both twins (Figure 1-7). We engaged 20 female nurses (ages 25-46) to estimate the twins' ages. The nurses were not informed beforehand about the age range of the twin pairs, and they assessed, based on the digital pictures, the ages of the first-born and the second-born twins on two different days. The mean of the nurses' age estimates for each twin was used as the twin's perceived age. The reliability of the mean age rating was estimated at .94 from a one-way analysis of variance. The correlation between real age and perceived age was 0.4 ($p < 0.001$), and the nurses' estimates regressed towards a mean of 77 years.

The correlation between the nurses' estimates for monozygotic twins ($r \approx 0.6$, $p < 0.01$) is about twice the correlation for dizygotic twins ($r \approx 0.3$, $p < 0.01$). This indicates an effect of additive genetic factors influencing perceived age. Biometrical models confirmed that the twin similarity is best explained by a model including additive genetic factors and nonfamily environment, and that the heritability (i.e., the proportion of the variance explained by genetic factors) of perceived age is about 60 percent, with no sex or age differences.

By January 2003, nearly two years after having been photographed, one of the pictured twins in 49 pairs had died. Among these 49 pairs, the longer surviving twin was rated as looking younger on average than his or her co-twin (mean of 1.15 years, SD = 3.63, t = 2.22 on 48 df, $p < 0.02$). This significant difference, however, owed entirely to those twin pairs who were perceived to be discrepant in age. Among the 26 pairs for which perceived age differed by two or more years, the oldest looking twin died first in 19 (73 percent) cases ($p < 0.01$), verifying that perceived age is associated with mortality. We are currently making a follow-up study of five years' survival, and the preliminary analyses strongly support the initial findings (also using raters of different age, gender, and education).

FIGURE 1-7 Longitudinal studies have demonstrated that perceived age is a predictor of survival among the elderly and that it is associated with a number of environmental exposures.

We have furthermore studied the determinants of perceived age and showed that when age and gender are controlled for factors like smoking, socioeconomic status, body mass index, marital status, and depression symptomatology, all affect perceived age in the expected direction (Rexbye et al., 2006). We did follow-up pictures in 2003 and 2005. We also made high-quality pictures of 100 selected pairs, including facial imprint of the facial structure, to shed light on which immediate and distant factors are the basis for perceived age and its association with health and survival.

OTHER BIOMARKERS

In the Danish surveys of twins and the oldest-old, we have used a number of validated biomarkers, such as grip strength, spirometry, and walk test, as well as molecular markers based on blood samples and cheek swabs. Here we report on what we consider the key findings, but the rapid development within the biomarker field is likely to provide new (combinations of) biomarkers, in particular from the neuroendocrine and the immune system (Goldman et al., 2006).

Finally, the ever-increasing possibilities in imaging and especially neuroimaging could yield major breakthroughs in the biomarker research in the coming years (Winterer et al., 2005), although neuroimaging is not likely to be included in large field surveys with 100+ interviewers in any foreseeable future.

Another promising research direction is the study of gene-environment interaction in aging, and how environmental factors, including social environment, can modify the gene effects and the biomarkers (Ryff and Singer, 2005).

CONCLUSION

The inclusion of biological markers in the Danish surveys of twins and the oldest-old has provided great leverage for these studies by making them highly interdisciplinary in nature and very productive in terms of results. Some 70 papers based on LSADT and the Danish 1905 Cohort Study have been published in international peer-reviewed journals, and half of the papers include biological indicators, even though their inclusion was not started until some years after the study had been launched. LSADT and the Danish 1905 Cohort Study aim at identifying the determinants of healthy aging and longevity. Several studies—including the traditional twin analyses in LSADT—have shown that there is indeed a genetic influence on human lifespan, longevity, and cognitive and physical functioning among the oldest, providing support for the search for candidate genes involved in aging (Hjelmborg et al., 2006; Frederiksen and Christensen, 2003; McGue and Christensen, 2001, 2002; Herskind et al., 1996; Hjelmborg et al., 2006).

We have so far examined several candidate genes for healthy aging and longevity and corroborated some findings and refuted others in large-scale studies with substantial power. In this area, our future studies will continue to focus on promising candidate genes identified in other studies or through biological insight. We intend to use haplotype-based analyses to thoroughly investigate the relevance of each candidate gene, including the regulatory areas surrounding it. Furthermore we intend to broaden our attention toward regulatory changes in the noncoding DNA, which could be functionally important. Also, we will extend our testing of biologically plausible gene-environment interaction (a research area in which we expected major findings five years ago but which has yielded few up to this point). Our studies so far have shown that small amounts of DNA from population-based studies are sufficient to perform simple genotypings on a large scale: at least 2,000 genotypings can be performed on a fully filled blood spot card. We expect that new techniques will evolve that may enable us to further widen the use of the collected material. Using the surveys as a screening instrument for ancillary studies has proved to be particular fruitful, and we intend to further explore this option. We feel that our most important research is a result of extending our surveys beyond the traditional items, and hence we feel that supplementing our epidemiological surveys with the collection of biological markers and indicators has brought us new insight and has helped us to think outside the box in our research field of healthy aging and longevity.

ACKNOWLEDGMENTS

This research was supported by a grant from the Danish National Research Foundation and the National Institute on Aging (P01-AG08761). Karen Andersen-Ranberg, Henrik Frederiksen, Frans Bødker, and Gitte Bay Christensen coordinated the surveys. Matt McGue, Bernard Jeune, and James W. Vaupel played a central role in planning and analyzing these studies.

REFERENCES

Andersen-Ranberg, K., Schroll, M., and Jeune, B. (2001). Healthy centenarians do not exist, but autonomous do: A population-based study of morbidity among Danish centenarians. *Journal of the American Geriatrics Society, 49,* 900-908.

Barbieri, M., Ferrucci, L., Ragno, E., Corsi, A., Bandinelli, S., Bonafe, M., Olivieri, F., Giovagnetti, S., Franceschi, C., Guralnik, J.M., and Paolisso, G. (2003). Chronic inflammation and the effect of IGF-I on muscle strength and power in older persons. *American Journal of Physiology, Endocrinology, and Metabolism, 284,* E481-E487.

Bathum, L., Christiansen, L., Tan, Q., Vaupel, J., Jeune, B., and Christensen, K. (2005). No evidence for an association between extreme longevity and microsomal transfer protein polymorphisms in a longitudinal study of 1651 nonagenarians. *European Journal of Human Genetics, 13,* 1154-1158.

Beekman, M., Blauw, G.J., Houwing-Duistermaat J.J., Brandt B.W., Westendorp, R.G, and Slagboom, P.E. (2006). Chromosome 4q25, microsomal transfer protein gene, and human longevity: novel data and a meta-analysis of association studies. *Journals of Gerontology Series A: Biological Sciences and Medical Sciences, 61*(4), 355-362.

Bischoff, C., Petersen, H.C., Graakjaer, J., Andersen-Ranberg, K., Vaupel, J.W., Bohr, V.A., Kølvraa, S., and Christensen, K. (2006). No association between telomere length and survival among the elderly and oldest-old. *Epidemiology, 17*(2), 90-194.

Bladbjerg, E.M., Andersen-Ranberg, K., de Maat, M.P., Kristensen, S.R., Jeune, B., Gram, J., and Jespersen, J. (1999). Longevity is independent of common variations in genes associated with cardiovascular risk. *Journal of Thrombosis and Haemostasis, 82,* 1100-1105.

Blanche, H., Cabanne, L., Sahbatou, M., and Thomas, G. (2001). A study of French centenarians: Are ACE and APOE associated with longevity? *Comptes Fendus de l'Académie des Sciences. Série III, 324,* 129-135.

Bonaccorso, S., Lin, A., Song, C., Verkerk, R., Kenis, G., Bosmans, E., Scharpe, S., Vandewoude, M., Dossche, A., and Maes, M. (1998). Serotonin-immune interactions in elderly volunteers and in patients with Alzheimer's disease (DAT): Lower plasma tryptophan availability to the brain in the elderly and increased serum interleukin-6 in DAT. *Aging (Milano), 10,* 316-323.

Borkan, G.A., Bachman, S.S., and Norris, A.H. (1982). Comparison of visually estimated age with physiologically predicted age as indicators of rates of aging. *Social Science and Medicine, 16,* 197-204.

Bruunsgaard, H., Pedersen, M., and Pedersen, B.K. (2001). Aging and proinflammatory cytokines. *Current Opinion in Hematology, 8,* 131-136.

Cawthon, R.M., Smith, K.R., O'Brien, E., Sivatchenko, A., and Kerber, R.A. (2003). Association between telomere length in blood and mortality in people aged 60 years or older. *The Lancet, 361,* 393-395.

Cherif, H., Tarry, J.L., Ozanne, S.E., and Hales, C.N. (2003). Ageing and telomeres: A study into organ- and gender-specific telomere shortening. *Nucleic Acids Research, 31,* 1576-1583.

Christensen, K. (2000). Biological material in household surveys: The interface between epidemiology and genetics. In National Research Council, *Cells and surveys: Should biological measures be included in social science research?* (pp. 42-63). Committee on Population, C.E. Finch, J.W. Vaupel, and K. Kinsella (Eds.). Commission on Behavioral and Social Sciences and Education. Washington, DC: National Academy Press.

Christensen, K., Iachina, M., Rexbye, H., Tomassini, C., Frederiksen, H., McGue, M., and Vaupel, J.W. (2004). Looking old for your age: Genetics and mortality. *Epidemiology, 15,* 251-252.

Christensen, K., Johnson, T.E., and Vaupel, J.W. (2006). The quest for genetic determinants of human longevity: Challenges and insights. *Nature Reviews Genetics, 7*(6), 436-448.

Christensen, K., Kristiansen, M., Hagen-Larsen, H., Skytthe, A., Bathum, L., Jeune, B., Andersen-Ranberg, K., Vaupel, J.W., and Ørstavik, K.H. (2000). X-linked genetic factors regulate hematopoietic stem-cell kinetics in females. *Blood, 95,* 2449-2451.

Christiansen, L., Bathum, L., Andersen-Ranberg, K., Jeune, B., and Christensen, K. (2004). Modest implication of interleukin-6 promoter polymorphisms in longevity. *Mechanisms of Ageing and Development, 125*(5), 391-395.

Cohen, H.J., Pieper, C.F., Harris, T., Rao, K.M., and Currie, M.S. (1997). The association of plasma IL-6 levels with functional disability in community-dwelling elderly. *Journals of Gerontology Series A: Biological Sciences and Medical Sciences, 52,* M201-M208.

Corder, E.H., Lannfelt, L., Viitanen, M., Corder, L.S., Manton, K.G., Winblad, B., and Basun, H. (1996). Apolipoprotein E genotype determines survival in the oldest old (85 years or older) who have good cognition. *Archives of Neurology, 53,* 418-422.

Costa, L.G., Vitalone A., Cole T.B., and Furlong C.E. (2005). Modulation of paraoxonase (PON1) activity. *Biochemical Pharmacology, 69*(4), 541-550.

de Craen, A.J., et al. (2005). Genetic influence of innate immunity: An extended twin study. *Genes and Immunity, 6,* 167-170.

de Maat, M.P., Bladbjerg, E.M., Hjelmborg, J.B., Bathum, L., Jespersen, J., Christensen, K., et al. (2004). Genetic influence on inflammation variables in the elderly. *Arteriosclerosis, Thrombosis, and Vascular Biology, 24,* 2168-2173.

Ershler, W.B, and Keller, E.T. (2000). Age-associated increased interleukin-6 gene expression, late-life diseases, and frailty. *Annual Review of Medicine, 51*, 245-270.

Ewbank, D.C. (2004). The APOE gene and differences in life expectancy in Europe. *The Journals of Gerontology Series A: Biological Sciences and Medical Sciences, 50*, B16-B20.

Ferrucci, L., Harris, T.B., Guralnik, J.M., Tracy, R.P., Corti, M.C., Cohen, H.J., Penninx, B., Pahor, M., Wallace, R., and Havlik, R.J. (1999). Serum IL-6 level and the development of disability in older persons. *Journal of the American Geriatrics Society, 47*, 639-646.

Finch, C.E., and Crimmins, E.M. (2004). Inflammatory exposure and historical changes in human life-spans. *Science, 305*, 1736-1739.

Fishman, D., Faulds, G., Jeffery, R., Mohamed-Ali, V., Yudkin, J.S., Humphries, S., and Woo, P. (1998). The effect of novel polymorphisms in the interleukin-6 (IL-6) gene on IL-6 transcription and plasma IL-6 levels, and an association with systemic-onset juvenile chronic arthritis. *Journal of Clinical Investigation, 102*, 1369-1376.

Franceschi, C., Bonafe, M., Valensin, S., Olivieri, F., De Luca, M., Ottaviani, E., and de Benedictis, G. (2000). Inflamm-aging. An evolutionary perspective on immunosenescence. *Annals of the New York Academy of Sciences, 908*, 244-254.

Frederiksen, H., and Christensen, K. (2003). The influence of genetic factors on physical functioning and exercise in second half of life. *Scandinavian Journal of Medicine and Science in Sports, 13*(1), 9-18.

Frederiksen, H., Gaist, D., Bathum, L., Andersen, K., McGue, M., Vaupel, J.W., and Christensen, K. (2003). Angiotensin I-converting enzyme (ACE) gene polymorphism in relation to physical performance, cognition, and survival: A follow-up study of elderly Danish twins. *Annals of Epidemiology, 13*, 57-65.

Gayagay, G., Yu, B., Hambly, B., Boston, T., Hahn, A., Celermajer, D.S., and Trent, R.J. (1998). Elite endurance athletes and the ACE I allele: The role of genes in athletic performance. *American Journal of Human Genetics, 103*, 48-50.

Geesaman, B.J., Benson, E., Brewster, S.J., Kunkel, L.M., Blanche, H., Thomas, G., Perls, T.T., Daly, M.J., and Puca, A.A. (2003). Haplotype-based identification of a microsomal transfer protein marker associated with the human lifespan. *Proceedings of the National Academy of Sciences, USA, 100*, 14115-14120.

Georges, J.L., Loukaci, V., Poirier, O., Evans, A., Luc, G., Arveiler, D., Ruidavets, J.B., Cambien, F., and Tiret, L. (2001). Interleukin-6 gene polymorphisms and susceptibility to myocardial infarction: The ECTIM study. (Etude Cas-Temoin de l'Infarctus du Myocarde). *International Journal of Molecular Medicine, 79*, 300-305.

Gerdes, L.U., Jeune, B., Ranberg, K.A., Nybo, H., and Vaupel, J.W. (2000). Estimation of apolipoprotein E genotype-specific relative mortality risks from the distribution of genotypes in centenarians and middle-aged men: Apolipoprotein E gene is a frailty gene, not a longevity gene. *Genetic Epidemiology, 19*, 202-210.

Goldman, N., Turra, C.M., Glei, D.A., Seplaki, C.L., Lin, Y., Weinstein, M. (2006). Predicting mortality from clinical and nonclinical biomarkers. *Journals of Gerontology Series A: Biological Sciences and Medical Sciences, 61A*(10), 1070-1074.

Haan, M.N., Shemanski, L., Jagust, W.J., Manolio, T.A., and Kuller, L. (1999). The role of APOE epsilon4 in modulating effects of other risk factors for cognitive decline in elderly persons. *Journal of the American Medical Association, 282*, 40-46.

Harris, T.B., Ferrucci, L., Tracy, R.P., Corti, M.C., Wacholder, S., Ettinger, W.H., Jr., Heimovitz, H., Cohen, H.J., and Wallace, R. (1999). Associations of elevated interleukin-6 and C-reactive protein levels with mortality in the elderly. *American Journal of Medicine, 106*, 506-512.

Herskind, A.M., McGue, M., Holm, N.V., Sorensen T.I., Harvald, B., and Vaupel, J.W. (1996). The heritability of human longevity: A population-based study of 2872 Danish twin pairs born 1870-1900. *American Journal of Human Genetics, 97*(3), 319-323.

Hjelmborg, J.B., Iachine, I., Skytthe, A., Vaupel, J.W., McGue, M., Koskenvuo, M., Kaprio, J., Pedersen, N.L., and Christensen, K. (2006). Genetic influence on human lifespan and longevity. *American Journal of Human Genetics, (119)3*, 312-321.

Hume, A.L. (2005). Safe and effective drug therapy in older adults. *Medicine and Health Rhode Island, 88*, 15-17.

Humphries, S.E., Luong, L.A., Ogg, M.S., Hawe, E., and Miller, G.J. (2001). The interleukin-6-174 G/C promoter polymorphism is associated with risk of coronary heart disease and systolic blood pressure in healthy men. *European Heart Journal, 22*, 2243-2252.

Ioannidis, J.P. (2003). Genetic associations: False or true? *Trends in Molecular Medicine, 9*(4), 135-138.

Jordan, B.D., Relkin, N.R., Ravdin, L.D., Jacobs, A.R., Bennett, A., and Gandy, S. (1997). Apolipoprotein E epsilon4 associated with chronic traumatic brain injury in boxing. *Journal of the American Medical Association, 278*, 136-140.

Keavney, B., McKenzie, C., Parish, S., Palmer, A., Clark, S., Youngman, L., Delepine, M., Lathrop, M., Peto, R., and Collins, R. (2000). Large-scale test of hypothesised associations between the angiotensin-converting-enzyme insertion/deletion polymorphism and myocardial infarction in about 5000 cases and 6000 controls. (International Studies of Infarct Survival (ISIS) Collaborators.) *Lancet, 355*, 434-442.

Kondo, S., Schutte, B.C., Richardson, R.J., Bjork, B.C., Knight, A.S., Watanabe, Y., Howard, E., de Lima, R.L., Daack-Hirsch, S., Sander, A., McDonald-McGinn, D.M., Zackai, E.H., Lammer, E.J., Aylsworth, A.S., Ardinger, H.H., Lidral, A.C., Pober, B.R., Moreno, L., Arcos-Burgos, M., Valencia, C., Houdayer, C., Bahuau, M., Moretti-Ferreira, D., Richieri-Costa, A., Dixon, M.J., and Murray, J.C. (2002). Mutations in IRF6 cause Van der Woude and popliteal pterygium syndromes. *Nature Genetics, 32*, 285-289.

Kritchevsky, S.B., Nicklas, B.J., Visser, M., Simonsick, E.M., Newman, A.B., Harris, T.B., Lange, E.M., Penninx, B.W., Goodpaster, B.H., Satterfield, S., Colbert, L.H., Rubin, S.M., and Pahor, M. (2005). Angiotensin-converting enzyme insertion/deletion genotype, exercise, and physical decline. *Journal of the American Medical Association, 294*, 691-698.

Lewis, S.J., and Brunner, E.J. (2004). Methodological problems in genetic association studies of longevity: The apolipoprotein E gene as an example. *International Journal of Epidemiology, 33*, 962-970.

Licastro, F., Pedrini, S., Caputo, L., Annoni, G., Davis, L.J., Ferri, C., Casadei, V., and Grimaldi, L.M. (2000). Increased plasma levels of interleukin-1, interleukin-6 and alpha-1-antichymotrypsin in patients with Alzheimer's disease: Peripheral inflammation or signals from the brain? *Journal of Neuroimmunology, 103*, 97-102.

Lorenz, D., Frederiksen, H., Moises, H., Kopper, F., Deuschl, G., and Christensen, K. (2004). High concordance for essential tremor in monozygotic twins of old age. *Neurology, 62*(2), 208-211.

Luft, F.C. (1999). Bad genes, good people, association, linkage, longevity, and the prevention of cardiovascular disease. *Clinical and Experimental Pharmacology and Physiology, 26*, 576-579.

Mackness, B., Mackness, M.I., Arrol, S., Turkie, W., and Durrington, P.N. (1998). Effect of the human serum paraoxonase 55 and 192 genetic polymorphisms on the protection by high-density lipoprotein against low-density lipoprotein oxidative modification. *FEBS Letters, 423*(1), 57-60.

Martin-Ruiz, C.M., Gussekloo, J., van Heemst, D., von Zglinicki, T., and Westendorp, R.G. (2005). Telomere length in white blood cells is not associated with morbidity or mortality in the oldest old: A population-based study. *Aging Cell, 4*, 287-290.

McGue, M., and Christensen, K. (2001). The heritability of cognitive functioning in very old adults: Evidence from Danish twins aged 75 and older. *Psychology and Aging, 16*, 272-280.

McGue, M., and Christensen, K. (2002). The heritability of level and rate-of-change in cognitive functioning in Danish twins aged 70 years and older. *Experimental Aging Research, 28*(4), 435-451.

Montgomery, H.E., Marshall, R., Hemingway, H., Myerson, S., Clarkson, P., Dollery, C., Hayward, M., Holliman, D.E., Jubb, M., World, M., Thomas, E.L., Brynes, A.E., Saeed, N., Barnard, M., Bell, J.D., Prasad, K., Rayson, M., Talmud, P.J., and Humphries, S.E. (1998). Human gene for physical performance. *Nature, 393,* 221-222.

Myerson, S., Hemingway, H., Budget, R., Martin, J., Humphries, S., and Montgomery, H. (1999). Human angiotensin I-converting enzyme gene and endurance performance. *Journal of Applied Physiology, 87,* 1313-1316.

Nauck, M., Winkelmann, B.R., Hoffmann, M.M., Bohm, B.O., Wieland, H., and Marz, W. (2002). The interleukin-6 G(-174)C promoter polymorphism in the LURIC cohort: No association with plasma interleukin-6, coronary artery disease, and myocardial infarction. *International Journal of Molecular Medicine, 80,* 507-513.

Nebel, A., Croucher, P.J., Stiegeler, R., Nikolaus, S., Krawczak, M., and Schreiber, S. (2005). No association between microsomal triglyceride transfer protein (MTP) haplotype and longevity in humans. *Proceedings of the National Academy of Sciences, USA, 102,* 7906-7909.

Panza, F., D'Introno, A., Colacicco, A.M., Capurso, C., Capurso, S., Kehoe, P.G., Capurso, A., and Solfrizzi, V. (2004). Vascular genetic factors and human longevity. *Mechanisms of Ageing and Development, 125,* 169-178.

Petersen, A.M., and Pedersen, B.K. (2005). The anti-inflammatory effect of exercise. *Journal of Applied Physiology, 98,* 1154-1162.

Pradhan, A.D., Manson, J.E., Rifai, N., Buring, J.E., and Ridker, P.M. (2001). C-reactive protein, interleukin 6, and risk of developing type 2 diabetes mellitus. *Journal of the American Medical Association, 286,* 327-334.

Puca, A.A., Daly, M.J., Brewster, S.J., Matise, T.C., Barrett, J., Shea-Drinkwater, M., Kang, S., Joyce, E., Nicoli, J., Benson, E., Kunkel, L.M., and Perls, T. (2001). A genome-wide scan for linkage to human exceptional longevity identifies a locus on chromosome 4. *Proceedings of the National Academy of Sciences, USA, 98,* 10505-10508.

Rauramaa, R., Vaisanen, S.B., Luong, L.A., Schmidt-Trucksass, A., Penttila, I.M., Bouchard, C., Toyry, J., and Humphries, S.E. (2000). Stromelysin-1 and interleukin-6 gene promoter polymorphisms are determinants of asymptomatic carotid artery atherosclerosis. *Arteriosclerosis, Thrombosis, and Vascular Biology, 20,* 2657-2662.

Ravaglia, G., Forti, P., Maioli, F., Chiappelli, M., Dolzani, P., Martelli, M., Bianchin, M., Mariani, E., Bolondi, L., and Licastro F. (2005). Associations of the-174 g/c interleukin-6 gene promoter polymorphism with serum interleukin 6 and mortality in the elderly. *Biogerontology, 6*(6), 415-423.

Reed, T., Dick, D.M., Uniacke, S.K., Foroud, T., and Nichols, W.C. (2004). Genome-wide scan for a healthy aging phenotype provides support for a locus near D4S1564 promoting healthy aging. *Journals of Gerontology Series A: Biological Sciences and Medical Sciences, 59,* 227-232.

Rexbye, H., Petersen, I., Johansen, M., Klitkou, L., Jeune, B., and Christensen, K. (2006). Influence of environmental factors on facial ageing. *Age and Ageing, 35,* 110-115.

Rigat, B., Hubert, C., Alhenc-Gelas, F., Cambien, F., Corvol, P., and Soubrier, F. (1990). An insertion/deletion polymorphism in the angiotensin I-converting enzyme gene accounting for half the variance of serum enzyme levels. *Journal of Clinical Investigation, 86,* 1343-1346.

Risch, N., and Zhang, H. (1995). Extreme sib pairs for mapping quantitative trait loci in humans. *Science, 268*(5217), 1584-1589.

Robak, T., Gladalska, A., Stepien, H., and Robak, E. (1998). Serum levels of interleukin-6 type cytokines and soluble interleukin-6 receptor in patients with rheumatoid arthritis. *Mediators of Inflammation, 7*, 347-353.

Ryff, C.D., and Singer, B.H. (2005). Social environments and the genetics of aging: Advancing knowledge of protective health mechanisms. *Journals of Gerontology Series B: Psychological Sciences and Social Sciences, 60*(Spec No I), 12-23.

Seres, I., Paragh, G., Deschene, E., Fulop, T., Jr., and Khalil, A. (2004). Study of factors influencing the decreased HDL associated PON1 activity with aging. *Experimental Gerontology, 39*(1), 59-66.

Silander, K., Komulainen, K., Ellonen, P., Jussila, M., Alanne, M., Levander, M., Tainola, P., Kuulasmaa, K., Salomaa, V., Perola, M., Peltonen, L., and Saarela, J. (2005). Evaluating whole genome amplification via multiply-primed rolling circle amplification for SNP genotyping of samples with low DNA yield. *Twin Research and Human Genetics, 8*(4), 368-375.

Skytthe, A., Pedersen, N.L., Kaprio, J., Stazi, M.A., Hjelmborg, J., Iachine, I., Vaupel, J.W., and Christensen, K. (2003). Longevity studies in GenomEUtwin. *Twin Research and Human Genetics, 6*(5), 448-454.

Swift, L.L., Zhu, M.Y., Kakkad, B., Jovanovska, A., Neely, M.D., Valyi-Nagy, K., Roberts, R.L., Ong, D.E., and Jerome, W.G. (2003). Subcellular localization of microsomal triglyceride transfer protein. *Journal of Lipid Research, 44*, 1841-1849.

Tan, Q., Kruse, T.A., and Christensen, K. (2006). Design and analysis in genetic studies of human ageing and longevity. *Ageing Research Reviews, (5)*4, 371-387.

Terry, C.F., Loukaci, V., and Green, F.R. (2000). Cooperative influence of genetic polymorphisms on interleukin 6 transcriptional regulation. *Journal of Biological Chemistry, 275*, 18138-18144.

Wierzbicki, A.S. (2003). New lipid-lowering agents. *Expert Opinion on Emerging Drugs, 8*, 365-376.

Williams, A.G., Rayson, M.P., Jubb, M., World, M., Woods, D.R., Hayward, M., Martin, J., Humphries, S.E., and Montgomery, H.E. (2000). The ACE gene and muscle performance. *Nature, 403*, 614.

Winterer, G., Hariri, A.R., Goldman, D., and Weinberger, D.R. (2005). Neuroimaging and human genetics. *International Review of Neurobiology, 67*, 325-383.

Yu, C.E., Oshima, J., Fu, Y.H., Wijsman, E.M., Hisama, F., Alisch, R., Matthews, S., Nakura, J., Miki, T., Ouais, S., Martin, G.M., Mulligan, J., and Schellenberg, G.D. (1996). Positional cloning of the Werner's syndrome gene. *Science, 272*, 258-262.

2

Whitehall II and ELSA: Integrating Epidemiological and Psychobiological Approaches to the Assessment of Biological Indicators

Michael Marmot and *Andrew Steptoe*

The Whitehall II study of British civil servants was set up with the explicit purpose of testing hypotheses as to the causes of the social gradient in cardiovascular and other diseases (Marmot and Brunner, 2005; Marmot et al., 1991). The scientific questions arose from key findings of the first Whitehall study: health followed a social gradient (Marmot, Shipley, and Rose, 1984). That is to say, the lower the position in the social hierarchy, the higher the mortality from cardiovascular disease and from a range of other major causes of death. Although the social gradient was defined in the Whitehall study by grade of employment, similar patterns emerge with classification on the basis of income or education (Mensah, Mokdad, Ford, Greenlund, and Croft, 2005). This gradient provided a particular challenge to explanation. One could not invoke the old notion of "executive stress" because people with more responsibility at work were at lower risk of disease. One could not look to poverty as an explanation because people second from the top of the hierarchy had worse health than those at the top—and so on all the way down the hierarchy—and it is an untenable hypothesis that higher executive officers in the British Civil Service, men (Whitehall II included women but the original Whitehall study was a women-free zone) with stable white-collar jobs, are somehow materially deprived.

Furthermore, the standard coronary risk factors—smoking, plasma cholesterol level, blood pressure, blood sugar level, and height (included as a marker of early life)—accounted for less than a third of the social gradient in cardiovascular disease (Marmot, Bosma, Hemingway, Brunner, and Stansfeld, 1997). Something else had to be going on. We hypothesized that the social gradient in disease occurrence could be attributed to psychosocial factors. In order to test this hypothesis, we set up the Whitehall II study: a longitudinal study of 10,308 men and women working in the British Civil Service ages 35-55 at baseline in 1985. Participants were recruited from the entire range of occupational grades, from senior civil servants responsible for large government programs to clerical workers, porters, and messengers.

In the biomedical world, the idea of social causation sounds mystical. How, a biomedical scientist wants to know, can someone's socioeconomic position get "under the skin" to cause disease (Adler and Ostrove, 1999)? Or, as a senior medical colleague put it to us, you will never convince medical scientists that people's social circumstances, and particularly psychosocial factors, influence health unless you have a biological pathway. An important part of the research agenda for Whitehall II is therefore to show how social and psychosocial factors influence biological pathways to cause social inequalities in disease.

Epidemiological studies are one, but not the only, research strategy to target this goal. Development and testing of hypotheses linking psychosocial factors to biological pathways come also from psychobiological studies (see below) in which smaller numbers of individuals are studied intensively, either in the laboratory or under naturalistic conditions, to link changes in emotion and behavior with changes in relevant biological markers. This is closely integrated with our overall scientific aim of understanding inequalities in health by studying civil servants from different levels of the hierarchy. There is thereby a conceptual link from Whitehall II to the psychobiology studies.

We also have a keen scientific interest in linking health, well-being, social participation, and economic and social circumstances in older people. To this end we set up ELSA, the English Longitudinal Study of Ageing (British spelling), very much influenced by, and modeled on, the Health and Retirement Study (HRS) in the United States (Marmot, Banks, Blundell, Lessof, and Nazroo, 2003). The sampling frame for ELSA was nationally representative surveys of the English population: the Health Surveys for England 1998, 1999, and 2001. The sample in ELSA comprises 11,392 men and women ages 50-72, with an additional 636 partners under 50, and 72 new partners were interviewed, leading to a total sample size of 12,100. The plan is to interview them at home every two years, and every second interview (i.e., every four years) will include a nurse visit in addition to the face-to-face interview. The nurse visit will allow physical function to be measured by a trained observer and blood samples to be drawn for biochemical analysis. One major difference from HRS is that the Health Survey data set contains biomarkers relevant to monitoring cardiovascular disease. The drawing of blood samples every four years means that these biomarkers can be assessed repeatedly and newer ones measured according to the scientific questions being addressed.

This chapter details the scientific insights that can be gained from measuring biological indicators in population studies and presents a rationale for complementing the observational epidemiological approach with more intensive psychobiological investigations.

BIOLOGICAL INDICATORS IN WHITEHALL II AND ELSA

The importance of biological markers can be illustrated by showing how they shed light on two issues of causation. Working with economists, we have found that the link between socioeconomic position and health has two major explanations. Public health scientists tend to start from the position that social circumstances associated with socioeconomic position lead to ill health. Economists tend to start from the position that health, or some resilience factor, leads to socioeconomic position. To sort between these competing positions, it is helpful to have further specification of a causal model. We, for example, posit that low social position is associated with increased exposure to psychosocial factors, which in turn activate the autonomic nervous system and the hypothalamic-pituitary-adrenal axis to influence metabolism and disease risk.

We do not, in other words, view the body as a black box with social conditions going in at one end and disease coming out the other or vice versa. Rather, we seek to gain evidence to support or refute our causal model. One major candidate to explain the social gradient in disease is the group of factors characterizing the metabolic syndrome: low high-density lipoprotein (HDL) cholesterol, high plasma triglyceride, high waist to hip ratio, and disturbances of insulin and glucose metabolism. To look at this in Whitehall II, we had to arrange for approximately 8,000 people to come to a clinic in a fasting state, have blood drawn, drink a standard glucose load, and have blood drawn again precisely two hours later—a mammoth undertaking. We performed this at the Phase 3 (5-year examination) of the Whitehall II cohort, repeated it at Phase 5 (10-year examination), and again at Phase 7 (15 years). We found that the metabolic syndrome showed a clear social gradient in men and women—lower grade, higher prevalence—making it a promising candidate to be a biological intermediary between social position and coronary heart disease (Brunner et al., 1997).

A further step in the causal model was to conduct a nested case-control study, the cases being people with the metabolic syndrome and the controls a sample from the Whitehall II cohort without. People with the metabolic syndrome had evidence of increased urinary output of metabolites of both cortisol and catecholamines, indicating more stress-related neuroendocrine activity (Brunner et al., 2002). We also examined heart rate variability to see if the metabolic syndrome was related to this indicator of adverse autonomic (sympathetic-parasympathetic) balance. We found that heart rate variability was less in cases of metabolic syndrome. Involvement of autonomic function was shown further by the finding that low employment grade was associated with low heart rate variability. The link to psychosocial factors was made explicit by the finding that people with low control at work had lower heart rate variability (high risk) than those with high control (Hemingway et al., 2005).

These findings shed light also on the question of confounding. When we showed, in Whitehall II, that low job control appeared to be an important link between low socioeconomic position and risk of coronary heart disease, we had to confront the question of whether low job control was confounded by some other causal exposure that was linked to socioeconomic position (Marmot et al., 1997). We have now shown that chronic work stress, measured as low control/high demands/low supports (known as isostrain), is related to the metabolic syndrome (Chandola, Brunner, and Marmot, 2006). This independent association completes an important link in the causal chain from chronic work stress to congenital heart disease and illustrates the strength of complementing observational epidemiological studies of social factors with biomarker assessment.

We are only now beginning to make use of the biological markers in ELSA. One example illustrates their usefulness. We performed a comparison of socioeconomic differences in morbidity in England (ELSA) and the United States (HRS). We were struck by the finding of a higher rate of morbidity in the United States, even among affluent groups in the two countries. The possibility of higher levels of diagnosis and reporting in the United States had to be considered. Our economist colleagues took the initiative to use the biological measures in ELSA and, for comparison as HRS has no biomarkers, took data from the National Health and Nutrition Examination Survey. The U.S. samples had higher levels of glycosylated hemoglobin, of plasma fibrinogen, of C-reactive protein, and of blood pressure. These results suggest that there is genuinely more illness in the United States. If we had an isolated finding of, say, higher levels of glycosylated hemoglobin in the United States, this would have raised the possibility of laboratory differences across countries. The facts that each of our biomarkers were adverse in the United States compared with England and each of the reported morbidity measures were higher make it likely that the intercountry differences are valid (Banks, Marmot, Oldfield, and Smith, 2006).

In our view these findings illustrate the value of working across disciplines and across "levels" of measurement, by which we mean economic, social, psychological, biological, and medical disease outcomes. Each discipline would have been impoverished without the input of the others in decisions of what to measure, how to measure them, how to analyze data, and interpretation.

A limitation of large-scale epidemiological studies that investigate biological pathways is the difficulty of examining the dynamics of biological markers. For example, if the interesting information will come from examining not the resting level of a biomarker but how it responds to a stress in the system, we may miss some relationships by using one measurement of a biomarker. It took a great deal of effort to measure how glucose and insulin responded to a glucose load administered in the fasting state. It would be difficult to repeat this by examining in 8,000 people how other biomarkers responded to a psychological stressor. To conduct this sort of investigation, smaller scale intensive studies are appropriate, although they have the drawback that they do not have the statistical power to examine disease end points.

WHITEHALL PSYCHOBIOLOGY STUDY

Psychobiological studies give us the opportunity to understand the connections between psychosocial factors and health-related biological responses in more detail than is possible in large-scale epidemiological research. There are two basic strategies used in psychobiological research. The first is psychophysiological or mental stress testing, in which biological responses are measured in response to standardized psychological or social stimuli (Steptoe, 2005). A wide range of challenging stimuli is employed, including cognitive and problem-solving tasks, simulated public speaking, and interpersonal conflict. These tests are usually carried out individually in the laboratory or clinic, with continuous monitoring of cardiovascular activity, sampling of saliva for the measurement of stress hormones like cortisol, and periodic blood sampling. The value of this experimental approach is that biological responses to psychosocial stimuli can be evaluated under environmentally controlled conditions, reducing many of the sources of confounding that might otherwise be present. Sophisticated measures can be employed that are difficult to assess in large-scale studies; for example, we have found that psychological stress stimulates transient disturbances of vascular endothelial function (thought to be important in the development of coronary atherosclerosis), up-regulation of the genes controlling release of inflammatory cytokines, and activation of blood platelets (Brydon et al., 2005; Ghiadoni et al., 2000; Steptoe et al., 2003c). The design of these studies does not assume that responses to these tasks are important per se, but rather that they represent the way in which people respond biologically to everyday challenges.

A more direct way of assessing responses in everyday life is to use a second strategy, namely ambulatory or naturalistic monitoring. The purpose of these methods is to investigate biological function in real life at work, at home, and in other situations, rather than when people are tested in the screening clinic or laboratory. We can also study the covariation between biology, activities, emotions, and behavior. The design depends on the availability of unobtrusive measures of biological activity in everyday life, with techniques depending on the health problem under investigation. Research into cardiovascular diseases primarily uses ambulatory monitoring devices for the measurement of blood pressure, heart rate, and heart rate variability. Research on stress often involves saliva sampling for the assessment of cortisol, while studies of musculoskeletal problems and pain utilize lightweight monitors of electromyographic activity (muscle tension).

The Whitehall psychobiology study used both mental stress testing and naturalistic monitoring to explore the dynamic relationships between the social gradient, psychosocial stress factors, and biological responses relevant to coronary heart disease. We focused on the hypothesis developed in the full Whitehall II study that biological stress responsivity is greater in lower than higher socioeconomic status groups and partly mediates differences in disease risk (Steptoe and Marmot, 2002). We recruited 129 middle-aged men and 109 women from the Whitehall II cohort to take part in this study. They were sampled from higher (n = 90), intermediate (n = 81), and lower (n = 67) grades of employment so that we could compare responses across the social gradient, and all the participants were in full-time paid employment. We excluded individuals with known coronary disease or diabetes, a task that was made simple by the detailed data collected from each person for more than a decade as part of the main Whitehall II study. The response rate was 55 percent. Participants attended a psychophysiological stress testing session and carried out ambulatory monitoring of blood pressure, heart rate, and salivary cortisol over a normal working day. Recruitment of the sample was relatively easy because participants were familiar with the Whitehall study setup. However, some of the people we asked did not take part since they thought it would be difficult to commit the time to this more intensive investigation.

The psychobiology study has generated interesting results relating psychosocial factors such as work stress, depression, and loneliness with biological function, but the main emphasis has been on socioeconomic position. Box 2-1 summarizes findings from the laboratory psychophysiological stress testing concerning socioeconomic position (Steptoe and

BOX 2-1
Psychophysiological Stress Testing and Socioeconomic
Position: Summary of Findings from the Whitehall
Psychobiology Study

Dynamic Effects
Prolonged activation or impaired poststress recovery in lower socioeconomic
status participants

- Systolic and diastolic blood pressure
- Heart rate
- Heart rate variability
- Factor VIII
- Plasma viscosity
- Blood viscosity

Larger stress reactions in lower socioeconomic status participants

- Interleukin-6
- Systolic and diastolic blood pressure (women only)

Static Effects
Higher stress levels in lower socioeconomic status participants, but no difference
in reactivity or recovery

- Plasma fibrinogen
- Von Willebrand factor
- Platelet activation
- Tumor necrosis factor α
- Interleukin-1 receptor antagonist
- Natural killer cell counts

Marmot, 2005). An intriguing result was that, by and large, the social gradient was not strongly related to the magnitude of biological responses to standardized tasks. However, lower socioeconomic status was characterized by delayed recovery or prolonged stress activation, so that biological responses remained elevated after tasks had been completed, rather than returning back to baseline levels promptly. For example, in comparison with high-status individuals, we calculated that the odds of impaired poststress recovery in the lower status group were 3.85 (95% CI, 1.48-10.0) for diastolic blood pressure, and 5.91 (CI, 1.88-18.6) for heart rate variability (Steptoe et al., 2002). As noted in Box 2-1, other biological measures did not show differential reactivity or recovery, but rather elevated levels in lower socioeconomic status participants throughout the protocol.

These findings have led us to think again about the role that biological stress responses might play in mediating social gradients in cardiovascular disease risk. Two simple models come to mind. The first is that there are differences in the magnitude and duration of biological responses between social groups that have developed through a lifetime of differential psychosocial challenge. The chronic allostatic load model formulated by McEwen (1998) suggests that the wear and tear imposed on biological regulatory systems as they are forced to respond repeatedly to life's demands ultimately provokes reduced adaptability and the failure of systems to operate within optimal ranges. The association between lower socioeconomic position and prolonged activation after tasks have ended is a manifestation of this pattern. The second possibility is that adverse health effects arise out of differential exposure to adversity. The biological systems of lower status groups receive more "hits" on a daily basis because of greater exposure to chronic stressors, failures of coping, or reduced psychosocial and material resources. The higher levels of activation both during stress and at rest observed in several biological parameters listed in Box 2-1 will contribute to increased disease risk.

Naturalistic monitoring studies complement the psychophysiological stress testing results while demonstrating how these processes play out in everyday life. We have found that ambulatory blood pressure was greater in the lower than higher occupational grade groups in the mornings of working days, even after controlling for physical activity levels, body weight, smoking, and other factors (Steptoe et al., 2003b). Early morning blood pressure levels have special significance for cardiovascular disease since there is a blood pressure surge soon after waking, which coincides with an increased incidence of cardiac events (Giles, 2006). We also found that the cortisol awakening response, the rise in cortisol over the first 20-40 minutes after waking in the morning, is greater in lower status groups (Kunz-Ebrecht, Kirschbaum, Marmot, and Steptoe, 2004). A heightened cortisol awakening response is also found in people experiencing work stress, financial strain, and depressive symptoms, and it appears to be an indicator of stress-related hypothalamic-pituitary-adrenocortical dysfunction (Steptoe, 2007).

There are two important limitations to the psychobiological studies that we have carried out so far. The first is that, although the study was quite large from the perspective of laboratory science, the sample size was still limited for addressing socioeconomic issues. Some of the findings summarized in Box 2-1 were graded across the three occupational status groups, while for others it was necessary to combine higher and intermediate groups and compare them with the lower occupational status group. Second, we have not shown associations with cardiovascular disease outcomes. Much larger samples need to be tracked over many years to document a relationship between disturbed stress responsivity and coronary heart disease end points, such as myocardial infarction or cardiac death. However, newer methods of imaging the coronary arteries have given us the opportunity to study the progression of subclinical coronary disease. A fresh study with a larger sample that began in 2006 is measuring psychobiological responses along with coronary artery calcification quantified using nuclear imaging. We will repeat this imaging of the coronary arteries after three years to provide direct evidence of the clinical significance of these processes.

INTERCHANGE BETWEEN EPIDEMIOLOGICAL AND PSYCHOBIOLOGICAL STUDIES

The nesting of psychobiological studies within the broader framework of the Whitehall II study has provided fertile ground for the exchange of methods and measures. Although the Whitehall II study is rooted in observational, population-based epidemiology and the psychobiology study has its origins in laboratory psychophysiology, each has gained from the association. For example, laboratory psychophysiological research has generally paid rather little attention to the sampling framework. Much work in this tradition continues to be carried out with select samples (notably college students), assuming that findings in such unique groups can be generalized to the rest of the population. The sampling framework provided by the Whitehall II cohort has allowed us to recruit participants into psychobiological testing who are much more representative of different social groups.

Findings from the Whitehall II study also give pointers to biological markers that we think would be interesting to study in more detail in laboratory work. For example, an association between elevated plasma fibrinogen and lower social class was identified in a pilot for the Whitehall II study, and this was subsequently confirmed in Whitehall II and other cohorts in the United States and Europe (Brunner et al., 1996; Markowe et al., 1985). Fibrinogen is involved both in thrombogenesis and in the stimulation of atherogenic cell proliferation, and elevated levels predict the development of coronary heart disease. We therefore assessed fibrinogen responses to acute stress in the psychobiology study and found that it increases acutely in response to stress, with higher levels in people from lower grades of employment (Steptoe et al., 2003a). In addition, we have seen that men who experience low levels of control at work show heightened fibrinogen stress reactivity, suggesting that it may be particularly relevant to the processes through which chronic work stress increases cardiovascular disease risk. We have also found that heightened fibrinogen stress reactivity predicts increases in ambulatory blood pressure over a three-year period (Brydon and Steptoe, 2005), a result that is consistent with emerging evidence for the role of inflammation in the development of hypertension.

Experience in the psychobiology study has also informed the measures in the Whitehall II and ELSA studies. An instance is cortisol assessment. Cortisol has long been recognized as an important stress hormone, but it has been difficult to measure in large-scale studies because of its marked diurnal variation. This means that it cannot be adequately assessed in single blood samples, and collecting urine samples over the night or day is practically inconvenient and potentially aversive. Sampling of saliva has revolutionized cortisol assessment, since repeated samples can be collected over the day, stored in domestic refrigerators, and returned to the investigators by post. We included saliva sampling over a working and weekend day in the psychobiology study, and it proved acceptable to participants as well as scientifically valuable (Kunz-Ebrecht, Kirschbaum, and Steptoe, 2004; Steptoe et al., 2003b). This experience has encouraged us to include salivary cortisol assessment in a recent phase of the Whitehall II study and in the ELSA cohort. We have successfully obtained saliva samples over the day from more than 5,000 participants in ELSA, with a response rate of 89 percent. In Whitehall II, we have data from about 4,600 people, representing 90 percent of those eligible for cortisol assessment. Each person was asked to collect six samples over the day, and 99.45 percent of these were collected. These response rates are very encouraging, given that participants had to take measures in their own time at home and at work. This indicates that even measurement of biological variables that require some effort from study participants may be feasible on a large scale with appropriate management of data collection.

The continuing interchange between epidemiology and psychobiology has also led us to analyze data in relation to emerging issues that were not near the top of the agenda when the studies were designed. A good example is the work we have been carrying out on the biological correlates of positive well-being. The notion that affective well-being or happiness has beneficial health effects over and above the mere absence of depression and distress has evolved over recent years with the development of the field of positive psychology (Pressman and Cohen, 2005). We realized that arguments concerning the role of positive states in health would be greatly strengthened by an understanding of the underlying biology. We utilized the data from the naturalistic monitoring phase of the psychobiology study to address this issue. As part of that study, we obtained ratings of happiness every 20 minutes over the working day, at the same time that blood pressure readings were taken by the automated monitor. We aggregated these over the working day, then tested associations with biological measures relevant to health.

We found that happiness was negatively associated with cortisol sampled over the day (Steptoe, Wardle, and Marmot, 2005). This relationship was independent of other factors known to influence cortisol, such as age, gender, smoking, body mass index, and grade of employment. It was also apparent on a weekend as well as a working day, and it was replicated in the same individuals three years later, suggesting a stable pattern (Steptoe and Wardle, 2005). Another interesting observation is that stress-induced fibrinogen responses were smaller among happier individuals. Effects all remained statistically significant when psychological distress (measured with the General Health Questionnaire) was included in the analysis. The results therefore suggest that affective well-being has favorable effects on biological responses relevant to health independently of the influence of negative psychological states.

LIMITATIONS TO THE USE OF BIOLOGICAL INDICATORS

The measurement of biomarkers is expensive, and collection of the data can be time-consuming and demanding both for the investigators and participants. The fact that a particular measure has been shown in animal research to relate to disease development, or that it predicts an outcome in clinical samples, is not a sufficient reason to embark on its assessment in a population cohort. Some biomarkers may be particularly significant at a specific point in the course of disease, while not being relevant at other stages. Other measures may be attractive from the scientific perspective but difficult to collect in practice. For example, there are currently no simple direct measures of sympathetic nervous system activity that are suitable for large-scale population studies. Measures such as heart rate or heart rate variability are stimulated by the sympathetic nervous system, but they are also influenced by other factors, while sweat gland activity indexes responses only in the cholinergic sudomotor component of the sympathetic nervous system. Urinary measures of catecholamine metabolites are possible, but these require 24-hour collections, as we have done in a subset of the Whitehall II cohort (Brunner et al., 2002). Such techniques are difficult to use on a large scale. The inclusion of biological indicators in a population study therefore requires thorough consideration of the literature to determine the relevance and potential value of the marker, consideration of the practicality of measurement, and careful thought about confounders.

Another fundamental issue concerns the dynamics of biological measures. Some indicators, such as heart rate, blood pressure, and cortisol, change rapidly over minutes or even seconds, while others, like C-reactive protein, are much more stable. This presents problems for both population and laboratory studies. In population studies, a very labile measure is difficult to capture with one or two assessments, as we have pointed out in the last section in relation to cortisol. Disappointing results have sometimes emerged from studies of psychosocial factors that have used one or two values to assess the biological marker. A good example comes from research on job stress, in which a relationship between job strain (high demands, low-control work) and blood pressure has emerged much more consistently from studies involving repeated measures over the day than from investigations that have used conventional clinical measures of blood pressure (Steenland et al., 2000).

From the perspective of psychobiological studies, important relationships between psychosocial experiences and biology may be missed if measures are taken at the wrong time. For instance, the cytokine interleukin 6 (IL-6) takes one to two hours to respond to psychological stress (Steptoe, Willemsen, Owen, Flower, and Mohamed-Ali, 2001). If blood samples are drawn before and immediately after a series of challenging stimuli lasting 15-20 minutes, then no IL-6 response will be observed. It is also becoming clear that stimuli of different intensities may be needed to elicit responses in different variables. An analysis of the experimental literature on cortisol responses to distress has shown that increases are much greater when social-evaluative tasks are used: situations like simulated public speaking or carrying out mental arithmetic in front of an audience (Dickerson and Kemeny, 2004). We have typically used much less challenging stimuli in our studies, since we have been interested in conditions that more closely reflect everyday life, and under these circumstances cortisol responses are very modest.

Another issue to be borne in mind is that the levels of biological markers are determined by multiple factors, of which current psychosocial experiences are only one set. Genetic factors, health status, demographic factors such as gender, age, and ethnicity, health behaviors, and clinical risk factor profiles may all be relevant. Even in the psychosocial domain, early life experience may exert an influence independently of current circumstances. For example, an analysis of the Dunedin birth cohort has recently shown that maltreatment in childhood is a predictor of adult inflammatory responses (including C-reactive protein and fibrinogen), independently of other early life risk factors, health behaviors like smoking, and adult stress and socioeconomic position (Danese, Pariante, Caspi, Taylor, and Poulton, 2007). In the light of these diverse influences, it is wise to expect that associations between biological markers and current social and economic experience will be moderate rather than strong.

PRACTICAL AND ETHICAL ISSUES

Data Collection and Participant Burden

Our main motive for carrying out both the Whitehall II and ELSA studies is to understand the role that social factors play in the development of disease. Data collection therefore involves face-to-face contact with participants and the assessment of physical measures, including blood sampling. There is danger that if these test sessions are too long or occur too frequently; participants and their employers may be discouraged from remaining in the studies. Data collection in the Whitehall II study involves assessment at a research clinic every five years, interspersed with postal questionnaire data collection approximately halfway through each quinquennium. The clinical assessment takes about half a day and involves each person moving through several research stations for the measurement of physical characteristics, blood pressure, lung function, cognitive ability, blood sampling, heart rate variability, and so on. The organization of these clinics is complicated, and the work is labor intensive, with different research assistants, technicians, and nurses involved in different tests. As the cohort ages, some participants find it difficult to travel to the research clinic, so home visits are carried out. The fieldwork for the ELSA study is conducted entirely through home visits, since many of the cohort are quite infirm, whereas the psychobiology study has involved for the participants half a day in the laboratory together with measurements in everyday life.

On one hand, the burden on participants is therefore rather greater than in demographic studies that do not involve biological sampling. On the other hand, participants benefit from these medical screening sessions, since health problems may be identified that would otherwise not be detected. Indeed, one of the major motives that volunteers have for continuing to take part is that they have a free health check every few years. In common with many cohort studies, there was a marked loss of about 16 percent of participants between the baseline of the Whitehall II study in 1985-1988 and the first clinical follow-up. Subsequently, attrition has been minimized by maintaining good relationships with the study cohort and keeping in contact between screening phases (Marmot and Brunner, 2005). All participants receive individual feedback on cardiovascular risk factors and other measures, and in addition we produce booklets containing results and research findings that are distributed to participants. Parenthetically, it should be noted that while the British National Health Service provides health care that is free at the point of access, routine health checks are not carried out, so the information we provide is useful to participants. Interestingly, we have discovered that involving cohort members in more intensive investigations does not put them off the whole enterprise or lead to sample attrition. Our analysis of the subgroup of Whitehall II participants who were involved in the psychobiology study suggests that, if anything, they were more rather than less likely to remain in the main study. This is partly because many people find these more intensive studies intrinsically interesting, and partly because they provide study members with even more detailed clinical information that helps them monitor their health status.

Ethical Issues

In Britain, there is a rather large anomaly in the issue of ethical approval for research. Although a social science survey does not need ethical approval, a health survey does—even if the health survey is limited to questionnaire data. One clear way to identify a social science survey as a health survey is to add biomarkers. It then needs ethical approval.

In general we have had little difficulty gaining approval for including biomarkers in population research. However, three general difficulties are worth highlighting. First, the entire process of ethical approval for research has become far more cumbersome and bureaucratic than in the past. We know of no researcher who fails to recognize the importance of independent ethical approval of research. Equally, we know of no researcher who thinks other than that the present system has become so tortuous that it is in danger of damaging the public good by making it needlessly difficult to gain ethical approval. It can never be ethical to engage the public in research if the science is of poor quality. But in our experience, panels are often composed of nonscientists or nonspecialists whose good intentions are not accompanied by the technical knowledge or relevant experience in the field necessary to make informed judgments. In such situations, approval of research is unnecessarily delayed.

Second, we have pursued the common practice of storing biological specimens. This means that when biological researchers develop a new hypothesis, it can be tested with stored specimens. It is, however, common practice for ethics committees to require that, as part of informed consent, participants be informed of *all* biomarkers that we plan to measure. This implies that if we decide that we could access stored data to test the hypothesis that, say, antibodies to the micro-organism *H. pylori* are associated with heart disease, we must go back to participants and ask permission for that specific test. It is a logistic nightmare with implications for participant cooperation.

Third is the specific and sensitive issue of genetic material. It is now common practice for this to be covered by a separate consent. There are unresolved issues that are still being debated as to how handle the sharing of genetic material within what is allowed by participants' informed consent.

Another ethical issue that arises with increasing frequency as the populations we study get older is that new information emerges about the health risks of individual participants. For example, during the course of clinical assessments, we have found that some people have high lipid levels of which they were not aware, or perhaps show disorders of glucose metabolism that indicate investigations for diabetes are warranted. In our new psychobiological study, we have identified some people with high levels of coronary calcification that suggest they are at particularly high risk for having a myocardial infarction. Even though our studies are observational and do not have an intervention component, it is clearly essential that we provide participants with risk data, so that they can decide in consultation with their general practitioners what steps should be taken. Such "interventions" pose knotty problems when it comes to statistical analysis of risk factors in relation to disease outcomes, but they cannot be avoided in clinically responsible research.

CONCLUSIONS: HARD AND SOFT DATA

There is a widespread belief that questionnaire data are "soft" and laboratory data and biomarkers are "hard." This is simplistic. First, they are neither hard nor soft but convey different information. Second, questionnaire data and biomarker data each have their own requirements for quality control. In the case of biomarkers, these requirements range from the circumstances under which the specimens are collected, to their mode of collection, to their handling, to the important issue of laboratory standardization. For all types of data, there are issues of reliability and validity, since all measurements have potential for error, bias, and contamination. In many ways, social scientists are more sensitive to these issues than some laboratory scientists, who appear to believe that the numbers produced by their instruments reflect the absolute truth. Social determinants of health are indeed hard to grasp, and there are many pitfalls in measurement and interpretation. But the integrated use of social epidemiological measures, biomarkers, and psychobiological approaches provides a means of triangulation on major problems of health and disease risk in an aging population.

ACKNOWLEDGMENTS

Michael Marmot is supported by a Medical Research Council professorship, and Andrew Steptoe is a British Heart Foundation professor. The Whitehall II study is supported by grants from the Medical Research

Council; the British Heart Foundation; the Health and Safety Executive; the National Heart, Lung, and Blood Institute; the National Institute on Aging; the Agency for Health Care Policy Research; and the John D. and Catherine T. MacArthur Foundation Research Networks on Successful Midlife Development and Socio-economic Status and Health. ELSA is supported by the National Institute on Aging and by several British government departments, including the Department for Education and Skills; the Department for Environmental, Food, and Rural Affairs; the Department of Trade and Industry; the Department for Work and Pensions; HM Treasury Inland Revenue; and the Office for National Statistics. The psychobiology studies are supported by the Medical Research Council and the British Heart Foundation. We thank all the participants in these studies and all the members of the study teams.

REFERENCES

Adler, N.E., and Ostrove, J.M. (1999). Socioeconomic status and health: What we know and what we don't. *Annals of the New York Academy of Sciences, 896*, 3-15.

Banks, J., Marmot, M., Oldfield, Z., and Smith, J.P. (2006). Disease and disadvantage in the United States and in England. *Journal of the American Medical Association, 295*, 2037-2045.

Brunner, E.J., Davey Smith, G., Marmot, M., Canner, R., Beksinska, M., and O'Brien, J. (1996). Childhood social circumstances and psychosocial and behavioural factors as determinants of plasma fibrinogen. *Lancet, 347*, 1008-1013.

Brunner, E.J., Marmot, M., Nanchahal, K., Shipley, M.J., Stansfeld, S.A., Juneja, M., and Alberti, K.G. (1997). Social inequality in coronary risk: Central obesity and the metabolic syndrome. Evidence from the Whitehall II study. *Diabetologia, 40*, 1341-1349.

Brunner, E.J., Hemingway, H., Walker, B.R., Page, M., Clarke, P., Juneja, M., Shipley, M.J., Kumari, M., Andrew, R., Seckl, J.R., Papadopoulos, A., Checkley, S., Rumley, A., Lowe, G.D., Stansfeld, S.A., and Marmot, M.G. (2002). Adrenocortical, autonomic, and inflammatory causes of the metabolic syndrome: Nested case-control study. *Circulation, 106*, 2659-2665.

Brydon, L., Edwards, S., Jia, H., Mohamed-Ali, V., Zachary, I., Martin, J.F., and Steptoe, A. (2005). Psychological stress activates interleukin-1 beta gene expression in human mononuclear cells. *Brain, Behavior, and Immunity, 19*, 540-546.

Brydon, L., and Steptoe, A. (2005). Stress-induced increases in interleukin-6 and fibrinogen predict ambulatory blood pressure at 3-year follow-up. *Journal of Hypertension, 23*, 1001-1007.

Chandola, T., Brunner, E., and Marmot, M. (2006). Chronic stress at work and the metabolic syndrome: Prospective study. *British Medical Journal, 332*, 521-525.

Danese, A., Pariante, C.M., Caspi, A., Taylor, A., and Poulton, R. (2007). Childhood maltreatment predicts adult inflammation in a life-course study. *Proceedings of the National Academy of Sciences, USA, 104*, 1319-1324.

Dickerson, S.S., and Kemeny, M.E. (2004). Acute stressors and cortisol responses: A theoretical integration and synthesis of laboratory research. *Psychological Bulletin, 130*, 355-391.

Ghiadoni, L., Donald, A., Cropley, M., Mullen, M.J., Oakley, G., Taylor, M., O'Connor, G., Betteridge, J., Klein, N., Steptoe, A., and Deanfield, J.E. (2000). Mental stress induces transient endothelial dysfunction in humans. *Circulation, 102,* 2473-2478.

Giles, T.D. (2006). Circadian rhythm of blood pressure and the relation to cardiovascular events. *Journal of Hypertension Supplement, 24,* S11-16.

Hemingway, H., Shipley, M., Brunner, E., Britton, A., Malik, M., and Marmot, M. (2005). Does autonomic function link social position to coronary risk? The Whitehall II study. *Circulation, 111,* 3071-3077.

Kunz-Ebrecht, S.R., Kirschbaum, C., Marmot, M., and Steptoe, A. (2004). Differences in cortisol awakening response on work days and weekends in women and men from the Whitehall II cohort. *Psychoneuroendocrinology, 29,* 516-528.

Kunz-Ebrecht, S.R., Kirschbaum, C., and Steptoe, A. (2004). Work stress, socioeconomic status, and neuroendocrine activation over the working day. *Social Science and Medicine, 58,* 1523-1530.

Markowe, H.L., Marmot, M., Shipley, M.J., Bulpitt, C.J., Meade, T.W., Stirling, Y., Vickers, M.V., and Semmence, A. (1985). Fibrinogen: A possible link between social class and coronary heart disease. *British Medical Journal, 291,* 1312-1314.

Marmot, M., Banks, J., Blundell, R., Lessof, C., and Nazroo, J. (Eds.) (2003). *Health, wealth, and lifestyles of the older population in England.* London, England: Institute of Fiscal Studies.

Marmot, M., Bosma, H., Hemingway, H., Brunner, E., and Stansfeld, S. (1997). Contribution of job control and other risk factors to social variations in coronary heart disease incidence. *Lancet, 350,* 235-239.

Marmot, M., and Brunner, E. (2005). Cohort profile: The Whitehall II study. *International Journal of Epidemiology, 34,* 251-256.

Marmot, M., Davey Smith, G., Stansfeld, S., Patel, C., North, F., Head, J., White, I., Brunner, E., and Feeney, A. (1991). Health inequalities among British civil servants: The Whitehall II study. *Lancet, 337,* 1387-1393.

Marmot, M., Shipley, M.J., and Rose, G. (1984). Inequalities in health: Specific explanations of a general pattern? *Lancet, I,* 1003-1006.

McEwen, B.S. (1998). Protective and damaging effects of stress mediators. *New England Journal of Medicine, 338,* 171-179.

Mensah, G.A., Mokdad, A.H., Ford, E.S., Greenlund, K.J., and Croft, J.B. (2005). State of disparities in cardiovascular health in the United States. *Circulation, 111,* 1233-1241.

Pressman, S.D., and Cohen, S. (2005). Does positive affect influence health? *Psychological Bulletin, 131,* 925-971.

Steenland, K., Fine, L., Belkic, K., Landsbergis, P., Schnall, P., Baker, D., Theorell, T., Siegrist, J., Peter, R., Karasek, R., Marmot, M., Brisson, C., Tuchsen, F. (2000). Research findings linking workplace factors to CVD outcomes. *Occupational Medicine, 15,* 7-68.

Steptoe, A. (2005). Tools of psychosocial biology in health care research. In A. Bowling and S. Ebrahim (Eds.), *Handbook of health research methods* (pp. 471-493). Maidenhead, Berkshire, England: Open University Press.

Steptoe, A. (2007). Cortisol awakening response. In G. Fink (Ed.), *Encyclopedia of stress, second edition* (Vol. 1, pp. 647-653). Oxford, England: Elsevier.

Steptoe, A., Feldman, P.M., Kunz, S., Owen, N., Willemsen, G., and Marmot, M. (2002). Stress responsivity and socioeconomic status: A mechanism for increased cardiovascular disease risk? *European Heart Journal, 23,* 1757-1763.

Steptoe, A., Kunz-Ebrecht, S., Owen, N., Feldman, P.J., Rumley, A., Lowe, G.D., and Marmot, M. (2003a). Influence of socioeconomic status and job control on plasma fibrinogen responses to acute mental stress. *Psychosomatic Medicine, 65,* 137-144.

Steptoe, A., Kunz-Ebrecht, S., Owen, N., Feldman, P.J., Willemsen, G., Kirschbaum, C., and Marmot, M. (2003b). Socioeconomic status and stress-related biological responses over the working day. *Psychosomatic Medicine, 65,* 461-470.

Steptoe, A., Magid, K., Edwards, S., Brydon, L., Hong, Y., and Erusalimsky, J. (2003c). The influence of psychological stress and socioeconomic status on platelet activation in men. *Atherosclerosis, 168,* 57-63.

Steptoe, A., and Marmot, M. (2002). The role of psychobiological pathways in socioeconomic inequalities in cardiovascular disease risk. *European Heart Journal, 23,* 13-25.

Steptoe, A., and Marmot, M. (2005). Socioeconomic position and coronary heart disease: A psychobiological perspective. In L. J. Waite (Ed.), *Aging, health, and public Policy.* New York: Population Council.

Steptoe, A., and Wardle, J. (2005). Positive affect and biological function in everyday life. *Neurobiology of Aging, 26* (Suppl 1), 108-112.

Steptoe, A., Wardle, J., and Marmot, M. (2005). Positive affect and health-related neuroendocrine, cardiovascular, and inflammatory processes. *Proceedings of the National Academy of Sciences, USA, 102,* 6508-6512.

Steptoe, A., Willemsen, G., Owen, N., Flower, L., and Mohamed-Ali, V. (2001). Acute mental stress elicits delayed increases in circulating inflammatory cytokine levels. *Clinical Science, 101,* 185-192.

3

The Taiwan Biomarker Project

Ming-Cheng Chang, Dana A. Glei,
Noreen Goldman, and *Maxine Weinstein*

It takes courage, optimism, and a remarkable degree of cooperation to field a study that engages colleagues from opposite ends of the earth. The Taiwan biomarker project—SEBAS (Social Environment and Biomarkers of Aging Study) for short—owes its success to dedicated teams in both Taiwan and the United States.

Our discussion in this chapter begins with an overview of the Taiwan biomarker project. We then review project logistics, summarize our substantive findings to date (and comment on the work we see in the near term), and offer some thoughts about how our experience might inform future research. Along the way, we engage in a candid discussion of the problems we have encountered.

OVERVIEW OF THE PROJECT

SEBAS builds on a longitudinal study of the elderly and near-elderly population of Taiwan. The Study of Health and Living Status of the Elderly in Taiwan was initiated by the Taiwan Provincial Institute of Family Planning (now the Bureau of Health Promotion, Department of Health) as a collaborative effort with Albert Hermalin at the University of Michigan. The first survey was conducted in 1989; 4,049 persons age 60 and older were interviewed (a response rate of about 92 percent) (Chang and Hermalin, 1989). The survey included eight modules on (1) marital history and other demographic characteristics; (2) household roster, social and economic networks, and exchanges; (3) health, health care utilization, and health-related behaviors; (4) occupational and employment histories; (5) activities and attitudes; (6) residential history; (7) economic and financial well-being; and (8) emotional and instrumental support (Weinstein and Willis, 2001). Since 1989, follow-up interviews have been conducted in 1993, 1996, 1999, and 2003. In both 1996 and 2003, the study drew a refresher sample to provide a sample of persons age 50 and older.

The initial impetus for the biomarker arm of the Taiwan study of the elderly grew out of a seminar on the cumulative effects of stress on health that was presented by Burton Singer at the Office of Population Research at Princeton University in 1995. The focus of his presentation was the MacArthur Study of Successful Aging—a study of predominantly high-functioning individuals drawn from community-based cohorts that were part of the Established Populations for Epidemiological Studies of the Elderly. The longitudinal study of the elderly in Taiwan seemed to offer an opportunity to do a population-representative study—albeit of persons middle-aged and older—that incorporated biomarkers. The study had been going on for some years, there was a strong base of sociodemographic data, the institute in Taiwan had a competent staff and substantial experience fielding surveys, we had a long and productive history of cooperative work with each other, and we knew that the study sample was cooperative and responsive.

We have already presented a (simplified) diagram of our basic theoretical model in the predecessor to this volume, *Cells and Surveys: Should Biological Measures Be Included in Social Science Research?* (Weinstein and Willis, 2001, p. 259). Underlying the study was our interest in exploring the (often) reciprocal relationships linking the social environment with stressful experience and with health outcomes, and in elaborating the physiological responses that lie between those links and between stressful experience and health outcomes. There are huge—and growing—literatures linking the social environment with exposure to challenge, linking the social environment with health outcomes, and some linking exposure to challenge with health outcomes. What we hoped to add to the discussion (primarily) were better data on the physiological pathways that lie between the environment and health outcomes and the physiological effects of exposure to challenge. Our original approach to incorporating physiological dysregulation was based on the concept of allostatic load. The idea behind allostatic load is that stressful experience causes a chain of physiological changes that interrupt normal processes; repeated or prolonged exposure to such stressors can result in physiological dysregulation (McEwen, 2002; McEwen and Stellar, 1993). Proponents of the framework would argue that allostatic load can be viewed as an index of the relative degree of failure at a physiological level—a marker of the cumulative physiological costs of efforts to cope with life's challenges.

The realization of allostatic load used in the few studies that had already been done by the time of the first biomarker collection in Taiwan was a simple index of 10 parameters (and our choice of biomarkers was driven by them) based on data from various domains of physiological regulation. Early in the MacArthur studies (and in some of our work that compares the Taiwan data with the MacArthur results), allostatic load was measured by summing the number of parameters for which individuals fell into the highest risk quartile over each of the components. In recognition of its limitations (e.g., equal weight given to each parameter, the definition of high risk as only a single tail of a given distribution, exclusion of parameters pertaining to immune function and, more generally, the limited number of markers), this formulation is evolving (Seplaki, 2004, 2005, 2006a) and undoubtedly will continue to evolve, as our understanding of physiological pathways increases. We have found, for example, that the simple score can be improved by incorporating additional biomarkers and allowing for two tails of risk (when relevant). At the same time, a simple index has the potential to mask important information disclosed by analysis of individual biomarkers. As discussed by Wachter (2001, p. 335), the measure is an example of "dimensionality reduction" that could be accomplished in a number of ways. Findings to date suggest that indicators over multiple systems provide greater power than single markers in understanding the costs and consequences of stressful experience (Cohen, 2000; Karlamangla, Singer, McEwen, Rowe, and Seeman, 2002; Seeman, McEwen, Rowe, and Singer, 2001). Still, the pre-2000 surveys provided data that allowed us to take a longitudinal look at the effects of the social environment on the physiological markers and the effects of the social environment on health. Our findings are discussed following a description of the project logistics.

PROJECT LOGISTICS

We did a pilot test of the biomarker collection in December 1997-January 1998. The idea was to assess feasibility: we obtained information regarding the logistics of collecting specimens and transporting them to the lab, training interviewers and staff, pretesting the instruments, and assessing participation rates (see Weinstein, Goldman, Hedley, Lin, and Seeman, 2003). SEBAS I—the first biomarker collection—was fielded between July and December 2000. We drew a random subsample of 1,713 persons ages 54 and older from those interviewed in 1999. Older persons (those age 71 and older in 2000) and persons in urban areas were over-sampled (for details see Goldman, Lin, Weinstein, and Lin, 2003 or Goldman, Glei, Turra, Glei, Lin, and Weinstein, 2006a). The study consisted of two parts: a face-to-face interview and a hospital-based examination. Our goal was to complete hospital examinations for 1,000 participants. The choice of a hospital setting for the tests was driven in large part by three factors. The first was the requirement in Taiwan that phlebotomy be performed by (or under the close supervision of) a physician. Second, we wished to provide some incentives for participation. The hospital setting allowed us to offer an abdominal ultrasound and provided the setting in which a physician could perform an examination similar to what the respondents would have received under National Health Insurance coverage. Third, our concern for the safety of the participants was the paramount consideration. We wanted to ensure that any problems could be addressed swiftly and effectively.

The face-to-face interviews were conducted by a local public health nurse who was well known and highly respected in the local area. A total of 1,497 persons responded to the home interview (93 percent of the elderly survivors and 91 percent of the near-elderly survivors). The in-home interview updated information on each respondent's living situation, employment, and marital status. It included questions on health (e.g., measures of depressive symptoms, cognitive function, activities of daily living, and instrumental activities of daily living), use of health care services, participation in social activities, stressful experiences and assessment of sources of stress and anxiety, and how respondents were affected by the 1999 earthquake (7.3 on the Richter scale and centered in Chi-Chi). The respondents also provided a subjective assessment of their position on a "social hierarchy ladder" separately for Taiwan as a whole and for their community. At the end of the interview, the interviewer evaluated whether the respondent's health would allow participation in the hospital protocol. Exclusion criteria included living in an institution, being seriously ill, needing a catheter or diaper, being on kidney dialysis, or having another health condition that would preclude blood drawing. About 7 percent of the respondents were ineligible for the examination. Eligible respondents were asked to participate in a health examination at a nearby hospital. The interviewer explained the protocol, scheduled the examination (for several weeks later), and arranged transportation if needed.

Hospitals were chosen by the Bureau of Health Promotion (BHP) based on reputation, accessibility, whether they had the interest—and capacity—to participate in the study, and whether they were within or close to a selected primary sampling unit (PSU). In the end, 24 hospitals were recruited; a few served more than one PSU. The night before the hospital appointment, a BHP staff member together with a public health nurse delivered a urine collection container (for an overnight 12-hour urine specimen) to the respondent's home, explained the proper procedures, provided written instructions for urine collection, and answered questions. They obtained informed consent for the health examination and reminded the participants not to eat anything from midnight until after the examination was completed. The next morning—the morning of the hospital examination—a member of the BHP met the participant at home, picked up the 12-hour urine specimen, and accompanied the participant to the hospital.[1]

During the hospital visit, a team of seven staff members coordinated the respondent's visit, processed the blood (collected at the hospital) and urine specimens, and processed the forms. Participants were asked about their health history, family disease history, health-related behaviors, and current long-term medications. A member of the team also confirmed that the participant had followed the urine collection procedure, had fasted since midnight, and did not have any contraindications to a blood draw. In addition, the team member reinterviewed the participant in order to obtain responses to questions that remained incomplete from the initial home interview and to resolve any inconsistencies in the home interview that were found when the staff reviewed and edited the questionnaire the previous evening.

At the hospital, the participant provided a spot urine specimen, a phlebotomist drew a blood specimen, and a nurse measured the participant's height, weight, waist and hip circumference, and blood pressure. Two blood pressure readings (about one minute apart with the respondent in a seated position) using a mercury sphygmomanometer on the right arm were taken at least 20 minutes after the participant arrived at the hospital. A physician performed a medical examination that included a third blood pressure reading, an abdominal ultrasound, and health counseling. Among participants in the medical exam, compliance with the clinical protocol was high: all but 10 individuals collected the 12-hour urine sample, provided a sufficient volume of blood for analysis, and completed the medical exam.

After the examination, participants were provided with breakfast, were given a small gift of nutritional supplements, and were accompanied back to their home. By noon on each day of the hospital visit, a staff member of Union Clinical Laboratories (UCL) based in Taipei collected the blood and urine specimens from the hospitals. UCL was responsible for transporting the specimens to Taipei. They followed standard lab procedures for the assays (details regarding the assays are provided in Seeman et al., 2004, and Goldman, Glei, Seplaki, Liu, and Weinstein, 2005) and returned the results to the BHP within two weeks.[2] Several weeks after the

---

[1]Some respondents chose to come to the hospital on their own or by taxi. In these cases, the respondents brought the urine collection container with them.

[2]In addition to the routine standardization and calibration tests performed by UCL, nine individuals (outside the target sample) contributed triplicate sets of specimens during the early stages of the fieldwork. In each case, two sets were submitted to UCL and a third was sent to Quest Diagnostics in the United States.

fieldwork, participants were sent the results of the standard tests based on the blood and spot urine specimens and the findings from the physical examination. Participants whose results were outside normal ranges were encouraged to see a physician for further examination and were informed about health counseling services available at the hospital.

Of the 1,497 respondents to the in-home interview, 1,023 participated in the hospital protocol (75 percent of the near-elderly and 61 percent of the elderly). Among the approximately 24 percent who declined to participate (as noted above, 7 percent were ineligible), the primary reasons were that the respondent felt that she or he was healthy and did not need an exam, that the exam was too much trouble, that she or he had just had a health exam, and that the respondent had no free time or was out of town during the several-day period that the exams were offered. Disproportionately high nonparticipation rates were found among the healthiest respondents and the least healthy. Overall, persons who received the medical exam reported the same average health status (on a five-point scale) as those who did not. Although respondents over age 70 were less likely than younger persons to participate, sex and measures of socioeconomic status were not significantly related to participation (Goldman et al., 2003).

A full list of the biomarkers and clinical measures that we obtained is presented in Box 3-1. Briefly, in addition to a hematology panel routinely collected as part of a health examination, we tested for total and high-density lipoprotein (HDL) cholesterol, glycosylated hemoglobin, dehydroepiandrosterone sulfate (DHEAS), IGF-1, and IL-6 from the blood specimens. We used the 12-hour urine specimens for assays of cortisol, epinephrine, norepinephrine, dopamine, and creatinine. At the suggestion of the study section that initially reviewed the proposal, we also added determination of apolipoprotein E (ApoE) genotype.

We stored three sets of aliquots of the specimens: one set at the BHP, one at UCL, and one at Georgetown University. Our experience with shipment to the United States was scary: one set of the reliability/validation specimens—fortunately not specimens from our sample—was held up in customs on the West Coast. This happened despite the fact that we had obtained all the necessary clearances. We also found that we had to ship the (sample) specimens from Taiwan to the United States in two separate shipments; evidently the amount of dry ice required for a single shipment exceeded the post-9/11 guidelines. The storage freezers are kept at −80 degrees C. All the storage facilities have emergency CO_2 back-up systems; the Georgetown specimens are housed at the Medical Center in a building that also has an emergency electrical generator. We have not yet received any requests for the use of the stored specimens (other than our own plans

BOX 3-1
Biomarkers and Clinical Measures Collected
During the SEBAS 2000

Measure

Physical Examination
  Anthropometry: height, weight, waist and hip circumference
  Systolic and diastolic blood pressure (3 readings)
  Examination of chest, heart rate, breathing, breasts, abdomen, arms, legs,
    lymph and thyroid glands for abnormalities (similar to National Health
    Insurance Exam)
  Abdominal ultrasound (liver, pancreas, gallbladder, kidneys)
Fasting Blood Sample
  Total and HDL cholesterol
  Glycosylated hemoglobin
  DHEA-S
  ApoE genotype
  Immune function and growth factor: IL-6, IGF-1,
  Other routine blood tests (e.g., blood cell counts, hemoglobin, glucose,
    triglycerides)
12-hour Urine Sample
  Cortisol
  Norepinephrine
  Epinephrine
  Dopamine
  Creatinine

SOURCE: Glei et al. (2006).

for additional assays); however, for the second round, we have made a provision for an advisory board to review such requests.

Moving forward to the second round of biomarker collection we have realized that one issue that we did not anticipate adequately is change in assay techniques. We recently discovered that standards for some assays have changed since 2000; indeed, they have changed to the extent that the reagents we used in 2000 are no longer available. UCL will be performing duplicate assays on the stored pilot study specimens with the new assays to see how well the results replicate the previous assays; however, at least for the assays that require fresh specimens, we will not be able to distinguish between the effects of sample deterioration and change in assay method.

An important question in adding biomarkers to general surveys relates to whether the additional burden imposed by the biomarker collection compromises participation in the parent study. We have no evidence that this has happened in Taiwan. The most recent round of household interviews, carried out by BHP as part of the Study of Health and Living Status, was done in 2003, based on a questionnaire similar to those in earlier waves of the survey. The response rate of 92 percent is consistent with rates from previous waves of the survey, suggesting that previous participation in the biomarker study did not affect participation rates in the main longitudinal study.[3] Our pretest for the second round of biomarker collection was conducted in September-October 2005. Of the 66 pretest participants in the household survey who were eligible for the hospital protocol, three reported that they were either tired of participating in the study as a whole or had a bad experience with the earlier (2000) biomarker study. It remains to be seen what will happen as we go into the field with the second round of the biomarker work.

Issues relating to human subjects and the process of informed consent are ongoing concerns for us. In particular, we are working with an elderly population, many of whom—particularly women—cannot read (31 percent overall, 50 percent among women). For the second round, we have consent procedures for respondents who are able to provide consent for themselves and separate procedures for those who require the assistance of a proxy. The low literacy rates affect what we can include in the protocol as well. A complex protocol, with regard to the urine collection for example, is difficult and time-consuming to explain. In the second round, we had hoped to collect three saliva specimens (at bedtime, next morning before getting out of bed, and 30 minutes after waking up), but our pretest suggested that participants, particularly illiterate ones, simply could not handle both sets of instructions (for urine and saliva).

The data from SEBAS I have been released through the Inter-university Consortium for Political and Social Research. The files include almost all the information that was collected at the time of the National Institutes of Health–funded 2000 study, along with some basic demographic data that were collected during the previous waves of the study (funded by the Taiwanese government). In order to protect the identity of the participants, some data were not included in the public release; we replaced identifiers for respondent, town of residence, PSU, hospital, examining physician, and town where respondent was during the earthquake with randomly generated identifiers, and we excluded other identifiers (i.e., for interviewer, type of organization in which the respondent works) and all of the

---

[3]Among the SEBAS participants, the response rate in 2003 was 97 percent. This very high response rate may be attributable, however, to the fact that it is based only on those who did not refuse the examination in 2000 and who were not lost to follow-up by 2000.

information regarding the arrangements for the hospital visit and reasons for nonparticipation. We did not anticipate the complex issues involved in responding to requests for cross-walks between the biomarker data and previous waves of the study. Apart from the 2000 wave of the study, which has been made publicly available as described above, the longitudinal data are the intellectual property of the BHP. Researchers who want to use the data can apply to the BHP for their use; upon approval, such researchers have been provided copies of (some) data.

The problem arises because the biomarker study does not have any contractual arrangements with these researchers regarding release or dissemination of the longitudinal data. If those data have already been shared with other researchers, and if we were to provide a cross-walk with the biomarker data, the possibility of deductive identification of the participants becomes very real indeed. A second problem—and we are working on this issue now—concerns the release of a second round of data. Again, the greatest concern is the need to protect our participants from deductive identification. We are working with Jim McNally, director of the National Archive of Computerized Data on Aging, and his staff in preparation for this release. There is also the need for ongoing support and funding by the research group—support that may not be available after the completion of the grants that fund the Taiwan project. This funding concern is also relevant to the storage and review of the use of the biological specimens. We have not yet identified a solution; however, the Behavioral and Social Research Branch of the National Institute on Aging is currently investigating these issues; we hope for some guidelines—and some support—as we go forward. The need for a curator (for the archived specimens and for the data) to ensure that these resources continue beyond the life of the grant (or the original researchers, for that matter) is very real indeed. Such an effort cannot be sustained indefinitely out of pocket of the research grant or the investigators.

This section on logistics would not be complete without acknowledging the work of our colleagues in Taiwan. We cannot overstate the contributions of the dedicated staff at the BHP. The three team leaders—Yu-Hsuan Lin, Shu-Hui Lin, and I-Wen Liu—and their supervisor, Yi-Li Chuang (now director of the Population and Health Research Center of the BHP) were remarkable. The interviewers, field workers, drivers, supervisors—absolutely everyone—pitched in to ensure the success of the project.

SUBSTANTIVE FINDINGS AND NONFINDINGS

Our study design integrated both biological and survey data from a population-representative sample; here we focus on results based on both sources, although our discussion includes findings from all aspects of the design. In Taiwan—as elsewhere—the social environment matters for an individual's health, but, unlike some previous work, we found that not all aspects of social ties or social support seem to matter. Specifically, contact with friends and participation in social or religious activities are associated with lower rates of functional disability, better survival, better self-rated health, fewer depressive symptoms, better cognitive function, and lower levels of allostatic load (Beckett, Goldman, Weinstein, Lin, and Chuang, 2002; Cornman, Goldman, Weinstein, and Chang, 2003; Glei et al., 2005; Goldman, Glei, and Chang, 2004a; Seeman et al., 2004; Yeager et al., 2006). Contrary to expectation—especially in light of the traditional importance of the extended family system and filial piety in Taiwan—coresidence patterns, the number of children, and contact with these children reveal little, if any, association with these health outcomes. In addition to highlighting the importance of measuring different components of the social environment, the results of these studies demonstrate the need to include a broad range of health outcomes. For example, whereas *perceptions* about social support matter little for measures of physical well-being in this population, these perceptions appear to be protective against depressive symptoms (Cornman et al., 2003). Not only are adverse aspects of the social environment associated with health decline and mortality, but also a socially rich environment is significantly related to maintaining health (Beckett et al., 2002).

Our research has also demonstrated that standard socioeconomic measures, most notably education and income, are related to health. A recent analysis, however, shows that the biomarkers measured in SEBAS do not account for the relationship between socioeconomic status (SES) and the two health outcomes considered—overall self-rated health and mobility difficulties (Dowd and Goldman, 2006). In particular, there is no evidence that sustained activation of neuroendocrine markers, including cortisol, is an important mediator in the relationship between SES and health. These results place an increased burden of proof on researchers who argue that psychosocial stress is an important link in the pathway linking low SES to poor health.

In an effort to gain new insights into social disparities in health, the Taiwan survey incorporated a recently developed instrument of subjective social position. This measure asks respondents to use the visual aid of a ladder to position themselves relative to other people in their community and society. An evaluation of this instrument reveals that the ladder captures diverse aspects of respondents' lives that extend beyond the conventional indicators of education, occupation, income, and wealth (Goldman, Cornman, and Chang, 2006). Cross-sectional analyses of the social determinants of health suggest that the ladder has a stronger association with health outcomes than the conventional measures of SES, but we await confirmation of these findings from longitudinal data (Hu, Adler, Goldman, Weinstein, and Seeman, 2005).

A question of interest—particularly to those of us doing comparative work—is whether biological profiles are relatively constant across broad populations or whether they reveal large variations, potentially attributable to environmental factors. A detailed comparison among SEBAS, the MacArthur study, and the Wisconsin Longitudinal Survey shows that with regard to risk factors for cardiovascular disease, men are at a clear disadvantage vis-à-vis women in the United States but not in Taiwan. The disparate findings for these two populations lead us to speculate that environmental factors, such as cultural practices or social roles (e.g., higher levels of stratification by sex in Taiwanese society), as well as inherent sex differences, affect these biomarkers (Goldman et al., 2004). In contrast, our analysis of the ApoE gene demonstrates that not all intercountry variation can be attributed to social factors. SEBAS provides the first estimates of the frequency of ApoE alleles for a national sample of the Taiwanese population. The results support earlier evidence based on select samples (Gerdes, Klausen, Sihm, and Foergeman, 1992) that Chinese populations have substantially lower frequencies of the ApoE-4 allele of this gene (a risk factor for Alzheimer disease and ischemic heart disease) than most other national and ethnic groups and underscore the need to obtain better information on the prevalence of dementia in Chinese populations.

When we look at the links between the social environment and physiology and between challenge and physiology, we find generally smaller effects than those found in the MacArthur study. Seeman et al. (2004), for example, used the longitudinal data in SEBAS on levels of social integration and extent of social support to predict allostatic load in 2000. Only a small set of significant links emerged from this analysis. The analysis by Goldman et al. (2005) of perceived stress and its relation to physiological dysregulation identified several biomarkers for which high or low values were significantly associated with perceptions of stress. At the same time, however, they found that, in general, the associations between indexes of perceived stress (at baseline and at earlier waves) and physiological dysregulation were small. We are still struggling with a generally unwieldy set of analyses exploring associations between stressful experiences (such as the death of family members, relocation, and financial difficulties) and physiological dysregulation, as well as the potential moderating role of "vulnerability," defined in terms of social position, social networks, and coping mechanisms. Still, our analyses identified the potential importance of traumatic experiences among the older population: questions that assess losses and reports of distress resulting from the 1999 earthquake show that damage to the home is significantly associated with an increase in depressive symptoms, most notably among middle-aged women (Seplaki, Goldman, and Weinstein, 2006b).

The generally weak findings regarding links between the biomarkers and the social environment (including environmental stressors) convinced us of the need to include a broader set of questions related to stressful experiences and perceptions of stress in the second round of the biomarker collection. We have added questions related to traumatic experience, perceptions of stress, caregiving, major life events, and daily hassles. Certainly, in terms of evaluating the framework of allostatic load, these are critical pieces. Our feeling is that much of the earlier work discussing allostatic load—which found associations between SES and allostatic load—*assumed* that lower SES was associated with greater exposure to challenge (or possibly that lower SES was associated with a poorer response to challenge).

It seems like a reasonable assumption, but it remains one that needs to be tested. The data in SEBAS II, along with the work of the National Survey of Midlife Development in the United States (MIDUS) II study, which includes a broad range of psychosocial measures along with similar biomarkers, may provide some insights into these questions. Of course, deficient measures of stressful experience may not be the only reason we are not finding strong associations. It may be that they simply aren't there, or—as we discuss below—we are not collecting the "right" biomarkers. It may also be unreasonable to expect to find the same relationships across societies. Taiwan, for example, may be a more socially integrated and less stratified society than the United States.

Finally, as noted above, we have recently been able to explore links between our measures of physiological dysregulation and downstream health outcomes. Based on mortality data for SEBAS I respondents between 2000 and 2003, Turra et al. (2005) demonstrated that biomarkers are predictive of survival even in the presence of extensive controls for sociodemographic factors and self-reported measures of physical and mental health. These biomarkers include both clinical markers—the cardiovascular and metabolic system measures customarily collected during physical examinations, which have well-defined clinical thresholds for normal function—and nonclinical markers—measures of neuroendocrine and immune dysfunction.

The findings of this and an earlier analysis suggest that individuals may underestimate their probability of dying because they have no information about risk factors that act silently on the body and are not detected by clinical exams (Goldman, Glei, and Chang, 2004a). Additional results demonstrate that the nonclinical markers appear to be better predictors of mortality than the clinical markers (although the small number of deaths in the three-year period makes us somewhat wary of pressing this claim), while the clinical markers generally have stronger relationships with (nonfatal) health outcomes. The findings also suggest that the physiological effects of the nonclinical measures, considered as primary mediators hypothesized to affect cardiovascular and metabolic outcomes in the allostatic load framework, are broader than those captured by the clinical markers in this analysis (Goldman et al., 2006a, 2006b).

The recent availability of longitudinal information on health outcomes (from the 2003 survey and from registered deaths between 2000 and 2003) has enabled us to compare some of our findings based on analyses of the 2000 data with similar analyses that use health outcome data assessed three years after the collection of the biomarkers and include health controls at baseline. We find some potentially important discrepancies between cross-sectional and longitudinal results. Whereas analyses by Seplaki et al. (2004) based on the 2000 data indicate that the neuroendocrine and immune measures have significant associations with physical and mental function, the more recent analysis using reports of health in 2003 does not find these effects (Goldman et al., 2006a); one caveat is that the two studies used different types of dysregulation scores to capture extreme values of these nonclinical markers (grade-of-membership analysis versus more conventional cumulative dysregulation scores).

In a separate set of analyses relating values of DHEAS to a range of health outcomes, estimates from statistical models based on 2000 data suggest stronger associations between high values of DHEAS and better health for women than for men (Glei et al., 2004). In contrast, statistical models that include measures of DHEAS and baseline health in 2000 to predict health outcomes in 2003 generally show that these associations exist for men, but not women. These two sets of comparisons underscore the limitations of cross-sectional analyses—in particular, the potential to overestimate associations between biomarkers and health because of possible reverse effects of poor health status on the biomarkers. The DHEAS results also raise intriguing questions as to why these types of biases may differ by sex.

LIMITATIONS

Of course, like all studies, our study has limitations. In terms of looking at the effects of stress on health, one might wonder whether an older population is the right target sample. Work by Crimmins and her colleagues on the data from the National Health and Nutrition Examination Survey (Crimmins, Johnston, Hayward, and Seeman, 2003) suggests that allostatic load flattens with age; more generally, some research suggests that SES gradients in health diminish with age (House et al., 1990, 1994). By using a sample age of 54 and older, have we missed the ages at which stressful experience and other social factors may have the greatest effect? We have some reassurance on this score: the probability of survival from ages 20 to 50 in 1970 was 0.90 for men and 0.94 for women (http://www. moi.gov.tw/stat/english/elife/1970.htm). Thus, while it seems reasonable to suppose that the effects of selective mortality are small, we cannot rule it out.

What about the biomarkers themselves? Did we collect the right biomarkers? Did we collect enough biomarkers? Are the generally small effects of our measures of stress on the biomarkers a result of having chosen markers that are relatively insensitive? Some biomarkers, cortisol for example, have substantial diurnal variation; our use of the overnight urine for this measure was intended to capture basal levels of production, but perhaps we are missing important information about more transient levels of response or recovery time.

At the outset of any project, it is difficult—perhaps impossible—to anticipate all the future uses of the data. Like all research, we began with a set of questions in mind and then discovered that we had additional questions. Even when we started, we knew that—ideally at least—some additional biomarkers would be desirable. The amount of blood we could collect was limited and assays are expensive; collection of cerebral spinal fluid is out of the question for a population-based study. Is there a better way of collecting a broad spectrum of markers? Dried blood spots are now being used successfully in a number of studies; it is a technique that allows collection in the home, and while we would have been unable to perform all the assays we wanted, our understanding is that additional assays based on blood spots are being developed. We are currently exploring the possibility of using tandem mass spectrometry on our stored serum specimens to identify peptides associated with the stress response; plasma would be an alternative compartment. Other approaches suggested by Singer and his colleagues (Singer, Ryff, and Seeman, 2004) include identifying time-related metabolic changes from, for example, urine specimens.

Clearly, a major limitation affecting our existing analyses is that we have measured the biomarkers only once; the second round in 2006 will be of great importance, although we suspect that it will raise many new questions and concerns. An important improvement in our second round is that we will also be measuring some markers in the home (blood pressure, lung function, grip strength, timed walks, and chair stands). These additional measurements should help us assess more accurately the magnitude of nonresponse bias affecting estimates derived from the hospital-based biomarker collection in the first round of the survey. In the case of blood pressure, the use of home assessments along with measurements in the hospital is likely to provide us with information about the degree of white coat hypertension (involving people's response to medical professionals), which we believe affected estimates in the first round.

We know, too, that many of our biomarkers are affected by medication use; we have some of the relevant information that would allow us to disentangle these effects, but it is a complicated undertaking, made more complex by the difficulties in obtaining complete lists of medications, documenting the extensive array of traditional Chinese medicines, and identifying the effects of both traditional and Western medicines on the biomarkers.

We remain concerned about the best way to measure physiological dysregulation; current scores—our own and those of other researchers—continue to have problems in terms of choice of cut points in the absence of clinical information, as well as potential biological interactions among these biomarkers. The problems with interactions among the biomarkers are exacerbated by limitations of conventional statistical models: they are not well designed to incorporate large amounts of information over successive waves, and we may not have the statistical power to get at some of the processes that interest us.

SUMMARY

Have we contributed to demographic knowledge? We would say yes. First, our analysis of SEBAS confirms earlier findings that the inclusion of biomarkers in household interview surveys improves the accuracy of the resulting health information (Goldman et al., 2003). Second, we demonstrate that the incorporation of biomarkers into statistical models of mortality substantially enhances the accuracy of the predictions, even in the presence of extensive control variables for self-reports of physical, mental, and cognitive health and sociodemographic information. Third, our exploratory analyses show that biological information can be used to identify biomarkers (and physiological systems) that do and do not account for demographic differentials in health and mortality (e.g., by sex), but that this may raise as many questions as it answers. Fourth, it is too early for us to answer the really big question: Does the inclusion of biomarkers in household surveys help us to understand SES differences in health, particularly with regard to the role of stressful experience? As described above, our research to date indicates that, at least among the older population in Taiwan, reports of stressful experience or perceptions of stress are only modestly correlated with the biological measures included in SEBAS and that these measures account for little of the association between SES and health outcomes.

Overall we think that, despite the limitations and occasional mistakes, and despite the logistical and financial constraints, the Taiwan study has been remarkably successful in achieving many of its objectives. Would we argue that all social surveys should collect biomarkers? Certainly not. Our experience has shown us that it is a complex task, not to be undertaken by the faint of heart. Because no single study can be all things to all people, we would suggest that considerable care be given to the choice of biomarkers and especially to archiving specimens for future use when the inevitable new questions arise and better technologies become available for the analysis of those specimens.

ACKNOWLEDGMENTS

The work on this project has been financed by the Office of Behavioral and Social Science Research of the National Institute on Aging under grant numbers R01AG16661 and R01AG16790. We gratefully acknowledge their support. The project has required a team with a wide range of skills across multiple disciplines. We particularly wish to acknowledge the contributions of Teresa Seeman, who has been an important collaborator throughout the project. Her advice and cooperation regarding criteria for our choice of laboratory and issues regarding choice of assay were invaluable, particularly during the early stages of the project. She generously provided the MacArthur protocols for blood and urine collection; these protocols served as the basis for the ones we used in Taiwan. Our technical decisions have also benefited from the accumulated wisdom of Chris Coe and Paul Aisen, who provided guidance on assays. Chris Peterson helped us through the difficulties of the public release of the first round of data. I-Fen Lin provided help with questionnaire design and has been indispensable for translation issues. Burt Singer and Germán Rodríguez have provided outstanding statistical guidance. Jennifer Cornman, Christopher Seplaki, and Cassio Turra have been wonderful collaborators at different stages of the project. Kaare Christensen has provided significant assistance with the protocols for the home-based assessments of function in the second round of the study. We extend our thanks to all of them.

REFERENCES

Beckett, M., Goldman N., Weinstein, M., Lin, I.-F., and Chuang, Y.-L. (2002). Social environment, life challenge, and health among the elderly in Taiwan. *Social Science and Medicine, 55*(2), 191-209.

Chang, M.-C., and Hermalin, A. (1989). The 1989 survey of health and living status of the elderly in Taiwan: Questionnaire and survey design. *Comparative Study of the Elderly in Four Asian Countries*. Research report 1. Ann Arbor: Population Studies Center, University of Michigan.

Cohen, J.I. (2000). Stress and mental health: A biobehavioral perspective. *Issues in Mental Health Nursing, 21,* 185-202.

Cornman, J.C., Goldman, N., Weinstein, M., and Chang, M.-C. (2003). Social ties and perceived support: Two dimensions of social relationships and health among the elderly in Taiwan. *Journal of Aging and Health, 15,* 616-644.

Crimmins, E.M., Johnston, M., Hayward, M., and Seeman, T. (2003). Age differences in allostatic load: An index of physiological dysregulation. *Experimental Gerontology, 38*(7), 731-734.

Dowd, J.B., and Goldman, N. (2006). Do biomarkers of stress mediate the relationship between socioeconomic status and health? *Journal of Epidemiology and Community Health, 60,* 633-639.

Gerdes, L.U., Klausen, I.C., Sihm, I., and Færgeman, O. (1992). Apolipoprotein E polymorphism in a Danish population compared to findings in 45 other study populations around the world. *Genetic Epidemiology, 9,* 155-167.

Glei, D.A., Chang, M.-C., Chuang, Y.-L., Lin, Y.-H., Lin, S.-H., Liu, I.W., Lin, H.-S., Goldman, N., and Weinstein, M. (2006, May). Results from the social environment and biomarkers of aging study (SEBAS) 2000: Survey Report (Chinese/English). *Taiwan Aging Study Series, 9.*

Glei, D., Goldman, N., Weinstein, M., and Liu, I.-W. (2004). Dehydroepiandrosterone sulfate (DHEAS) and health: Does the relationship differ by sex? *Experimental Gerontology, 39,* 321-331.

Glei, D., Landau, D.A., Goldman, N., Chuang, Y.-L., Rodríguez, G., and Weinstein, M. (2005). Participating in social activities helps preserve cognitive function: An analysis of a longitudinal, population-based study of the elderly. *International Journal of Epidemiology, 34,* 864-871.

Goldman, N., Cornman, J., and Chang, M.-C. (2006). Measuring subjective social status: A case study of older Taiwanese. *Journal of Cross-Cultural Gerontology, 21,* 71-89.

Goldman, N., Glei, D., and Chang, M.-C. (2004). The role of clinical risk factors in understanding self-rated health. *Annals of Epidemiology, 14,* 49-57.

Goldman, N., Glei, D., Seplaki, C., Liu, I.-W., and Weinstein, M. (2005). Perceived stress and physiological dysregulation. *Stress, 8,* 95-105.

Goldman, N., Lin, I.-F., Weinstein, M., and Lin, Y.-H. (2003). Evaluating the quality of self-reports of hypertension and diabetes. *Journal of Clinical Epidemiology, 56*(2), 148-154.

Goldman, N., Turra, C.M., Glei, D.A., Lin, Y.-H., and Weinstein, M. (2006a). Physiological dysregulation and changes in health in an older population. *Experimental Gerontology, 41,* 862-870.

Goldman, N., Turra, C., Glei, D., Seplaki, C., Lin, Y.-H., and Weinstein, M. (2006b). Predicting mortality from clinical and non-clinical biomarkers. *Journal of Gerontology: Medical Sciences, 61*(10), 1070-1074.

Goldman, N., Weinstein, M., Cornman, J., Singer, B., Seeman, T., and Chang, M.-C. (2004). Sex differentials in biological risk factors for chronic disease: Estimates from population-based surveys. *Journal of Women's Health, 13,* 393-403.

House, J.S., Kessler, R.C., Herzog, A.R., Mero, R.P., Kinney, A.M., and Breslow, M.J. (1990). Age, socioeconomic status, and health. *The Milbank Quarterly, 68*(3), 383-311.

House, J.S., Lepkowski, J.M., Kinney, A.M., Mero, R.P., Kessler, R.C., and Herzog, A.R. (1994). The social stratification of aging and health. *Journal of Health and Social Behavior, 35*(3), 213-234.

Hu, P., Adler, N., Goldman, N., Weinstein, M., and Seeman, T. (2005). Relations between subjective social status and measures of health in older Taiwanese persons. *Journal of the American Geriatrics Society, 53,* 483-488.

Karlamangla, A.S., Singer, B.H., McEwen, B.S., Rowe, J.W., and Seeman, T.E. (2002). Allostatic load as a predictor of functional decline. *Journal of Clinical Epidemiology, 55*(7), 696-710.

McEwen, B.S. (2002). Sex, stress, and the hippocampus: Allostasis, allostatic load, and the aging process. *Neurobiology of Aging, 23,* 5, 921-939.

McEwen, B.S., and Stellar, E. (1993). Stress and the individual: Mechanisms leading to disease. *Archives of Internal Medicine, 153,* 2093-2101.

Seeman, T.E., Glei, D., Goldman, N., Weinstein, M., Singer, B., and Lin, Y.-H. (2004). Social relationships and allostatic load in Taiwanese elderly and near elderly. *Social Science and Medicine, 59,* 2245-2257.

Seeman, T.E., McEwen, B.S., Rowe, J.W., and Singer, B.H. (2001). Allostatic load as a marker of cumulative biological risk: MacArthur studies of successful aging. *Proceedings of the National Academy of Sciences, USA, 98*(8), 4770-4775.

Seplaki, C., Goldman, N., Glei, D., and Weinstein, M. (2005). A comparative analysis of measurement approaches for physiological dysregulation in an older population. *Experimental Gerontology, 40,* 438-449.

Seplaki, C., Goldman, N., Weinstein, M., and Lin, Y.-H. (2004). How are biomarkers related to physical and mental well-being? *The Journals of Gerontology Series A: Biological Sciences and Medical Sciences, 59,* 201-217.

Seplaki, C., Goldman, N., Weinstein, M., and Lin, Y.-H. (2006a). Measurement of cumulative physiological dysregulation in an older population. *Demography, 43,* 165-183.

Seplaki, C., Goldman, N., Weinstein, M., and Lin, Y.-H.. (2006b). Before and after the 1999 Chi Chi earthquake: Traumatic events and depressive symptoms in an older population. *Social Science and Medicine, 62,* 3121-3132.

Singer, B., Ryff, C.D., and Seeman, T. (2004). Operationalizing allostatic load. In: J. Schulkin (Ed.), *Allostatis, homeostasis, and the costs of physiological adaptation* (pp. 113-149). Cambridge, England: Cambridge University Press.

Turra, C.M., Goldman, N., Seplaki, C.L., Weinstein, M., Glei, D.A., and Lin, Y.-H. (2005). Determinants of mortality at older ages: The role of biological markers of chronic disease. *Population and Development Review, 31,* 677-701.

Wachter, K.W. (2001). Biosocial opportunities for surveys. In National Research Council, *Cells and surveys: Should biological measures be included in social science research?* (pp. 329-338). Committee on Population, C.E. Finch, J.W. Vaupel, and K. Kinsella, Eds. Commission on Behavioral and Social Sciences and Education. Washington, DC: National Academy Press.

Weinstein, M., Goldman, N., Hedley, A., Lin, Y.-H., and Seeman, T. (2003). Social linkages to biological markers of health among the elderly. *Journal of Biosocial Science, 35,* 433-453.

Weinstein, M., and Willis, R. (2001). Stretching social surveys to include bioindicators: Possibilities for the Health and Retirement Study, experience from the Taiwan Study of the Elderly. In National Research Council, *Cells and surveys: Should biological measures be included in social science research?* (pp. 250-276). Committee on Population, C.E. Finch, J.W. Vaupel, and K. Kinsella, Eds. Commission on Behavioral and Social Sciences and Education. Washington, DC: National Academy Press.

Yeager, D.M., Glei, D.A., Au, M., Lin, H.-S., Sloan, R.P., and Weinstein, M. (2006). Religious involvement and health outcomes among older persons in Taiwan. *Social Science and Medicine, 63,* 2228-2241.

4

Elastic Powers:
The Integration of Biomarkers into the Health and Retirement Study

David Weir

The title of this chapter, "Elastic Powers," appeared as the last words of *Cells and Surveys: Should Biological Measures Be Included in Social Science Research?*, the influential 2001 volume from the Committee on Population (National Research Council, 2001). In his closing reflections on the biosocial opportunities for surveys, Kenneth Wachter quoted from a Matthew Arnold poem that described the aging process as exhaustion from the cumulation of shocks and change—a literary anticipation of the concept of allostatic load that Wachter in turn stretched from a model of individual physiology in a social context into a model of paradigm shifts in the history of science (Wachter, 2001). He warned of the challenges to established modes of thinking that the integration of biology and social surveys would pose, calling for a renewal of what Arnold had termed "the elastic powers."

Beginning the story of the integration of biomarkers into the Health and Retirement Study (HRS) with the closing words of *Cells and Surveys* is more than mere symbolism. Several contributors to that volume went on to be contributors to the HRS effort: Robert Wallace and Robert Willis as investigators, Eileen Crimmins and Douglas Ewbank as members of advisory groups, Teresa Seeman as an adviser to the study's sponsor, the National Institute on Aging (NIA), and Jeffrey Halter as a consultant. Richard Suzman's considerable role in organizing the first volume was acknowledged by Jane Menken in her preface, and his leadership on the NIA side of the HRS cooperative agreement is unrelenting. Together, their own elastic powers in picking up where *Cells and Surveys* left off have been considerable, and through their influence the elasticity of the entire field of population research has been renewed.

The scientific rationale for including biomarkers in HRS is not fundamentally different from the rationale for including them in any population survey concerned with health. They validate and add nuance to self-reports of health, they allow richer modeling of pathways of influence between the socioeconomic and the physical, and they may capture aspects of health unknown to survey participants. This chapter gives examples of how each of these are realized in the HRS. The development of biomarker data in other studies of older populations in the United States, such as the National Survey of Midlife Development in the United States (MIDUS) and the National Social Life, Health, and Aging Project (NSHAP), and outside the United States in the English Longitudinal Study of Ageing (ELSA) and the Mexican Health and Aging Study (MHAS), has both provided models of what can be done and created great potential for comparative work with the addition of such data to the HRS.

Because of the unique place of the HRS in population surveys of aging, however, the decision to add biology to the HRS involved a number of other considerations, several of which were clearly anticipated by Weinstein and Willis in their chapter of *Cells and Surveys* (Weinstein and Willis, 2001). The HRS is a large longitudinal study that serves a large constituency of researchers from many different disciplines. At last count, there were over 6,000 registered users of the data, and over 1,000 unique authors of written research using the data. Putting its traditional aims at risk through attrition of respondents or elimination of critical established content would have been unacceptable. Similarly, the confidentiality of respondents had to be protected, as well as the integrity of a longitudinal observation study not be transformed into an intervention study.

The ethical issues were considered carefully by the HRS investigators as well as the institutional review board (IRB) governing the study. Notifying respondents of the results of well-established and commonly available diagnostic tests was deemed an ethical responsibility that overrides any concern that the information might alter future behavior. Because the tests contemplated by HRS assess familiar risk factors and do not identify, for example, life-threatening cancers, the ethical conflict is not particularly difficult at this time. Biological material stored in repository for future use is governed by a separate IRB review. Respondents were asked to consent to having this material stored anonymously for future research. Ethical issues arising from any particular future test will need to be addressed at that time. For example, it is conceivable that some test of scientific value might not be permitted if it carried with it the ethical obligation to notify children or other nonparticipants of the possibility of an inherited disease, and if that notification were considered detrimental to the confidentiality of respondents.

Through the use of supplemental studies, some funded through peer review as competing supplements to the HRS, most of the elements of the HRS biomarker expansion were piloted on subsamples well before their introduction to the main survey. Those pilot efforts from 2001 through 2004 are thus a crucial part of the story.

Adding biomarkers without subtracting other things from an ongoing panel study leads to the question of cost. After the baseline interviews, the primary mode of interview in the HRS has been telephone. Although some biological material can be collected by mail, and the HRS had some success with this in a study of diabetes described at some length below, it was clear that, for a thorough integration of biology, some form of personal contact would be needed: clinic visits, nurse visits to the home, or in-person interviewing. Because of the high cost of clinics or nurses, and because of technical innovations that have expanded what can be done by interviewers in the home, the HRS has designed its biomarker effort around conventional in-person interviewing.

First described in its 2005 renewal proposal and now fully implemented in the ongoing 2006 data collection, the HRS has developed an integrated package of new content for a new model of in-home interview we describe as "enhanced" face-to-face. It includes anthropometrics, physical measures, blood spots, salivary DNA, and a self-administered psychosocial questionnaire. Most of the elements of this dramatically new development for the HRS were piloted in one way or another in smaller supplemental studies.

AGING, DEMOGRAPHICS, AND MEMORY STUDY

The first effort at collecting biological specimens from HRS respondents came in a supplemental study of dementia, known as the Aging, Demographics, and Memory Study (ADAMS). The primary aim of the ADAMS study was to establish the prevalence of dementia in the population over 70 years of age from a nationally representative sample (Langa et al., 2005). Because dementia is not a common condition even at that age, a simple random sample of the population would need to be fairly large to derive reliable estimates. The great virtue of using the HRS as a sampling frame for the ADAMS was the ability to sample at higher rates from persons with higher likelihoods of dementia based on the cognitive assessments conducted by the HRS.

The stratification of sample selection for ADAMS was based on five cognitive categories from the HRS interview. These had to be established separately for persons who did their own cognitive assessments and respondents for whom interviews were taken by proxy. In the case of proxies, the proxy reporter provides to the HRS interviewer an assessment of the cognitive function of the proxied respondent from the Jorm Informant Questionnaire on Cognitive Decline in the Elderly (IQCODE). The cognitively normal group was further stratified by age (ages 70-79 versus 80 or older) and sex in order to ensure adequate numbers in each of these subgroups. Finally, because of the long anticipated field period for the in-home ADAMS assessments, the recruitment was split between the 2000 wave of HRS and the 2002 wave, with some geographic areas drawn in one and some in the other, on a randomly sampled basis.

Because the rigorous in-home dementia assessments were conducted by collaborators at Duke University Medical School, the consent process for ADAMS was in two parts. First, an interviewer from the Institute for Social Research (ISR) at the University of Michigan contacted the respondent and an informed caregiver to obtain consent for Duke to contact them about the study. Duke then established contact and obtained consent for the home visit. This two-stage process worked against response rates in two ways. It extended the time interval between the HRS interview and the ADAMS visit, which resulted in loss due to mortality, and it provided respondents two opportunities to say no.

Overall, the response rate among survivors was 56 percent, yielding a final sample size of 856. Mortality rates (14 percent overall) were higher among the more cognitively impaired, but response rates conditional on survival were slightly higher for those groups. There were few significant predictors of nonresponse to the ADAMS study. Racial minorities participated at slightly higher rates than whites.

There are no established blood tests for dementia or cognitive impairment. Consequently, the ADAMS protocol relied primarily on extensive neuropsychological testing and did not include any blood sampling. There is one well-established genetic risk factor for dementia, and that is the E-4 allele of the ApoE gene. The ADAMS protocol did include collecting a cheek swab. Nearly everyone who participated in ADAMS agreed to provide this sample; only 11 of the 856 refused (1.3 percent). The samples were sent to the pathology lab of the University of Michigan for extracting and typing of ApoE. Only three of the samples could not be genotyped.

The (weighted) distribution of genotypes found in the ADAMS respondents replicates fairly closely the expected population distributions (Table 4-1). As has been found elsewhere, the presence of any ApoE-4 allele is a risk factor for dementia, but not a particularly powerful one (Hyman et al., 1996). Preliminary analyses show that presence of the ApoE-4 allele (either homozygous 4/4 or heterozygous 3/4) was associated with approximately twice the odds of dementia compared with those with the 3/3 genotype, which is in the range of previously reported values

TABLE 4-1 Distribution of ApoE Genotypes in ADAMS and Other Studies

Genotype	ADAMS[a] %	IOWA 65+ Population[b]	Framingham Population[b]
ApoE e2/2	1	1	1
ApoE e2/3	12	15	12
ApoE e2/4	2	2	2
ApoE e3/3	60	58	63
ApoE e3/4	22	22	19
ApoE e4/4	2	2	3

[a]ADAMS percentages are weighted.
[b]Data from Hyman et al. (1996).

(Breitner, Jarvik, Plassman, Saunders, and Welsh, 1998; Skoog et al., 1998). In the ADAMS data, the odds ratio for dementia associated with residing in a rural area compared with urban or suburban areas is nearly as high.

The ADAMS study demonstrated that HRS respondents would be willing to provide samples of DNA for research purposes. In the group of respondents willing to participate in a three-hour home interview to assess dementia, cooperation with the DNA request was nearly universal. Combined with the 56 percent response rate to the ADAMS study overall, however, only 55 percent of the HRS respondents approached for ADAMS ultimately provided a DNA sample.

DIABETES STUDY

In contrast to the ADAMS study, for which biomarker collection was a relatively small part of the overall assessment, the diabetes study was motivated in large part by the idea of collecting a clinically meaningful biomarker of the disease. Diabetes can result from a variety of underlying conditions (pancreatic failure to produce insulin, cellular resistance to absorbing insulin), but it is always characterized by excessive levels of glucose in the blood. High levels of blood glucose cause damage to both large and small blood vessels and to nerves, potentially leading to many severe consequences (including cardiovascular disease). Among persons with diagnosed diabetes, the management of the disease targets the maintenance of lowered glucose levels (and, increasingly, the management of other cardiovascular disease risk factors, especially hypertension). Thus, while there is considerable interest in understanding how people manage and cope with the disease, further study without a clinical marker for glucose levels seemed of relatively low priority.

A study on the scale of ADAMS, with an in-home assessment and blood draw, would have been quite expensive. Much of the nonbiological information about diabetes could easily be collected by a self-administered mail survey, which is far less expensive than even telephone interviews. The innovative aspect of the diabetes study was the attempt to gather dried blood spots (DBS) through the mail for the analysis of glycoslylated hemoglobin (HbA1c). A1c is an ideal measure for this study, as it is for the medical management of diabetes, because it summarizes the average levels of blood glucose over a two- or three-month period. It also does not require fasting and can be done from blood collected at any time of the day. Glucose levels vary widely over the course of a day and in response to the intake of food, making standard point-in-time readings very difficult to interpret in isolation.

Because of the reliance on A1c measures in the medical management of diabetes, commercial laboratories have developed assays for A1c that can be done in DBS. This allows patients to take their own samples and mail them to a lab from which the results can be reported to their doctors, saving time and money. DBS assays for A1c require special proprietary pretreatment of filter paper and utilize proprietary laboratory methods for analysis. Working with a commercial partner that has developed a DBS assay is therefore essential. Flexsite Diagnostics was the laboratory that did the HRS diabetes study, and their support and cooperation were outstanding. They designed a specimen collection card specifically for the study. This allowed the respondents to mail their specimens directly to the lab with only an arbitrary numeric identifier, so that the laboratory would not know the name or address of the respondent. Results were reported to HRS by numeric identifiers and then merged with the questionnaire data and the usual HRS identifications.

The diabetes sample was selected from respondents to the 2002 wave of HRS. Only persons reporting a doctor diagnosis of diabetes were eligible. About 20 percent of the eligible sample was excluded because of their participation in another HRS mail survey (the Consumption and Activities Mail Survey). The eligible sample numbered 2,518. Of that group, 133 (5.3 percent) died prior to the beginning of the diabetes study in late 2003.

The diabetes study proceeded in two stages. First, respondents were sent a self-administered questionnaire, along with a check for $40 and an explanation that they would be receiving a second request to send a blood sample later (the usual HRS incentive for a mail survey is $20). Blood test kits were sent out to respondents when questionnaires were returned. A standard protocol of reminders was followed. After about six weeks, duplicate questionnaires and blood kits were mailed to persons who had not responded. After about eight weeks, follow-up telephone calls were placed to some respondents.

Questionnaires were returned by 1,897 sample members (79.7 percent). In contrast to both the core HRS interviews and the ADAMS study, but quite consistent with other HRS mail surveys, there were substantial racial and ethnic differences in participation. Blacks and Hispanics had response rates about 10 percent lower than those of whites. Blood kits were returned by 1,233 respondents, which is 65 percent of those who returned the questionnaire. There was not much difference between Hispanic and other respondents on the blood test response rate conditional on participation in the questionnaire, but there was again a lower response rate among blacks. Combined with the questionnaire response rate, the net biomarker rate was 52 percent of the eligible surviving sample.

The quality of the A1c data collected seems to be quite satisfactory. Figure 4-1 shows the level of A1c according to the type of treatment regime: 7.9 for those on insulin, 7.2 for those taking oral medication only, and 6.5 for those not taking medication (F-statistic = 53.8, p < .0001). The

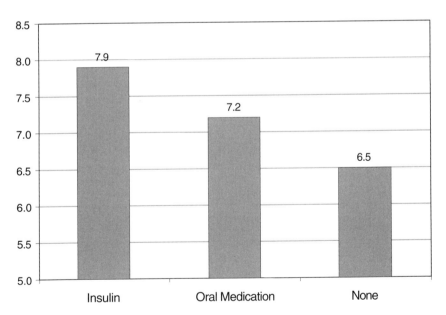

FIGURE 4-1 Mean HbA1c score by type of medication regime.
SOURCE: HRS Diabetes Study.

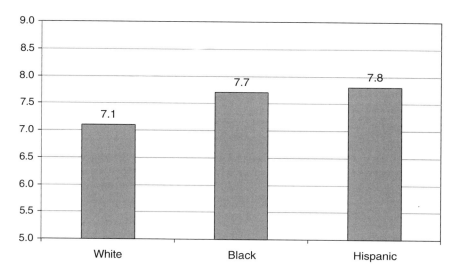

FIGURE 4-2 Mean HbA1c score by race and ethnicity.
SOURCE: HRS Diabetes Study.

corresponding numbers for the population ages 50 and older from the National Health and Nutrition Examination Survey (NHANES) study for 2003-2004 are 7.8, 7.0, and 6.2. Figure 4-2 shows the differentials by race and ethnicity: 7.1 for white non-Hispanics, 7.7 for blacks, and 7.8 for Hispanics (F-statistic = 35.1, $p < .0001$). That again differs only slightly from the comparable NHANES figures of 7.1, 7.9, and 7.8. Finally, Figure 4-3 shows that A1c also varies according to the respondent's self-assessed performance at managing the disease. Respondents giving themselves an "A" had A1c scores of 7.0, compared with 7.3 for "B" grades, and 7.8 for "C" or lower (F-statistic = 15.1, $p < .0001$).

Thus, in comparison with ADAMS, the diabetes study had a much higher overall participation rate but a fairly comparable net completion rate on the biomarker. Taken together, these two experiments suggested several important guidelines for future work on biomarkers in the HRS. First, multistage requests, in which the biomarker request is conditional on agreeing to one or more prior request, are bad for response rates. Second, self-administration and mailback of blood spots, while inexpensive, is unlikely to yield high response rates and seems particularly ill-suited to maintaining high response rates of minorities. In-home requests, with a trained person present to take the sample, seemed to provide the best basis for administering biomarkers.

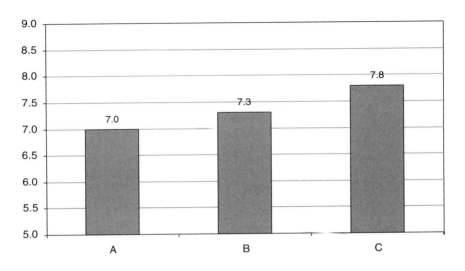

FIGURE 4-3 Mean HbA1c score by self-rated assessment (letter grade) of self-management of diabetes.
SOURCE: HRS Diabetes Study.

FACE-TO-FACE INTERVIEWS

In 2004, the HRS was given additional funding from the Social Security Administration to use in-person interviewing to improve consent rates for linkage to Social Security records for two groups: all of the original 1992 HRS cohort (born 1931-1941, plus spouses) and members of the 1998 war baby cohort (born 1942-1947, plus younger spouses) who had not yet given consent. This effort was successful. Seizing the opportunity created by in-person interviewing to pilot some other measures, the HRS obtained administrative supplements from NIA to conduct in-person interviews with samples of the other cohorts to create a representative sample of the whole. From the combined set of in-person interviews, samples of about 100 persons from each single year of birth were assigned to do physical performance measures, with a subset getting height and weight measures.

Although the 2004 interviews did not include any blood or DNA work or blood pressure testing, they were an important step in developing the 2006 strategy for biomarkers. We observed that there was a fairly high loss of sample due to respondents declining the in-person interview in favor of telephone—about 10 percent of those assigned. Thus, addressing "mode switches" is important for the HRS, given its history as a telephone survey. We also observed that failure to complete the physical performance measures (timed walk, puff test, grip strength) was related to self-reported physical limitations. Having good self-report indicators of those abilities would aid in understanding that censoring.

2006 Enhanced Face-to-Face Interview

All the work of the various supplemental studies and pilot projects were brought together in the design of the enhanced face-to-face interview for 2006. The key elements of the 2006 HRS enhanced face-to-face interview are

- measured height and weight and waist circumference,
- blood pressure,
- timed walk, grip strength, puff test, balance test,
- dried blood spots for HbA1c, total cholesterol, high-density lipoprotein cholesterol, C-reactive protein and repository,
- salivary DNA for repository, and
- self-administered mailback psychosocial questionnaire.

Selection of Measures

As a multidisciplinary population survey serving a wide community of researchers, the decision process in HRS about any survey content, including biomarkers, must consider a wide range of potential uses and not focus narrowly on specific hypotheses or interests. Input was sought from a large number of experts. The choice of measures attempted to balance scientific value against cost and respondent and interviewer burden. There are two primary foci of the measures: the first is obesity and metabolic syndrome, for which the main goal is obtaining assessments now to model risks of future events, and the second is frailty, for which the main goal is improving our characterization of the dynamics of disability and care needs of the elderly.

As an example of this selection process, the Quetelet body mass index (BMI) is obviously a critical measure for understanding obesity, and direct measures of height and weight should help to resolve any doubts about the accuracy of self-reports. But BMI is far from a perfect measure because of variability in muscle mass and other nonfat components of body weight. Waist circumference adds valuable complementary information about fatness and in particular about central adiposity. Waist-hip ratio was considered to add relatively little to waist alone, and hip measurement is more intrusive for respondents and difficult for interviewers. Grip strength is a somewhat expensive measure because of the high price of the dynamometer devices, but when this is factored over an average of 50-60 enhanced face-to-face (EFTF) interviews per interviewer in this wave, and the potential for reuse in other waves, its value in assessing loss of muscle strength clearly outweighs its cost. A more difficult set of choices had to be made regarding the physical performance measures. There are a number of well-known assessments of lower body mobility, such as chair stands and "get-up-and-go." We determined in 2004 that timed walk was the best single measure for our needs. For 2006, after consultation with other experts, we decided that what was needed to complement timed walk was not another measure focused on lower body strength, but rather measures that directly assessed balance, because that can also be useful not only for modeling falls, but also in understanding cognition. We therefore added a 10-second semitandem stand, followed by a side-by-side stand, or a full tandem stand, depending on performance.

The most significant restriction imposed by cost constraints was on blood testing. Drawing of whole blood in the home or in a clinic would be extremely expensive in a dispersed national sample like the HRS. At the same time, the laboratory technologies for using dried blood spots are advancing rapidly, making the scientific potential of this relatively inexpensive field collection protocol extremely attractive. At present, the HRS blood spots will be used to assay for HbA1c, total cholesterol, HDL cholesterol, and C-reactive protein. All these measures are of course important in metabolic syndrome and cardiovascular risk. Having established the protocol for DBS collection, other assays can be added as the technology improves and as scientific interest and funding develop.

Despite the restrictions imposed by both cost concerns and scientific focus, the new measures added to HRS cover a lot of ground. In their paper in *Cells and Surveys*, Eileen Crimmins and Teresa Seeman outlined a table of 17 measures on 8 different physiological systems that were related to social and behavioral influences and health outcomes (Crimmins and Seeman, 2001, p. 20). Of these, the HRS covers eight measures in four of the systems. The two most significant systems not covered are the sympathetic nervous system and the hypothalamic pituitary adrenal axis. Good measures of functioning in these systems require either whole blood, urine, or multiple measures during the course of a day or over several days (e.g., cortisol). The HRS will continue to follow research using these measures and technologies for assessing them.

Sample Design

The 2005 renewal proposal called for spreading the EFTF interviews over the next three waves of the HRS—randomly assigning one-third of the sample to each. Following the successful review of the proposal, NIA recommended that this be accelerated to assign one-half of the sample in 2006 and the other half in 2008, creating the possibility of a four-year interval between biomarker collections rather than six. That recommendation was adopted. The assignments were made randomly at the household level. That means that both persons in a two-person household will get the biomarker interview in the same year. It also means that all sample clusters will get a mix of conventional telephone follow-ups and enhanced face-to-face interviews, and therefore that all interviewers must be trained for both types of interviews. While this increases training costs slightly, it allows for operational efficiencies when interviewers can mix the two types of activities, and it allows for complete geographical representation in each wave.

Interviewer Training

The HRS was fortunate to follow, with about a year's lag, the development of in-home biomarker interviews in the National Social Life, Health, and Aging Project at the National Opinion Research Center and the University of Chicago. Their success set a high standard. Critical to that success is the successful training of interviewers in both persuading respondents to participate and in conducting the various measures successfully. The HRS had already developed protocols for height and weight measurement and physical performance measures. The ADAMS study had used cheek swabs to collect DNA. The switch to mouthwash samples offered both better quantities of genetic material and an easier mode of administration. The two main areas for which new training materials had to be developed were blood pressure and blood spots. For the latter we were advised by Professor Thomas McDade of Northwestern University, as well as the commercial laboratory conducting the assays. For both we had input from Robert Wallace and Kenneth Langa, the two medical doctors on the HRS investigative team.

The HRS survey operations group developed a DVD that demonstrated the protocols for all aspects of the new content. This video was sent in advance to prospective interviewers interested in working on HRS (most of whom had worked for the study in previous years). It helped to screen out interviewers who were too uncomfortable with the methods to do the work. It also served as a training vehicle and continues to serve as a refresher for interviewers in the field.

Consent and Reporting

The HRS developed a booklet for the administration of physical measures and biomarkers. Respondents of course consent to participate in the

HRS itself before interviewing begins. The biomarker assessments occur around the middle of the interview. For each of three sections—physical measures, blood spots, and DNA sample—the respondents are shown a printed information form and asked to read it and sign a consent before proceeding. In addition, respondents are asked after signing the consent whether they feel it is safe for them to perform each measure immediately before doing it. Blood pressure results can be reported during the interview, and any respondent exceeding a specified threshold is given a card recommending that they see a doctor about their blood pressure. Respondents are also told that their blood test results for HbA1c and cholesterol will be reported to them by mail. Both the blood test and the DNA consents permit future analysis to be done for HRS-related research purposes without reporting back to the respondent.

Early Results

At this writing, the HRS is about 10 weeks from the end of its 2006 field period, and about 90 percent of the expected interviews are completed. Early indications are that cooperation with the new EFTF interview is going well. Relatively few respondents have refused the face-to-face mode. Of those who have been interviewed, consent rates are over 94 percent for the physical measures, 82 percent for the DNA sample, and 81 percent for the blood spots. At present, older respondents are somewhat more likely to give DNA and less likely to give blood spots than younger respondents.

In the 2004 pilot work with physical performance measures, we noted a significant correlation between noncompletion of the measures and self-reported physical limitations. The distribution of measured scores thus does not represent the true population distribution of abilities. In the case of timed walk, we have several good self-reports of lower body function that allow one to assess the function level of persons who decline to do the task and potentially to impute a physical performance score. For the other measures, we do not. In 2006 we therefore added self-rating questions on hand strength and on lung function to aid in understanding the functional abilities of those who do not do the measures.

Table 4-2 shows that self-rated hand strength correlates very strongly with measured grip strength, and that persons reporting weakness in the hand had substantially lower completion rates on the grip strength test. For lung function, shown in Table 4-3, the question on frequency of breathlessness is not quite as good at predicting response rates to the puff test. This may be due to the fact that (unwarranted) fear of infection from the device leads some well-functioning respondents to decline this test. Self-rating does correlate well with performance on the test.

TABLE 4-2 Self-Rated Grip Strength, Response Rate to Grip Test, and Measured Grip Strength

Self-Rated Hand Strength	Measured Grip Strength	Response Rate (%)	N Measured
Very strong	36.7	86.5	1,440
Somewhat strong	31.9	88.1	4,149
Somewhat weak	24.7	78.8	1,213
Very weak	20.6	48.9	174
F-statistic	351.7	151.3	
p-value	< .0001	< .0001	

SOURCE: Preliminary HRS 2006 production data, unweighted.

TABLE 4-3 Self-Rated Shortness of Breath, Response Rate to Lung Test, and Measured Lung Function

Self-Rated Shortness of Breath	Expiratory Force	Response Rate (%)	N Measured
Often	272.2	72.3	391
Sometimes	313.2	85.3	1,221
Rarely	367.2	87.7	2,134
Never	378.0	87.6	3,411
F-statistic	135.6	33.7	
p-value	< .0001	< .0001	

SOURCE: Preliminary HRS 2006 production data, unweighted.

There is some controversy about the quality of self-reports of height and weight, although the general finding seems to be that there is a general tendency to overstate height among the elderly (Ezzati et al., 2006; Gunnell et al., 2000). The most plausible explanation for this is that older people report their maximum adult height, not their current height after shrinkage due to age-related compression. Weight tends to be underreported by the overweight, and overreported by the underweight, leaving a relatively small bias on average. In preliminary results from 2006, as well as in a very small sample from 2004, the HRS tends also to find rather small errors in reported weight, and systematic overreporting of height. The self-reports of height and weight are obtained before respondents are told that they will be measured.

Figure 4-4 shows the pattern of heights found in 2006, graphing the mean measured height and mean self-reported height against the self-

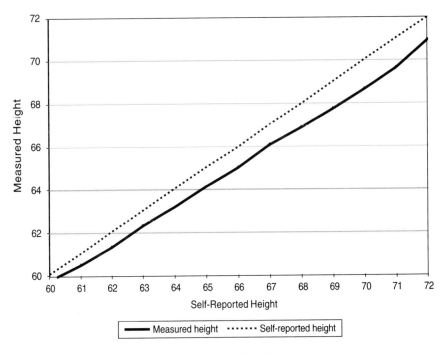

FIGURE 4-4 Measured versus self-reported height.
SOURCE: Preliminary HRS 2006 production data.

reported height. If self-reports were unbiased (equal to measured) on average, the graph should show a perfect 45-degree line. Instead, measured heights are lower than self-reports at every level of self-report, and more so at taller self-reported heights. The average differential is just under one inch. Measured heights are recorded to the nearest quarter-inch, and self-reports are in round inches. In addition to the bias, there is some random error, as shown by the correlation coefficient of .89 between measured and self-reported height.

Figure 4-5 shows a similar graph for weights, grouping self-reports in 10-pound ranges on the horizontal axis and graphing on the vertical axis the mean of measured and self-reported weight for each of those groups. The average error is about three pounds (self-reports below measured weight). The correlation is also impressively high at .97.

To put this in perspective, the average HRS respondent has a BMI of 29.1 using the measured data. Using instead the self-reported height lowers this to 28.2, while using instead the self-reported weight lowers it much less, to 28.6. Both together lower it to 27.8. Based on this evidence,

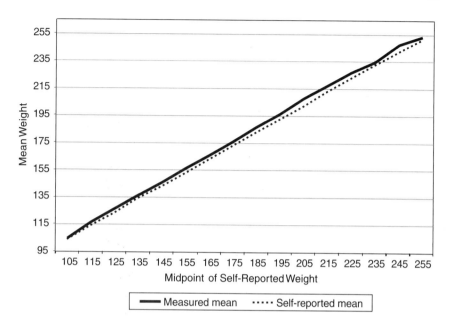

FIGURE 4-5 Measured versus self-reported weight.
SOURCE: Preliminary HRS 2006 production data.

the real scientific value from measuring weight in an older population, as opposed to relying on self-report, does not appear to be as great as the gain from measuring height. Height measurements are also less costly and less burdensome than using scales to measure weight.

Blood pressure data appear to be of good quality. The HRS protocol calls for three repeated measures, using an upper arm cuff with an automatic inflation device. These three measures are correlated at about .95,

TABLE 4-4 Biomarkers and Self-Reports: Measured Blood Pressure by Self-Reported High Blood Pressure Diagnosis and Control (mean of 3 measurements)

	Systolic	Diastolic
No high blood pressure	126.4	78.1
Under control no meds	137.4	84.5
Under control using meds	134.2	79.9
Not under control	148.4	87.6
F-statistic	138.7	71.2
p-value	< .0001	< .0001

SOURCE: Preliminary HRS 2006 production data, unweighted.

indicating good reliability. Table 4-4 shows that the mean of the three measures is reasonably well correlated with self-reported status. Interestingly, the mean blood pressure of persons who report a diagnosis and say it is under control is not much higher than those who say they do not have hypertension. Those who report their blood pressure is not under control do indeed have substantially higher measured levels.

CONCLUSION

The integration of biomarkers into the HRS is very much a work in progress. The first big steps have been taken to transform a primarily telephone study into one using in-person interviewing to obtain direct physical measurements and collect biological samples in the home, challenging the "elastic powers" of the survey's designers and its funders. HRS respondents have shown themselves willing to participate in this new survey experience, and the data they have provided appears to be of high quality. The HRS investigators hope to continue to expand and innovate in the inclusion of biomarkers as appropriate to the overall aims of the HRS. While this provides valuable new content to the HRS and new points of contact with clinical and lab-based studies, a large population survey like HRS cannot replace the vastly greater biological detail attainable in small clinical studies.

Soon the challenge to the elastic powers will shift from the design and implementation of the measures to their integration into longitudinal analyses using the data. It is this crucial intellectual transformation of how researchers conceive of problems that Wachter saw as the real challenge. The effort to collect such measures in population surveys will be warranted only by the new research insights they support. To support that challenge, we will need to seek ways to encourage researchers to develop models to make use of them. And that in turn will stimulate new ideas and new measures for future waves of data collection.

REFERENCES

Breitner, J.C.S., Jarvik, G.P., Plassman, B.L., Saunders, A.M., and Welsh, K.A. (1998). Risk of Alzheimer disease with the epsilon-4 allele for apolipoprotein E in a population-based study of men aged 62-73 years. *Alzheimer Disease and Associated Disorders, 12*(1), 40-44.

Crimmins, E., and Seeman, T. (2001). Integrating biology into demographic research on health and aging (with a focus on the MacArthur Study of Successful Aging). In National Research Council, *Cells and surveys: Should biological measures be included in social science research?* (pp.9-41). Committee on Population, C.E. Finch, J. Vaupel, and K. Kinsella, Eds. Commission on Behavioral and Social Sciences and Education. Washington, DC: National Academy Press.

Ezzati, M., Martin, H., Skhold, S., VanderHoorn, S., and Murray, C. (2006). Trends in national and state-level obesity in the USA after correction for self-report bias: Evidence from health surveys. *Journal of the Royal Society of Medicine, 99*, 250-257.

Gunnell, D., Berney, L., Holland, P., Maynard, M., Blane, D., Frankel, S., and Davey Smith, G. (2000). How accurately are height, weight, and leg length reported by the elderly and how closely are they related to measurements recorded in childhood? *International Journal of Epidemiology, 29*(3), 456-464.

Hyman, B.T., Gomez-Isla, T., Briggs, M., Chung, H., Nichols, S., Kohout, F., and Wallace, R. (1996). Apolipoprotein E and cognitive change in and elderly population. *Annals of Neurology, 40*(1), 55-66.

Langa, K.M., Plassman, B.L., Wallace, R.B., Herzog, A.R., Heeringa, S.G., Ofstedal, M.B., Burke, J.R., Fisher, G.G., Fultz, N.H., Hurd, M.D., Potter, G.G., Rodgers, W.R., Steffens, D.C., Weir, D.R., and Willis, R.J. (2005). The aging, demographics, and memory study: Design and methods. *Neuroepidemiology, 25*, 181-191.

National Research Council. (2001). *Cells and surveys: Should biological measures be included in social science research?* Committee on Population, C.E. Finch, J.W. Vaupel, and K. Kinsella, Eds. Commission on Behavioral and Social Sciences and Education. Washington, DC: National Academy Press.

Skoog, I., Hesse, C., Aevarsson, O., Landahl, S., Wahlstrom, J., Fredman, P., and Blennow, K. (1998). A population study of ApoE genotype at the age of 85: Relation to dementia, cerebrovascular disease, and mortality. *Journal of Neurology, Neurosurgery, and Psychiatry, 64*(1), 37-43.

Wachter, K. (2001). Biosocial opportunities for surveys. In National Research Council, *Cells and surveys: Should biological measures be included in social science research?* (pp. 329-336). Committee on Population, C.E. Finch, J. Vaupel, and K. Kinsella, Eds. Commission on Behavioral and Social Sciences and Education. Washington, DC: National Academy Press.

Weinstein, M., and Willis, R. (2001). Stretching social surveys to include bioindicators: Possibilities for the health and retirement study, experience from the Taiwan study of the elderly. In National Research Council, *Cells and surveys: Should biological measures be included in social science research?* (pp. 250-276). Committee on Population, C.E. Finch, J.W. Vaupel, and K. Kinsella, Eds. Commission on Behavioral and Social Sciences and Education. Washington, DC: National Academy Press.

5

An Overview of Biomarker Research from Community and Population-Based Studies on Aging

Jennifer R. Harris, Tara L. Gruenewald, and *Teresa Seeman*

The goal of this chapter is to provide an overview of findings from community-based studies that have profitably incorporated biomarkers along with more traditional interview data to address important questions regarding factors that affect health risks at older ages. The focus on older age stems from a series of activities (National Research Council, 1997, 2001a) funded by the National Institute on Aging that promoted a new era of population aging research predicated on the importance of integrating biomarkers into survey research.

Prior efforts to include biomarkers in community and population studies have yielded a wealth of knowledge regarding the role of biological systems and processes in cognitive and physical functioning, mental and physical disease development, and mortality outcomes. The biomarker initiative under the National Institute on Aging was particularly concerned with identifying biomarkers of physiological age (Butler et al., 2004). In contrast, biomarker research in the social and behavioral sciences has emphasized elucidating the interplay of biological systems with sociodemographic, behavioral, psychosocial, pharmacological, and genetic factors in health outcomes (Institute of Medicine, 2006; National Research Council, 2001b).

This chapter highlights research contributions from a selected group of community and population studies that include various biomarker measurements in study assessments. It is not intended to review completely, catalogue, or present in-depth results, but rather to provide an overview of the range of biomarker research conducted in these studies and highlight some key findings and research directions stemming from this integrative research.

For the purposes of this review, we have limited our consideration of biomarkers mainly to DNA and physiological biomarkers collected from blood (e.g., glycosylated hemoglobin, cholesterol), saliva (cortisol) or urine (cortisol and catecholamines); not included are more functional parameters that are also considered to be biomarkers, such as hand grip strength, measures of vision or hearing, and assessments of cognitive or physical functioning. This chapter highlights the types of questions that can be addressed when social and behavioral studies are supplemented with biomarker data, and therefore we also exclude descriptions of biomarker research derived from postmortem studies of brain tissue. It is important to note, however, the value of postmortem biomarkers for elucidating biological pathways and mechanisms, including those linking social and behavioral measures with disease outcomes. This is illustrated by research from, for example, the Nun's Study (Snowdon, Kemper, Mortimer, Wekstein, and Markesbery, 1996), the Religious Order Study (Wilson, Bienias, and Evans, 2004), and the Memory and Aging Project (Bennett et al., 2005).

Our decision to focus on physiological parameters and genes was driven in large part by a central goal of the current volume, namely, to help inform social scientists about the potential value of incorporating biomarkers into their projects through exposition of prior research in which analyses of such biomarkers or candidate genes have provided insights into processes and mechanisms affecting healthy aging. In that context, the focus taken herein seems warranted, as such biomarkers have generally not been included in social surveys (whereas assessments of functioning have), so that evidence based on physiological biomarkers is much less well known to the social science community.

Much of the information presented in this chapter derives from the studies summarized in Table 5-1. These were selected to represent a sampling of ongoing community or population-based studies on aging that have collected DNA or other biological or physiological biomarkers and that are *not* reviewed elsewhere in this volume. We have organized the wide range of findings generated from the studies reviewed according to a number of critical thematic areas that emerged during our review. These include biomarkers and aging, genetic and environmental influences on risk factors for cardiovascular disease (CVD), social and psychological factors, behavior genetics, biomarkers of cognitive aging, biomarkers of physical function and aging, indices of cumulative biological risk, and the relationship between biomarkers and genetic pleiotropy.

TABLE 5-1 Description of Selected Community and Population-Based Studies Conducting Biomarker Research

Study Name (year started)	Sample	Design and General Purpose	Specimen Type and How Collected	Biomarkers Ascertained
MacArthur Study of Successful Aging (Berkman et al., 1993)	1,189 men and women, ages 70-79 at baseline. Selection criteria employed to enroll a cohort representing the top third of this age group in terms of physical and cognitive functioning.	Longitudinal cohort study to elucidate the factors (sociodemographic, behavior, psychosocial, biological) associated with more successful aging (i.e., maintenance of higher cognitive and physical functioning). Data collected at baseline and 3- and 7-year follow-up exams via in-person and phone assessments.	Biomarker data were collected at baseline and three-year follow-up. Home specimen collections—phlebotomist collected blood and 12-hour (overnight) urine sample.	Total/HDL cholesterol, glycosylated hemoglobin, albumin, IL-6, C-reactive protein, fibrinogen, complete blood count (CBC), blood chemistry tests (SMAC-24), antioxidants, homocysteine, vitamin B and folate, urinary norepinephrine, epinephrine, cortisol and dopamine, resting and postural blood pressure, waist/hip ratio, peak flow rate (Mini-Wright meter), ApoE genotyping.

Normative Aging Study (1963) (Bell, Rose, and Damon, 1972)	Sample of 2,280 initially healthy men, ages 21–80 at enrollment. Study conducted through Department of Veteran Affairs in Boston (96% of participants are veterans).	Longitudinal cohort study to examine biomedical and psychosocial characteristics of aging and development of disease.	Clinical examinations every 3–5 years with collection of blood and 24-hour urines.	CBC, creatinine, albumin, fasting glucose, insulin, 2-hour glucose tolerance test, lipid profiles (LDL, HDL, triglycerides), serum glutamic-oxaloacetic transaminase (SGOT), calcium, blood urea nitrogen (BUN), blood lead levels, urinary catecholamines, blood pressure, heart rate, EKG reading, pulmonary function, anthropometry, homocysteine, vitamin B and folate.
Cardiovascular Health Study (CHS, 1989; Fried et al., 2001)	Representative sample of 5,000 adults, ages 65+, sampled from Medicare listings for 4 communities (N = 1,250 each; Forsyth County, ND; Sacramento, CA; Washington County, MD; Pittsburgh, PA).	Longitudinal cohort study of risk factors for coronary heart disease and stroke in older adults. Assessments at baseline and one-, two-, and three-year follow-ups.	Clinic-based collection of blood for wide array of biomarkers, EBCT, ultrasound, and MRI.	Resting and postural blood pressure, ankle-arm index, fasting glucose, insulin, 2-hour oral glucose tolerance test, lipid profile (LDL, HDL, triclycerides), albumin, CBC, left ventricular ejection fraction, markers of inflammation and coagulation, ApoE, body fat (bioelectric impedance), height, weight, waist/hip ratio, 12-lead ECGs (24-hour ambulatory ECGs on subset of 600), forced vital capacity, forced expiratory volume, grip strength, ultrasonography of carotid arteries, m-mode, Doppler echocardiography (for left ventricular mass, ejection fraction, stroke volume and end-systolic stress, regional and segmental wall motion, % fractional shortening).

Continued

TABLE 5-1 Continued

Study Name (year started)	Sample	Design and General Purpose	Specimen Type and How Collected	Biomarkers Ascertained
Later Life Resilience Study (Ryff, Singer, and Dienberg Love, 2004)	135 older women, ages 61-91, recruited from a prior longitudinal investigation of move to an assisted living facility from personal residence in Madison or Milwaukee, WI (approximately half of the larger sample participated in biomarker substudy).	All participants took part in a prior 4-wave longitudinal study of community relocation. Participants in the biomarker study participated in a fifth wave of data collection, including assessment of biomarkers.	Biological data collected through overnight visit to General Clinical Research Center, including blood, 12 hour (overnight) urines and saliva.	Blood pressure, heart rate, urinary norepinephrine, epinephrine, and cortisol, glycosylated hemoglobin, total/HDL cholesterol, DHEAS, salivary cortisol, waist/hip ratio, BMI.

Women's Health and Aging Studies (WHAS) I and II. WHAS I (Simonsick et al., 1997); WHAS II (Fried et al., 2000)	WHAS I: cohort of women (n = 1,002) ages 65+ sampled to represent the one-third most disabled Medicare enrollees from the Baltimore, MD, area (based on 12 zip code areas in eastern Baltimore and parts of Baltimore County). WHAS II = cohort of women, age 65+, sampled to represent the remaining two-thirds least disabled women in that age range.	Longitudinal cohorts, followed to elucidate factors associated with trajectories of functional decline or lack thereof.	Home-based specimen collection.	Fasted glucose, markers of inflammation, growth factors (IGF-1), genetic information (e.g., IL-6 haplotype), antioxidants, ambulatory ECG, resting 12-lead ECG, BP (resting, ankle-arm), height, weight, grip strength, spirometry.

Continued

TABLE 5-1 Continued

Study Name (year started)	Sample	Design and General Purpose	Specimen Type and How Collected	Biomarkers Ascertained
Swedish Adoption Twin Study of Aging (SATSA) 1984–present (Pedersen et al., 1991)	Population-based subset of 958 twin pairs from the Swedish Twin Registry. The sample includes twins reared apart and a matched sample of twins reared together.	Longitudinal, currently in 7th wave, questionnaires on all and in-person testing on a subsample. Combines twin and adoption design to study genetic and environmental influences affecting variation in normal aging.	Blood samples were drawn during in-person testing (12-hour fasting) drawn on subsample of twins at central testing sites or in home.	Serum lipoproteins: total cholesterol, total triglycerides, HDL cholesterol, LDL cholesterol, apolipoproteins A-1 and B, MaoB H pylori, metals, creatinine, gamma-glutamyltransferase (gamma-GT), potassium, sodium, urea and uric acid, creatinine, electrolytes, telomere length. Whole blood for DNA bank. Plasma at in-person testing occasion 3 for evaluation of a variety of coagulation factors.
Origins of Variance in the Oldest-Old: Octogenarian Twins (OCTO-Twin) 1991-2002 (McClearn et al., 1997)	Swedish twins ages 80 and older. Original sample identified 351 pairs for the first wave of in-person testing.	5-wave longitudinal study, measurements conducted at 2-year intervals, to explore the origins of individuality in the oldest-old.	Blood samples for clinical assessment and DNA extraction for molecular genetic analyses. Collected in home.	Albumin, calcium, total cholesterol, HDL cholesterol, creatinine, gamma-glutamyltransferase (gamma-GT), potassium, sodium, urea and uric acid, creatinine, electrolytes, cobalamin, free thyroxin, folic acid, prostate-specific antigen (PSA), and thyroid-stimulating hormone (TSH), homocysteine.

Study	Sample	Study Design/Measures	Blood Collection	Biological Samples
Screening Across the Lifespan Study (SALT) 1998-2002	Pilot included a random sample of 2,000 twins ages 5-85. Followed by full-scale screening of all twins born 1958 or earlier, population-based.	Computer-assisted telephone interview. Multidimensional health, demography, health behaviors, disease identification diagnostic items.	Blood samples collected at local clinics.	Biological samples only collected in pilot project include clinical biochemistries, zygosity plus blood stored for future analyses. Through recent efforts, DNA, serum, TG, HDL, LDL, apolipoproteins, glucose, HbA1C, and DNA collected on all pairs alive (goal 16,000 individuals, current at 12,000).
Men and Women's Aging (GENDER)	486 unlike-sexed twins born 1906-1925, population-based.	3 wave longitudinal study with measures 1995-1997, 1999-2001, and 2001-2005.	Collected during in-person testing in home.	Whole blood, serum, DNA, lipids, clinical panel.
Study of Dementia in Swedish Twins (HARMONY) (Gatz et al., 1997)	Twins born in or before 1935, population-based.	Cross-sectional, 1998-2001. Genetic and environmental factors for Alzheimer disease and other dementias. Diagnostic assessments for dementia, cognitive testing, and risk factor information. Three-year follow up for small subsample.	Collected in home.	Whole blood, serum, plasma, DNA, lipids, clinical panel.

NOTES: HDL = high-density lipoprotein; IL-6 = interleukin 6; DHEAS = dehydroepiandrosterone sulfate; ApoE = apolipoprotein E; LDL = low-density lipoprotein; ECG or EKG = electrocardiogram; EBCT = electron beam computed tomography; MRI = magnetic resonance imaging.

BIOMARKERS IN COMMUNITY- OR POPULATION-BASED STUDIES

Scientists have long been interested in obtaining measures of biomarkers to understand the role of biological systems in functioning and disease processes. Much of this work has been conducted via well-controlled experiments in the laboratory with nonhuman animals or in small-scale human studies focused on specific physiological processes in the lab (e.g., physiological responses to a specific challenge or interaction of specific physiological systems). More recently, the value of collecting biomarker measurements in large-scale community and population studies has been recognized (National Research Council, 2001a) and greater emphasis is being placed, by researchers and by funding agencies, on banking biological samples.

The impetus for the collection of biomarkers is to gain a better understanding of the role of specific biological systems in health conditions, including an understanding of the role of biological systems in association with other sociodemographic, behavioral, pharmacological, psychosocial, and genetic contributions to health outcomes. There is a growing literature in the behavioral sciences literature linking social and behavioral factors (ranging from sociocultural and neighborhood influences to interpersonal relations) to biomarkers and health (Berkman and Kawachi, 2000; Cacioppo, Hughes, Waite, Hawkley, and Thisted, 2006; Hawkley, Masi, Berry, and Cacioppo, 2006; House, Landis, and Umberson, 1988; Kiecolt-Glaser et al., 2005; Ryff and Singer, 2001; Uchino, Cacioppo, and Kiecolt-Glaser, 1996; Wen, Hawkley, and Cacioppo, 2006). These findings, in conjunction with methodological advances, are fostering integrative lines of research to study health and pathways to disease. For example, the recent Institute of Medicine report *Genes, Behavior, and the Social Environment* (2006) focuses on social environments in the study of gene by environment interactions and health.

A wide variety of biomarkers have been assessed in community- or population-based studies. Biomarkers of cardiovascular, metabolic, endocrine, and immune systems or processes are most commonly assessed. However, other types of biomarker measurements have also been obtained, including exposure indices (e.g., bone and blood lead levels or pesticide blood levels to assess environmental exposure), measurements of vitamin or antioxidant levels, anthropometric measures (e.g., bone length, height, weight), bone density scans, measurements of brain activity (e.g., functional magnetic resonance imaging, fMRI, or electroencephalogram, EEG), as well as markers of the functional status of a bodily system (e.g., forced expiratory volume to assess lung function).

Biomarker values are typically assessed for biological systems at a resting state (e.g., resting blood pressure, blood levels of glucose), but values are also sometimes assessed under conditions of challenge to a system (e.g., blood pressure levels after standing, levels of glucose after a glucose challenge test). Blood, urine, and saliva samples are typical sources for biomarker assessments, although for some of the biomarkers described above, mechanical or electrical devices (e.g., MRI machine) are also used to obtain measurements. The measurement of biomarkers in community- or population-based studies presents a methodological challenge for researchers, as they must determine how and where to obtain biomarker measurements on a large number of people. Most studies have used home-based or clinic-based protocols for the collection of biological specimens, with some investigations utilizing both approaches.

The primary advantage of clinic-based specimen collection is the ability to implement protocols requiring greater temperature control (e.g., samples must be kept cold or iced) as well as meeting more restricted processing requirements (e.g., within minutes or several hours at most). However, sample representativeness can sometimes suffer as certain subgroups are less able or willing to come to a clinic for reasons related to health, transportation, or unfamiliarity with the location. Examples of studies that have used clinic-based protocols successfully include the Women's Health and Aging Studies I and II, the Cardiovascular Health Study, and the Health, Aging and Body Composition Study. Studies using home-based protocols include the MacArthur Study of Successful Aging (Berkman et al., 1993), the Later Life Resilience Study (Ryff, Singer, and Dienberg Love, 2004), and the Swedish Twin Studies (Lichtenstein et al., 2002).

BIOMARKERS AND AGING

Most established biomarkers indices reflect age norms in disease-free samples from which individuals with known risks and diseases are excluded from study. This poses a challenge for aging studies because these criteria make it difficult to define "normal" values among groups of elderly individuals for whom morbidity is common. Furthermore, this approach could mask age changes in many biomarkers and values of routine biochemical blood tests. A study of clinical biochemical values in a population-based sample of twins ages 82 and older from the Swedish study of Origins of Variance in the Oldest-Old: Octogenarian Twins (OCTO-Twin) found few participants without clinical diagnoses; therefore, subsequent survival for six years was used as a marker of overall health in late life. Results revealed an association between mortality and higher serum levels of urea, urate, gamma-glutamyltransferase (gamma-GT), free thyroxin, and plasma homocysteine. In women, increased mortality was associated with low serum values for albumin and total cholesterol. The authors propose that these results could provide guidelines for clinical practice and general health examinations (Nilsson et al., 2003a, 2003b).

Further study of the association of biochemical values with morbidity, drug therapy, and anthropometry was examined. In addition to expected findings showing that biochemical values deviate under disease states common among the elderly, a number of biological risk factors exhibit patterns of increasing risk with age, including blood pressure, glucose, and markers of inflammation and homocysteine, each of which is associated with risks for one or more common diseases of aging, such as cardiovascular disease, osteoporosis, hip fracture, depression, and dementia (Nilsson et al., 2003a, 2003b). These findings indicate that morbidity and health-related factors common in aging populations substantially influence routine biochemical values.

Analyses of community-based studies, such as the MacArthur Study of Successful Aging, also point to potentially important age-related reductions in risks associated with some biological factors (e.g., a reduction in the apparent risks associated with elevated total cholesterol—Karlamangla, Singer, Reuben, and Seeman, 2004), although other major risk factors continue to exhibit strong effects with respect to risks for major outcomes, such as physical function (Reuben et al., 2002), cognitive function (Weaver et al., 2002), and longevity (Hu et al., 2005). Analyses of biomarker data from the Cardiovascular Health Study (CHS) have confirmed that lipid profiles (specifically total and low-density lipoprotein [LDL], cholesterol) are not significant predictors of myocardial infarction (MI), stroke, or mortality among older adults (Psaty et al., 2004).

CHS data also point to the continued importance of such biomarkers as high blood pressure, fasted glucose, low albumin, elevated creatinine, and low forced vital capacity as significant, independent risk factors for mortality, along with additional measures of subclinical disease, including aortic stenosis, abnormal left ventricular ejection fraction, major electrocardiographic abnormalities, and stenosis of the internal carotid artery (Fried et al., 1998) and markers of inflammation (Jenny et al., 2006). Analyses from the Health, Aging and Body Composition Study (Health ABC) (another cohort study of adults ages 70 and older) suggest that the presence of metabolic syndrome (based on a complex of risk factors, including cholesterol, blood pressure, and glucose) in older adults does continue to predict subsequent coronary events, heart failure, myocardial infarctions, and cardiovascular-related mortality (Butler et al., 2006). Like the MacArthur and other studies of aging, Health ABC data indicate that inflammatory markers such as interleukin-6 (IL-6) and tumor necrosis factor (TNF)-α are associated cross-sectionally with the presence of subclinical or clinical cardiovascular disease (Cesari et al., 2003a), and the presence of peripheral artery disease (McDermott et al., 2005) and that high levels of IL-6, TNF-α, and C-reactive protein (CRP) also prospectively predict incident coronary events, including coronary heart disease, stroke, and congestive heart failure (Cesari et al., 2003b).

A recent analysis of time trends in biological markers of health risks in participants ages 65 and older from the National Health and Nutrition Examination Surveys (NHANES) III (1988-1994) and NHANES IV (1999-2000) found significant reductions in the levels of high-risk indices of total cholesterol and homocysteine. These findings are consistent with the hypothesis that lipid-lowering medications and folate supplementation, respectively, have been effective in reducing the prevalence of "high-risk" values for these biological parameters. However, other changes indicate an increased burden associated with obesity-related measures and high-risk levels of CRP (Crimmins et al., 2005).

Analyses based on available biomarker data in the Women's Health and Aging Studies (WHAS I and II) have yielded insights into relationships of novel and potentially modifiable biomarkers, such as serum selenium and carotenoids and mortality—low levels in each case predicting higher mortality risk (Ray et al., 2006). Analyses also confirmed mortality risks associated with gradations of diabetic hyperglycemia (Blaum et al., 2005a).

Another example of age effects is illustrated by studies of mortality and the apolipoprotein E (ApoE) genotype. The ApoE gene is a cholesterol transporter involved in brain repair mechanism and is among the most widely studied genes in aging research. It has three common alleles, ApoE-2, ApoE-3, and ApoE-4, with ApoE-4 conferring a higher risk for developing Alzheimer disease (AD). Demographic models investigating the association between ApoE genotype and mortality rates using data from several countries revealed diminishing risks with age and that the ApoE genotypes are associated with little variation in mortality among centenarians (Ewbank, 2002).

GENETIC AND ENVIRONMENTAL INFLUENCES ON RISK FACTORS FOR CARDIOVASCULAR DISEASE

In certain respects, the strategy behind biomarker research reflects an endophenotype (intermediary phenotype) approach, described in psychiatric genetics (Gottesman and Gould, 2003). In polygenic systems, in which environmental influences also figure prominently, biomarkers often represent "downstream" traits in the pathways between genes and a measured behavior or outcome. Endophenotype-based strategies are becoming more common in gene detection studies, and it is critically important to recognize that many biomarkers are themselves complexly determined.

As illustrated by studies of CVD risk factors, genetic and environmental analyses of biomarkers can provide insights regarding biomarker effects in complex causal pathways.

CVD risk factors are among the most studied biomarkers. They are well known, easy to measure, and routinely assayed in clinical practice. Heritable and environmental factors have long been known to play an important role in cardiovascular disease, although quantification of these risks in the population at large is not straightforward. Twin and family studies provide ample evidence that biological risk factors (biomarkers) for cardiovascular disease, such as serum lipids and apolipoproteins, are also influenced by multiple genetic and environmental factors. Developmental trends in lipid and apolioproteins have been described (e.g., early life, adolescence, menopause in women, and older age) and age-sex differences emerge in lipid profiles.

A key question is how variation in these age and sex patterns is differentially influenced by genes and environment. Research on CVD risk factors from the Swedish Adoption Twin Study of Aging (SATSA) have helped to elucidate this. Genetic differences account for more than half the variation in plasma factor VII levels (Hong, Pedersen, Egberg, and deFaire, 1999). Analyses of total cholesterol, high-density lipoprotein (HDL), apolipoproteins A-I and B, and triglycerides revealed substantial heritabilities for these measures ranging from 0.63 to 0.78 in the younger group (ages 52-65) and from 0.28 to 0.55 in the older group (ages 66-86) for these measures. Heritabilities were consistently lower in the older age groups and significantly lower for apolipoprotein B and triglyceride levels (Heller, de Faire, Pedersen, Dahlen, and McClearn, 1993).

Furthermore, the unique design involving twins reared apart and reared together revealed that shared rearing environmental effects contributed to variation in total cholesterol levels. Most of the nongenetic variation for the other biomarkers was explained by unique environmental factors. This research was expanded on in a further twin study using augmented data that included SATSA, the Screening Across the Lifespan Twin (SALT) pilot study, and the Men and Women's Aging (GENDER) study. Sex and age differences in lipid and apolipoprotein levels were investigated in three age groups (17-49, 50-69, and 70-85). Heritabilities ranged from 35 to 74 percent and were consistently greater in women across all age groups. This age effect is almost entirely explained by greater variance associated with environmental factors (Iliadou, Lichtenstein, deFaire, and Pedersen, 2001). Age dependency in genetic effects for plasma lipids was also investigated in Dutch data, which showed some evidence for partially different sets of genes to influence lipid levels at different ages (Sneider, Van Doornen, and Boomsma, 1997).

Numerous other studies have confirmed the importance of genetic and environmental factors on cardiovascular risk factors in different countries (see Iliadou and Sneider, 2004, for a review), illustrating that genetic influences differentially affect various risk factors. For example, heritabilities for lipoprotein a Lp(a) levels are exceptionally high (.90) but may be affected by the use of sex hormones among women (Hong et al., 1995). Future research that incorporates measured environments into these models will help identify the nature of nonshared influences that impinge on CVD risk factors and become more important with age and, potentially, impact CVD outcomes.

SOCIAL AND PSYCHOLOGICAL FACTORS

A large and rapidly growing literature documents links among social, behavioral, and psychosocial measures with physiological, biological, and genetic factors (Harris, 2007; Ryff and Singer, 2005). A key goal of many of the studies reviewed here is to understand the nature of some of these links and to elucidate the interplay between these factors and their relationships to biology as these latter relationships affect healthy aging. This research has investigated an array of factors spanning different spheres of influence, including socioeconomic status, social engagement or social relationships, and various psychological factors.

Socioeconomic Status

Although previous research has suggested that socioeconomic status (SES) is less strongly related to health outcomes at older ages (House et al., 1994), data from multiple population-based studies point to the continued impact of lower SES on health risks in later life, including links to greater biological dysregulation in various major regulatory systems. Multiple studies based on data from the Health ABC Study have documented relationships between indicators of SES and biomarkers and also provide evidence that lower SES confers a greater risk for poorer cognitive and physical functioning outcomes. Lower SES (assessed by levels of education, income, and assets) was associated with higher levels of CRP, TNF-α, and IL-6 (Koster et al., 2006). Behavioral factors, such as smoking, alcohol consumption, and obesity accounted for a greater proportion of these associations than prevalent disease (e.g., heart disease, diabetes).

Lower SES participants were also at greater hazard of developing mobility disability (Koster et al., 2005a). Body mass index and an index of the number of inflammatory biomarkers (CRP, IL-6, TNF-α) for which participants' values were in the highest tertile were significant covariates in analyses, suggesting that these biomarkers may be important mediators of relationships between SES and mobility disability.

Lower SES has also been shown to predict the likelihood of cognitive decline in Health ABC participants (Koster et al., 2005b). Although a number of biomarkers were associated with both SES and cognitive decline in this investigation, biomarkers explained only a small portion of the relationship between SES and cognitive decline.

Data from the MacArthur Study of Successful Aging document similar SES gradients in major biomarkers (lower SES being associated with increased prevalence of higher risk levels), with a significant, negative gradient seen for a summary index of allostatic load (Seeman et al., 2004a). Analyses also indicate that these SES differences in allostatic load mediate 35 percent of the SES gradient in mortality (Seeman et al., 2004a). Parallel evidence for such SES gradients in cumulative biological risk has been reported from the Wisconsin Longitudinal Study (Singer and Ryff, 1999). And low levels of educational attainment were also shown to be associated with higher levels of allostatic load in the Normative Aging Study (NAS), with much of the association apparently mediated by higher levels of hostility (Kubzansky, Kawachi, and Sparrow, 1999). Recent analyses of NHANES III data also show consistent SES gradients in levels of biological risks with respect to blood pressure, cholesterol, metabolic profiles, and levels of inflammation (Seeman, Merkin, Crimmins, Koretz, and Karlamangla, no date).

Social Engagement and Social Relationships

A consistent body of evidence links greater social integration to better health and longevity (Seeman, 1996). A growing literature further indicates that health effects of social relationships persist importantly into older age, with evidence linking greater social integration or engagement to lower risks for cognitive decline (Bassuk, Glass, and Berkman, 1999; Seeman, Lusignolo, Albert, and Berkman, 2001a) and dementia (Fratiglioni, Wang, Ericsson, Maytan, and Winblad, 2000), as well as lower risks for physical disability (Seeman, Bruce, and McAvay, 1996) and greater longevity (Seeman, Kaplan, Knudsen, Cohen, and Guralnik, 1987; Seeman et al., 1993).

More recently, attention has shifted to the identification of potential biological pathways through which social relationships affect health and aging. Greater social integration and higher levels of reported emotional support have been linked to lower levels of major stress hormones (e.g., cortisol, norepinephrine, epinephrine) among older men, with weaker and nonsignificant trends among women (Seeman, Berkman, Blazer, and Rowe, 1994), and similar findings have been reported more recently for patterns of association with IL-6 (Loucks, Berkman, Gruenewald, and Seeman, 2006). Examination of more cumulative measures of biological risk has yielded evidence of an inverse relationship between levels of positive social engagement and cumulative risk for both men and women (Seeman, Singer, Ryff, and Levy-Storms, 2002). Importantly, data on levels of social conflict point to the increased health risks associated with greater exposure to such interactions (see Burg and Seeman, 1994, for a review).

Paralleling these earlier data, more recent data provide a growing body of evidence indicating that such negative social interactions are associated with heightened physiological activity and reactivity, resulting in significantly increased levels of biological risk (Seeman and McEwen, 1996; Taylor, Repetti, and Seeman, 1997; Seeman et al., 2002). These findings are consistent with the fact that man has evolved as a social animal. As such, it is not surprising that qualitative aspects of our social interactions should be associated with underlying and parallel patterns of physiological activation reflecting reduced versus heightened biological risk profiles.

Psychological Factors

The Later Life Resilience Study is another investigation that has focused on links between biomarkers and a number of social and psychological factors. Using a sample of women drawn from a larger study of the process of relocating to "senior housing," Ryff and colleagues (2006) examined biomarker correlates of positive and negative well-being in study participants to address the question of whether positive and negative well-being are opposite sides of a single mental health continuum or whether these forms of well-being are separate and independent dimensions. They hypothesized that if positive and negative well-being are unique constructs, they should have distinct biomarker correlates, and this premise was largely supported by study findings.

For example, salivary cortisol, norepinephrine, waist-hip ratio, HDL cholesterol, and total/HDL cholesterol showed significant associations with some measures of well-being (e.g., purpose in life, personal growth, perceived autonomy, positive affect, positive relations) but not with measures of ill-being, while dehydroepiandrosterone sulfate (DHEAS) and systolic blood pressure showed significant associations with negative well-being (e.g., depressive symptoms, negative affect, trait anxiety, trait anger) but not positive well-being. Only two biomarkers, weight and glycosylated hemoglobin, showed significant associations with indicators of both positive and negative well-being.

Other analyses have examined links between biomarkers and two dominant forms of positive well-being, eudaimonic (e.g., self-actualization, purpose in life) and hedonic (e.g., happiness, pleasure, satisfaction) well-being. A number of biomarkers, including salivary cortisol, urinary norepinephrine, the soluble receptor for IL-6, weight, waist-hip ratio, HDL cholesterol, total/HDL cholesterol, glycosylated hemoglobin, and indicators of better sleep quality were found to relate to scores on measures of eudaimonic well-being, while only HDL cholesterol was associated with hedonic well-being (Ryff et al., 2004).

Women with more positive social relations have also been found to have lower levels of IL-6 and better sleep efficiency (Friedman et al., 2005). Interestingly, social relations and sleep quality showed an interaction in their association with IL-6, such that those with both poor social relations and sleep efficiency had the highest levels of IL-6, while those women with poor scores on only one measure showed more moderate levels. This finding suggests that social relationships and sleep may act to buffer one another for deficits in one domain in terms of potential impacts on inflammatory activity. Evidence that poor sleep may be related to CVD risk, particularly in older people under chronic stress, has also been reported. Poor sleep, as measured using polysomnography, was associated with higher plasma IL-6 and procoagulant marker fibrin D-dimer in a study of older community-dwelling caregiver and noncaregiver adults, and these effects were more pronounced in caregivers of patients with Alzheimer disease (von Kanel et al., 2006).

Data from the Normative Aging Study also point to the potential importance of hostility as a factor affecting patterns of biological risk as well as major health outcomes. For example, high hostility has been linked to high levels of insulin and triglycerides and low HDL levels, with much of these associations mediated by links between higher hostility and higher body mass index and waist-hip ratio (Niaura et al., 2000). As indicated earlier, high levels of hostility have also been found to be associated with high scores on an allostatic load index, which represented a summary score of high-risk values on a range of cardiovascular, endocrine, and metabolic biomarkers (Kubzansky, Kawachi, and Sparrow, 1999). Investigations of NAS participants have also examined the interaction of both biological and psychosocial risk factors in the development of disease. For example, one investigation found that high levels of hostility and the presence of metabolic syndrome (high-risk scores on measures of triglycerides, glucose, HDL cholesterol, blood pressure, and body mass index [BMI]) independently predicted the likelihood of myocardial infarction over a 14-year follow-up period, but the concomitant presence of both high hostility and metabolic syndrome was associated with the greatest risk for subsequent myocardial infarction (Todaro et al., 2005).

Chronic stress has also been linked to a myriad of effects, including poor health, reduced immune function, premature aging, less robust responses to treatment, and earlier age of disease onset. Epel and colleagues (2004) analyzed peripheral blood mononuclear cells in premenopausal women to investigate the relationship between life stressors (chronic caretaking of an ill child) and cellular aging. Although there was no relationship between caregiver status and the indices of cellular aging, results revealed that perceived stress and duration of stress were associated with higher oxidative stress, lower telomerase activity, and shorter telomere length (TL), even after controlling for age. This research provides insight into potential mechanisms that link psychological stress to biological aging and suggests that perceived or chronic stress may play a role in premature senescence. Longitudinal research that tracks changes in telomere length in association with other psychosocial and health indices is needed to provide further insights into the factors affecting telomere shortening across the lifespan. Although this research is not based upon a community sample, measures of telomere length and telomerase activity are becoming more common biomarkers in larger cohort studies. For example, in the MacArthur Study of Successful Aging, stored DNA has been used to assess telomere length, and analyses now under way are examining both relationships to subsequent health risks and predictors of TL.

BEHAVIOR GENETICS

Following the human genome project there has been a tremendous shift in the focus of behavioral genetic research that has moved this field more squarely into the realm of biomarker research. A large body of research unequivocally documents the importance of anonymous genetic influences for an array of behaviors that affect healthy aging, including symptoms of depression, physical functioning, personality, health-related behaviors, and cognition (see special issue of *Behavior Genetics on Aging*, 33(2), March 2003).

Candidate genes, and their protein products, represent biomarkers that are becoming more routinely catalogued and studied. For example, based on information regarding relevant biological pathways, projects from the Swedish twins (SATSA, OCTO-Twin, GENDER, HARMONY, and SALT) have identified a variety of candidate genes and gene markers to study in association with aging-related behaviors and health outcomes. An example finding from this approach and combining data from several studies (SATSA, OCTO-Twin, and GENDER) is the reported association between depressed mood in the elderly and the AA gene variant in the serotonin receptor gene (5-HTR2A) among men but not among women (Jansson et al., 2003). These results raise the question whether different genes or genetic mechanisms contribute to the development of depressed mood in the elderly.

Methodological advancements have added a new focus to behavior genetic research, and an increasing number of studies are interested in the localization and identification of functional genetic variants influencing individual differences in human behavior. For example, loneliness is common among the elderly and is central to a cluster of socioemotional states and affects behavior, psychological health (Cacioppo et al., 2006), and physical health (Tomaka, Thompson, and Palacios, 2006). Social isolation and loneliness are implicated in the pathogenesis of multiple diseases, responses to therapy, and mortality (Hawkley and Cacioppo, 2003). Putative disease pathways through which loneliness exerts an influence include health behaviors, excessive stress reactivity, and deficiencies in physiological repair and maintenance. Research indicates that genetic differences among people explain about half of the variation in loneliness among children (McGuire and Clifford, 2000) and adults (Boomsma, Willemsen, Dolan, Hawkley, and Cacioppo, 2005).

Data from the Netherlands Twin Registry (Boomsma et al., 2002) were analyzed to investigate the molecular-genetic basis for these findings. Genotypic marker (400 microsatellite markers) data were collected in 682 sibling pairs and their parents. Linkage and association analyses were conducted to elucidate candidate regions that may contain genes that influence variation in loneliness. Results pointed to a region on chromosome 12q23-24, and follow-up association tests showed significant association to two neighboring markers, D12S79 and D12S395. Linkage results in this region have been reported previously for a number of psychiatric disorders and neuroticism. Although the collective linkage results are not definitive or consistent, they provide enough evidence for the authors to postulate that this region on chromosome 12 may contain genes involved in affective and social regulation and dysregulation (Boomsma, Cacioppo, Slagbom, and Posthuma, 2006).

Although linkage studies per se do not constitute strict biomarker research, the effort to identify biomarkers of complex behaviors is becoming more and more common, as illustrated by a recent special issue of *Behavior Genetics, 36*(1), January 2006, dedicated to genetic linkage studies for behavioral traits, including emotionality, depression, loneliness, cognition, addictive behaviors, health behaviors, and their endophenotypes. Such an approach may potentially generate numerous useful biomarkers that may become standard covariates in large cohort studies. Aided by publicly available databases, the value of these data for exploring complexly determined phenotypes could be greatly enhanced through cross-study comparisons and data pooling possibilities.

Recently, attention has focused on the importance of studying gene-environment interactions, including how behaviors may modify gene expression (Harris, 2007; Rutter, Moffit, and Caspi, 2006). One of the most extensively studied common genetic variants in human studies of social behavior is the 5'-promoter polymorphism of the serotonin transporter gene (5HTT). A wave of studies has explored interactions between functional polymorphisms of this gene with stressful life events and depression. Evidence supporting such an interaction effect is reported by some (Caspi et al., 2003; Eley et al., 2004; Grabe et al., 2005; Kaufman et al., 2004; Kendler, Kuhn, Vittum, Prescott, and Riley, 2005) but not confirmed by all (Gillespie, Whitfield, Williams, Heath, and Martin, 2005; Surtees et al., 2006) studies. Although these findings do not derive from aging studies per se, they generate important aging-related questions regarding the role of functional polymorphisms in the serotonin transporter gene and late life depression.

New findings from the Swedish SATSA study report evidence for genotype by environment interactions affecting change in a semantic memory task. This work investigated the role of several candidate genes including those coding for ApoE and estrogen receptor alpha (ESR1) and serotonin candidates (HTR2A and 5HTT). Further investigation aimed at identifying the nature of the environmental influences involved in this interaction examined social and stress factors, including social support, life events, and depressive symptoms. Results suggested that influences associated with depressive symptoms may moderate the gene-environment interaction observed for ESR1 and ApoE and longitudinal semantic memory change. The authors explain that noncarriers of putative risk alleles may be relatively more sensitive to depression-evoking environmental contexts than carriers of the risk allele. This suggests that the contexts that facilitate or reduce depressive symptoms could affect resiliency in semantic memory dependent on genotype (Reynolds, Gatz, Berg, and Pedersen, 2007).

BIOMARKERS OF COGNITIVE AGING

Cognition represents one of the two major domains of functioning, the other being physical functioning (see below). Among the biomarkers known to affect risks for cognitive decline, the ApoE genotype, more specifically the ApoE-4 allele, has received by far the most attention to date. After the initial discovery of the role of ApoE-4 for dementia, numerous works have examined the role of ApoE and the ApoE-4 variant in trajectories of normal aging as well as in the pathogenesis of other diseases. For example, analysis of stored samples in the MacArthur study confirmed that ApoE status is predictive of cognitive declines in this initially high-functioning cohort (Bretsky, Guralnik, Launer, Albert, and Seeman, 2003). Studies of diverse populations reveal that the risk conferred by the ApoE-4 allele does not pertain to all populations. Results from a study of healthy Medicare recipients in the Washington Heights–Inwood community of

New York City (WHICAP) revealed that ApoE-4 conferred a greater risk for Alzheimer disease among non-Hispanic whites, but that blacks and Hispanics were at increased risk regardless of their ApoE genotype (Tang et al., 1998). These results indicate that other, unknown genetic or environmental risk factors contribute to the increased risk of Alzheimer disease in blacks and Hispanics.

A particularly exciting opportunity is offered by the growing availability of more population-based genetic information in studies that include a rich array of other information. These data offer extraordinary new possibilities to investigate the ways in which risks associated with genotype may be importantly modified by other characteristics (i.e., to test for interactions). Such research, particularly as it relates to potentially modifiable characteristics of individuals (or environments), may offer important insights regarding protective factors that can compensate for and reduce ultimate risks associated with particular genotypes—factors that could be the focus of interventions to reduce health risks. One such question has been whether acquisition of higher levels of education might provide a protective effect against the known risks for cognitive decline usually associated with the ApoE-4 allele. Results to date are mixed, although comparisons across studies are hampered by differences in study populations and, perhaps more importantly, by lack of comparability in measures used to assess cognitive aging. Findings are nonetheless illustrative of the potential for use of growing genotypic data in the context of population studies with information on other characteristics of the individual, the environment, or both. For example, analysis from the MacArthur Successful Aging Study revealed that the presence of the ApoE-4 allele was associated with greater declines in cognitive performance (based on detailed assessments of major domains of cognitive function, including naming, spatial recognition, praxis, and executive function) over a seven-year follow-up among the more educated but not among those with less than high school education; risks for cognitive decline were highest and comparable for those with and without the ApoE-4 allele (Seeman et al., 2005).

Another epidemiological study from Washington state has examined this question using a community-dwelling (n = 2,168) sample of nondemented elderly who were followed prospectively for six years using the cognitive abilities screening instrument. With their larger sample, analyses of gene-dose effect were possible and provided evidence for biological effects of the ApoE-4/ApoE-4 genotype compared with the heterozygous and other homozygous configurations. Analyses of a possible interaction with education yielded evidence for education modification of effects only among those with the ApoE-4/ApoE-4 genotype; cognitive decline was greater among those with less education (Shadlen et al., 2005). Differences in analytic approach and operationalization of ApoE status preclude direct comparisons to the MacArthur findings. However, the findings do appear to contrast with trends seen in the MacArthur study, in which greater overall declines were associated with the presence of the ApoE-4 allele among those with higher education (though they remained at higher levels of function than carriers of the ApoE-4 allele who had lower education throughout the follow-up). However, comparisons between these studies are hampered by noncomparability on outcome measures as well as study populations. Nonetheless, their findings suggest that further attention to the joint effects of ApoE and education with respect to cognitive aging are clearly merited.

Along similar lines to the foregoing studies are efforts to investigate the potential of early life socioeconomic environment to modify the relationship between ApoE status and risk for development of Alzheimer disease. Census data were used to index socioeconomic risk based on a number of parental and demographic measures. In addition to the increased risk conferred by genetic predisposition and early life environment, risk for AD was found to be greatly elevated (OR = 14.8; 95% CI, 4.9-46) when both the genetic and the environmental risk factors were present (Moceri et al., 2001). Other data from the CHS also point to possibly important interactions with dietary fatty fish consumption in relation to dementia risk: dietary intake had a significant effect only for those without the ApoE-4 allele (Huang et al., 2005).

Findings from community-based studies on the effect of ApoE on memory performance and memory change are mixed, with some finding deficits in performance or quicker rates of decline and others reporting no effect. However, ApoE-4 influences are more consistently reported for episodic versus working memory.

Two other genes with mixed support as genetic risk factors for Alzheimer disease, A2M (alpha-2-macroglobulin) and low-density LRP (lipoprotein receptor-related protein), have now been studied in relation to memory among nondemented adults. Variation in these three genes was analyzed in latent growth models measuring memory performance over a 13-year period in SATSA. Polymorphisms of ApoE and A2M (but not low-density LRP) were associated with memory performance and change in memory in this nondemented sample. Specifically, ApoE status affected ability levels of working and recall memory, the ApoE-4/ApoE-4 genotype was associated with worst performance across all ages. Furthermore, this study provided evidence of within-locus interactions (genetic dominance deviations) because memory performance among the heterozygotes was better than among the noncarriers of the ApoE-4 allele. Finally, the rare del/del genotype of A2M was associated with a more rapid rate of decline on figural recognition than the other two genotypes (Reynolds et al., 2006).

The role of other candidate genes has recently been explored in relation to cognitive aging. Age-related loss of serotonin receptors 2A (5-HT2A) is associated with a loss in brain regions including the hippocampus and posterior medial prefrontal cortex. A functional variant (H452Y) of the gene coding for the 5-HT2A serotonin receptor (HTR2A) has been associated with recall tests in young adults (de Quervain et al., 2003). This was further investigated using a new approach that explored allelic associations and trajectories of change in memory performance over a 13-year period in the SATSA data. Findings suggested that the 5-HT2A serotonin receptor is involved in the formation of episodic memories in older adults. Performance on figural memory at age 65 and change in figural memory were associated with the HTR2A genotype. Genotype-dependent effects comparing the AG, AA, and AG configurations revealed the steepest declines associated with AG heterozygotes. Performance over time was consistently worse among those with the AA compared with the GG genotypes, with trajectories differing by 2-6 percent per year (Reynolds, Jansson, Gatz, and Pedersen, 2006).

Other biomarkers also show an association with cognitive decline. Data from the OCTO-Twin study revealed lower homocysteine values among those with intact cognitive function, a finding that contrasts results regarding dementia. These findings overlap somewhat with those from the MacArthur Study of Successful Aging and the NAS, both of which found that high levels of homocysteine and low levels of vitamin B and folate were associated with cognitive decline (Kado et al., 2005; Tucker, Qiao, Scott, Rosenberg, and Spiro, 2005). Elevations in various stress hormones (i.e., urinary free cortisol and/or epinephrine) have also been shown to predict increased risks for cognitive decline (Seeman, McEwen, Singer, Albert, and Rowe, 1997a; Karlamangla, Singer, Chodosh, McEwen, and Seeman, 2005b; Karlamangla, Singer, Greendale, and Seeman, 2005a). Additional biomarkers found to predict cognitive and mental health status, independent of other standard sociodemographic and lifestyle risk factors, include low serum thyroxine (a marker of thyroid function) as a risk factor for cognitive decline (Volpato et al., 2002) and vitamin B12 deficiency as a risk factor for depression (Penninx et al., 2000). High levels of lead in bone and blood have also been shown to be associated with cognitive impairment (Payton, Riggs, Spiro, Weiss, and Hu, 1998). Greater vitamin E intake or vitamin E levels have also been associated with less cognitive impairment and dementia in the InChianti study (Cherubini et al., 2005).

More detailed neuropsychological protocols have also provided evidence linking early cognitive test performance to risks for Alzheimer disease (Saxton et al., 2004), and more detailed MRI data show that levels of inflammation (e.g., fibrinogen) and forced vital capacity are positively and significantly related to white matter disease (Ding et al., 2003).

BIOMARKERS OF PHYSICAL FUNCTION AND AGING

Physical functioning represents the second major functional domain and has been a focus of considerable research attention. Among the most detailed of this work has been that by Fried and colleagues as part of the WHAS. A particular strength of the WHAS I and II studies is their detailed evaluation of both lower and upper extremity function, including various assessments of balance, walking speed, timed chair stands and knee extension force for lower extremity function and hand grip, pegboard, and putting-on-blouse tests for upper extremity function. Analyses based on these data have documented the significant contribution of muscle weakness to risks for disability (Rantanen et al., 1999) and the more general contributions of physical performance across the various domains of upper and lower extremity functioning to both progressive and catastrophic disability (Onder et al., 2005). In the WHAS I and II, availability of detailed performance measures, plus information on biological processes thought to impact such performance, such as inflammation or growth factors, has afforded important opportunities to examine the joint and independent contributions of performance abilities and biological processes to actual levels of functional disability.

Several analyses have highlighted the significant, negative impact of higher burdens of inflammation in terms of reduced muscle strength and declines in physical function (Ferrucci et al., 2002) as well as parallel negative associations between low insulin-like growth factor-1 (IGF-1) and declining IGF-1 with slower walking speed and reported difficulty with mobility tasks, respectively (Cappola, Bandeen-Roche, Wand, Volpato, and Fried, 2001). These findings on the relationship between muscle strength and inflammation mirror results from Health ABC showing that high levels of oxidized LDL (oxLDL) and IL-6 predict incident mobility disability, with those individuals with high levels of both biomarkers being at greatest risk. Loss of muscle strength and muscle mass may be one pathway involved in relationships between inflammation and physical disability, as high circulating levels of inflammatory biomarkers are associated with low muscle strength and muscle mass in study participants (Visser et al., 2002; Yende et al., 2006).

Inflammatory markers have also been linked to both levels of physical activity and risks for functional decline. Data from the InChianti (Invecchiare in Chianti, meaning "aging in the Chianti area") Study, a prospective investigation of over 1,000 community-dwelling French adults, indicate that compared with sedentary individuals, physically active men have lower fibrinogen, CRP, IL-6, and TNF-α, as well as lower uric acid and a lower erythrocyte sedimentation rate, while physically active women have lower CRP, IL-6, and uric acid (Cherubini et al., 2005; Elosua et al., 2005). Similar findings have been reported from the MacArthur Study of Successful Aging in the United States (Reuben, Judd-Hamilton, Harris, and Seeman, 2003). Greater inflammatory burden has also been associated with increased risks for physical disability (Reuben et al., 2002) and with poorer physical performance (Cesari et al., 2005).

A number of investigations in InChianti have also examined biomarker correlates of anemia and links between anemia and indicators of physical health or functioning. Anemia is typically defined by low levels of blood hemoglobin (< 12 g/dL in women, < 13 g/dL in men). Hemoglobin is found in red blood cells, which contain iron and are responsible for carrying oxygen to bodily tissues. Anemia or low blood hemoglobin levels are associated with low muscle density, low skeletal muscle strength, a low muscle/total area ratio, and low bone mass and density (Cesari et al., 2005). Consistent with these bone and muscle correlates of anemia, anemic persons have also been found to be more likely to have physical disabilities, poor physical performance, lower hand grip strength, and lower knee extensor strength, compared with nonanemic individuals (Penninx et al., 2005).

Data from the InChianti study also point to the importance of additional biomarkers, including antioxidants such as vitamin E (alpha-tocopherol), which has been associated with a number of indicators of health and functioning in study participants. Those with lower vitamin E levels were less likely to be frail (as assessed by an index of weight loss, low energy, slow gait, low grip strength, and low physical activity; Ble et al., 2006), had higher conduction velocity in peripheral nerves (a slowing of conduction velocity is thought to contribute to decline in muscle strength; Di Iorio et al., 2006), and were less likely to have peripheral arterial disease (Antonelli-Incalzi et al., 2006).

The concept of frailty has become the focus of a growing body of research, stimulated in good measure by Linda Fried, a leader in developing and testing of an operational definition of the concept (Fried et al., 2001). Using data from the WHAS I and II studies, Fried and colleagues have provided confirmation of the internal validity of the component measures, which include poor extremity strength (low grip strength), slow gait, low levels of physical activity, exhaustion and weight loss, and the independent risks for functional disability, institutionalization, and mortality associated with such frailty (Bandeen-Roche et al., 2006). WHAS I and II data have also contributed to understanding of the various physiological processes that appear to contribute to frailty, including documenting the contribution of anemia (Chaves et al., 2005) and, perhaps most surprisingly, of obesity, which was found to be associated with prefrailty and frailty despite the fact that a defining characteristic of frailty is weight loss (Blaum, Xue, Michelon, Semba, and Fried, 2005b). This obesity-frailty association remained significant even with adjustments for multiple conditions associated with frailty (e.g., inflammation burden). Data from the CHS provide parallel evidence linking increased burdens of inflammation to risks for frailty (Walston et al., 2002). Research is needed to determine whether frailty represents an important pathway that underlies known links between a number of the biomarker correlates of frailty (e.g., inflammation, anemia, vitamin E) and disability, morbidity, and mortality outcomes (see review above).

INDICES OF CUMULATIVE BIOLOGICAL RISK

Using available biomarker data, investigators associated with the MacArthur study have been among the leaders in efforts to develop operational indices of allostatic load (AL)—that is, a multisystems measure of physiological dysregulation. Beginning with initial work using a simple count of the number of available biomarkers for which an individual had a value placing them in the top risk quartile, Seeman, McEwen, Singer, Albert, and Rowe (1997b) examined health risks associated with differences in such cumulative AL and demonstrated that higher levels of baseline AL were associated with significantly increased risks for cardiovascular disease, cognitive and physical decline, as well as mortality (Seeman, Singer, Horwitz, and McEwen, 1997b; Seeman Singer, Rowe, and McEwen, 2001). Subsequent work by Karlamangla, Singer, McEwen, Rowe, and Seeman (2002) using canonical correlation techniques demonstrated improved prediction of cognitive and physical decline when the full range of scores was used for each biomarker, and unequal weighting of the different biomarkers was incorporated into the scoring of overall AL. Recent work has also documented that measured change in AL predicts subsequent mortality risk (Karlamangla, Singer, and Seeman, 2006). Stimulated by Singer's work on recursive partitioning (Zhang and Singer, 1999; Singer, Ryff, and Seeman, 2004), recent analyses by Gruenewald, Seeman, Ryff, Karlamangla, and Singer (no date) have demonstrated that particular combinations of these biomarkers (e.g., inflammation and neuroendocrine biomarkers) predict higher versus lower mortality risks over a 12-year period.

A common thread seen in all of the work on AL has been the confirmation of contributions to cumulative health risks from multiple biological systems and the value of taking account of this range of contributions in understanding population variations in burdens of morbidity, disability, and mortality. As illustrated in Seeman et al. (2004a), analyses of AL as a mediator of education effects on mortality risk, the more comprehensive AL index accounted for the largest percentage reduction in the education effect on mortality, with subsets of biomarkers representing cardiovascular, inflammatory, and sympathetic nervous system/hypothalamic-pituitary-adrenal activity, each contributing to this overall effect (Seeman et al., 2004a).

As noted earlier, data from the Wisconsin Longitudinal Study (a cohort of men and women, ages 58-62, approximately a decade younger than the MacArthur study cohort) have provided evidence for relationships between SES and levels of positive social engagement and a similar "count" index of allostatic load (Singer and Ryff, 1999; Seeman et al., 2002). Using the MacArthur Study of Successful Aging protocols, the Social Environment and Biomarkers of Aging Study (SEBAS) (see Chapter 3) collected parallel biological data and provides another comparison to the MacArthur study. Analyses of SEBAS data have yielded intriguing similarities and differences with respect to relationships and allostatic load (weaker in Taiwan; Hu, Wagle, Goldman, Weinstein, and Seeman, 2006), and between social integration and allostatic load (nonsignificant in Taiwan; Seeman, Glei, Goldman, Weinstein, Singer, and Lin, 2004b). The possible importance of sociocultural differences between Asia and the United States represents one area of potentially fruitful future research to better understand how aspects of the social environment impact health.

RELATIONSHIP BETWEEN
BIOMARKERS AND GENETIC PLEIOTROPY

Two other recent areas of interest in WHAS I and II include examination of possible interactions among biomarkers, with initial work showing cross-sectional relationships between higher serum levels of antioxidants and lower levels of IL-6 (as a marker of inflammation) and longitudinal relationships between initially low antioxidant levels and subsequent increases in IL-6 levels (Walston et al., 2006). A second area of investigation has been to incorporate consideration of genetic information in tracking the factors contributing to observed profiles of biological activity. Initial analyses focused on IL-6 alleles and their relationship to serum IL-6 levels and to decreased muscle strength and frailty. No significant relationships were found for any of these outcomes with any single IL-6 single nucleotide polymorphism (SNP) or any IL-6 haplotype (Walston et al., 2005).

Genetic pleiotropy (in which a gene or set of genes influences multiple traits) could explain the association between biomarkers. A series of studies from the Swedish twin projects have investigated the variance architecture explaining the clustering between biomarkers for cardiovascular disease. Genetic and environmental correlations among the following five serum lipid measures—total cholesterol, HDL cholesterol, triglycerides, and apolipoproteins A-I and B—were analyzed in two different age groups from SATSA. Substantial genetic correlations were found in each age group, although there is no evidence for a single genetic factor common to all five lipids. There were significant age differences in the heritabilities for the various serum lipid levels, and genetic factors seemed to be more important for explaining the covariation between the lipid levels in the younger compared with the older group (Heller, Pedersen, de Faire U, and McClearn, 1994). Further research focused on the sources of clustering among five principal components (BMI, insulin resistance, triglycerides, HDL cholesterol, and systolic blood pressure) of the insulin resistance syndrome (IRS). Results suggest a single set of genetic factors is common to all five components; of particular note was the strong genetic association between BMI and insulin resistance. In contrast, the relationship between only three of the IRS components—triglycerides, insulin resistance, and HDL cholesterol—could be explained by shared sources of individual environmental influences. These findings demonstrating a strong genetic correlation between BMI and insulin resistance raises the question whether behavioral factors that affect both of these phenotypes, such as overeating, act through genetic pathways, perhaps related to control and sensations of satiation, rather than through such environmental factors as availability of food (Hong, Pedersen, Brismar, and deFaire, 1997).

SUMMARY AND FUTURE DIRECTION

Recent growth in the number of studies incorporating biomarkers into larger population-based surveys has yielded a rapidly growing body of evidence linking various aspects of biological functioning not only to major health outcomes, including cognitive and physical functioning and longevity, but also importantly to individual differences in socioeconomic and other social, psychological, and behavioral characteristics. Findings linking aspects of life situations to major biological risk factors provides important evidence on two fronts. First, it provides validation for various biopsychosocial models of aging and helps elucidate biological pathways through which social, psychological, and behavioral factors affect trajectories of aging and risks for various health outcomes. Second, such evidence provides further support for the potential value of interventions targeting such social, psychological, and behavioral factors as a means of altering underlying biological risk profiles.

A number of ongoing studies will soon provide even richer databases for use in elucidating the complex pathways through which individuals' life experiences and situations affect their health and aging and the biological pathways through which these effects are mediated. For example, current data collection for the MIDUS (Midlife Development in the United States) study will result in a rich set of data on life experiences, including longitudinal data covering the past decade, along with detailed biomarker data and daily diary data for subsets of some 1,500 MIDUS participants. These data will offer significantly improved opportunities to examine a wide variety of hypotheses regarding the role of life experiences (both current and past) in shaping patterns of biological risk and health trajectories. As also outlined in other chapters, data collection for other studies, such as the Health and Retirement Study and the reassessments currently under way for the Taiwan SEBAS study, will offer yet additional national and cross-national data, including socioeconomic, psychosocial, and biological data, that can be used to replicate and extend findings outlined here.

The data and research potential generated from these studies will be greatly enhanced by coordinating efforts such as those undertaken by the Chicago Core on Biomarkers in Population-Based Aging Research (CCBAR) at the University of Chicago-NORC Center on Aging. Through a number of activities, CCBAR provides a central resource to help foster collaboration, exchange information, and promote interdisciplinary research related to biomarker collections in population-based health research and aging, including an interactive website, http://biomarkers.health-studies. org/studydemo.php.

It is instructive to consider the factors that affect the selection of biomarkers for inclusion in these studies. These factors include at least three common and critical considerations—two scientific and one logistical or financial. First, from a scientific standpoint, biomarkers were selected to include those needed to address primary substantive, scientific questions of interest to the study investigators. Second, also from a scientific standpoint, in cases in which there is a presumption from the beginning that the wider community of health researchers will ultimately use these databases to address other, as yet unspecified, questions, there is clearly the additional question of whether additional biomarkers (i.e., other than those already selected based on the substantive interests of study investigators) should be included. Here, considerations generally relate to whether there are major biological systems or processes that are likely to affect health for which data would otherwise be missing.

Perhaps most challenging are a third set of nonscientific considerations that relate to issues of feasibility both with respect to requirements for implementation of protocols to collect and process needed biospecimens and with financial considerations. Perhaps foremost are the logistical constraints imposed by time and handling requirements for obtaining many biological measurements. Examples include (1) the need for phlebotomy to collect venous blood (e.g., when dried blood spots cannot be used), (2) the need for fasted blood samples, which can constrain collection to morning hours, (3) the need for sample collection at specific times due to diurnal rhythms of parameters such as cortisol, necessitating collection at the same time of day for everyone and collection of multiple samples over time, (4) the need for blood or urine samples to be processed within a limited time frame (usually within a couple of hours). The selection of biomarkers for inclusion in a given study will thus necessarily be heavily influenced by what is possible in the context of specific study designs and logistical parameters. For example, the national scope of the HRS study precludes collection of venous blood so dried blood spots are being collected. Selection of biomarkers is thus restricted to those for which assays are available that can use dried blood spots rather than venous blood. By contrast, the MIDUS study is collecting a wide array of biomarkers because participants are being brought to regional clinical research centers (each in a hospital setting) where venous blood can be drawn first thing in the morning (allowing for fasted samples) and where these samples can be processed immediately.

Thus, the final set of biomarkers included in any studies will reflect both the underlying scientific questions investigators seek to address and what is possible given their logistical and financial constraints. Despite the demands and challenges inherent in incorporating biomarkers into social science surveys, continued research development and efforts in the directions presented herein are critical to understanding better the factors that affect patterns of biological aging and trajectories of health at older ages.

REFERENCES

Antonelli-Incalzi, R., Pedone, C., McDermott, M.M., Bandinelli, S., Miniati, B., Lova, R.M., Lauretani, F., and Ferrucci, L. (2006). Association between nutrient intake and peripheral artery disease: Results from the InCHIANTI study. *Atherosclerosis, 186*(1), 200-206.

Bandeen-Roche, K., Xue, Q.L., Ferrucci, L., Walston, J., Guralnik, J.M., Chaves, P., Zeger, S.L., and Fried, L.P. (2006). Phenotype of frailty: Characterization in the Women's Health and Aging Studies. *Journals of Gerontology Series A: Biological Sciences and Medical Sciences, 61*(3), 262-266.

Bassuk, S.S., Glass, T.A., and Berkman, L.F. (1999). Social disengagement and incident cognitive decline in community-dwelling elderly persons. *Annals of Internal Medicine, 131,* 165-173.

Bell, B., Rose, C.L., and Damon, A. (1972). The normative aging study: An interdisciplinary and longitudinal study of health and aging. *Aging and Human Development, 3*(1), 5-17.

Bennett, D.A., Schneider, J.A., Buchman, A.S., et al. (2005). The rush memory and aging project: Study design and baseline characteristics of the study cohort. *Neuroepidemiology, 25,* 163-175.

Berkman, L.F., and Kawachi, I. (Eds.) (2000). *Social epidemiology.* New York: Oxford Press.

Berkman, L.F., Seeman, T.E., Albert, M.A., et al. (1993). High, usual and impaired functioning in community-dwelling older men and women: Findings from the MacArthur Foundation Research Network on Successful Aging. *Journal of Clinical Epidemiology, 46,* 1129-1140.

Blaum, C.S., Volpato, S., Cappola, A.R., Chaves, P., Xue, Q.L., Guralnik, J.M., and Fried, L.P. (2005a). Diabetes, hyperglycaemia, and mortality in disabled older women: The Women's Health and Ageing Study I. *Diabetic Medicine, 22*(5), 543-550.

Blaum, C.S., Xue, Q.L., Michelon, E., Semba, R.D., and Fried, L.P. (2005b). The association between obesity and the frailty syndrome in older women: The Women's Health and Aging Studies. *Journal of the American Geriatrics Society, 53*(6), 1069-1070.

Ble, A., Cherubini, A., Volpato, S., Bartali, B., Walston, J.D., Windham, B.G., Bandinelli, S., Lauretani, F., Guralnik, J.M., and Ferrucci, L. (2006). Lower plasma vitamin E levels are associated with the frailty syndrome: The InCHIANTI study. *Journals of Gerontology Series A: Biological Sciences and Medical Sciences, 61*(3), 278-283.

Boomsma, D.I., Vink, J.M., van Beijsterveldt, T.C., de Geus, E.J., Beem, A.L., Mulder, E.J., Derks, E.M., Riese, H., Willemsen, G.A., Bartels, M., van den, Berg, M., Kupper, N.H., Polderman, T.J., Posthuma, D., Rietveld, M.J., Stubbe, J.H., Knol, L.I., Stroet, T., and van Baal, G.C. (2002). Netherlands Twin Register: A focus on longitudinal research. *Twin Research and Human Genetics, 5*(5), 401-406.

Boomsma, D.I., Willemsen, G., and Dolan, C.V., Hawkley, L.C., and Cacioppo, J.T. (2005). Genetic and environmental contributions to loneliness in adults: The Netherlands twin register study. *Behavioral Genetics, 35*(6), 745-752.

Boomsma, D.I., Cacioppo, J.T., Slagboom, P.E., and Posthuma, D. (2006). Genetic linkage and association analysis for loneliness in Dutch twin and sibling pairs points to a region on chromosome 12q23-24. *Behavioral Genetics, 36*(1), 137-146. Epub 2005 Dec 24.

Bretsky, P., Guralnik, J., Launer, L., Albert, M.A., and Seeman, T.E. (2003). The role of APOE-ε4 in longitudinal cognitive decline in an elderly cohort: MacArthur studies of successful aging. *Neurology, 60,* 1077-1081.

Burg, M.M., and Seeman, T.E. (1994). Families and health: The negative side of social ties. *Annals of Behavioral Medicine, 16,* 109-115.

Butler, J., Rodondi, N., Zhu, Y., Figaro, K., Fazio, S., Vaughan, D.E., Satterfield, S., Newman, A.B., Goodpaster, B., Bauer, D.C., Holvoet, P., Harris, T.B., de Rekeneire, N., Rubin, S., Ding, J., Kritchevsky, S.B, and Health ABC Study. (2006). Metabolic syndrome and the risk of cardiovascular disease in older adults. *Journal of the American College of Cardiology, 47*(8), 1595-1602.

Butler, R.N., Sprott, R., Warner, H., Bland, J., Feuers, R., Forster, M., Fillit, H., Harman, M., Hewitt, M., Hyman, M., Johnson, K., Kligman, E., McClearn, G., Nelson, J., Richardson, A., Sonntag, W., Weindruch, R., and Wolf, N. (2004). Biomarkers of aging: From primitive organisms to humans. *Journal of Gerontology: Biological Sciences, 59A,* 560-567.

Cacioppo, J.T., Hughes, M.E., Waite, L.J., Hawkley, L.C., and Thisted, R.A. (2006). Loneliness as a specific risk factor for depressive symptoms: Cross-sectional and longitudinal analyses. *Psychology and Aging, 21*(1), 140-151.

Cappola, A.R., Bandeen-Roche, K., Wand, G.S., Volpato, S., and Fried, L.P. (2001). Association of IGF-I levels with muscle strength and mobility in older women. *Journal of Clinical Endocrinology & Metabolism, 86*(9), 4139-4146.

Caspi, A., Sugden, K., Moffitt, T.E., et al. (2003). Influence of life stress on depression: Moderation by a polymorphism in the 5-HTT gene. *Science, 301,* 386-389.

Cesari, M., Penninx, B.W., Newman, A.B., Kritchevsky, S.B., Nicklas, B.J., Sutton-Tyrrell, K., Tracy, R.P., Rubin, S.M., Harris, T.B., and Pahor, M. (2003a). Inflammatory markers and cardiovascular disease (The Health, Aging, and Body Composition [Health ABC] Study). *The American Journal of Cardiology, 92*(5), 522-528.

Cesari, M., Penninx, B.W., Newman, A.B., Kritchevsky, S.B., Nicklas, B.J., Sutton-Tyrrell, K., Rubin, S.M., Ding, J., Simonsick, E.M., Harris, T.B., and Pahor, M. (2003b). Inflammatory markers and onset of cardiovascular events: Results from the Health ABC study. *Circulation, 108*(19), 2317-2322.

Cesari, M., Penninx, B.W., Pahor, M., Lauretani, F., Corsi, A.M., Rhys Williams, G., Guralnik, J.M., and Ferrucci, L. (2004a). Inflammatory markers and physical performance in older persons: The InCHIANTI study. *Journals of Gerontology Series A: Biological Sciences and Medical Sciences, 59*(3), 242-248.

Cesari, M., Penninx, B.W., Lauretani, F., Russo, C.R., Carter, C., Bandinelli, S., Atkinson, H., Onder, G., Pahor, M., and Ferrucci, L. (2004b). Hemoglobin levels and skeletal muscle: Results from the InCHIANTI study. *Journals of Gerontology Series A: Biological Sciences and Medical Sciences, 59*(3), 249-254.

Cesari, M., Pahor, M., Lauretani, F., Penninx, B.W., Bartali, B., Russo, R., Cherubini, A., Woodman, R., Bandinelli, S., Guralnik, J.M., and Ferrucci, L. (2005). Bone density and hemoglobin levels in older persons: Results from the InCHIANTI study. *Osteoporosis International, 16*(6), 691-699.

Chaves, P.H., Semba, R.D., Leng, S.X., Woodman, R.C., Ferrucci, L., Guralnik, J.M., and Fried, L.P. (2005). Impact of anemia and cardiovascular disease on frailty status of community-dwelling older women: The Women's Health and Aging Studies I and II. *Journals of Gerontology Series A: Biological Sciences and Medical Sciences, 60*(6), 729-735.

Cherubini, A., Martin, A., Andres-Lacueva, C., Di Iorio, A., Lamponi, M., Mecocci, P., Bartali, B., Corsi, A., Senin, U., and Ferrucci, L. (2005). Vitamin E levels, cognitive impairment, and dementia on older persons: The InCHIANTI study. *Neurobiology of Aging, 26*(7), 987-994.

Crimmins, E.M., Alley, D., Reynolds, S.L., Johnston, M., Karlamangla, A., and Seeman, T. (2005). Changes in biological markers of health: Older Americans in the 1990s. *Journals of Gerontology Series A: Biological Sciences and Medical Sciences, 60*(11), 1409-1413.

de Quervain, D.J.F., Henke, K., Aerni, A., Coluccia, D., Wollmer, M.A., Hock, C., et al. (2003). A functional genetic variation of the 5-HT2A receptor affects human memory. *Nature Neuroscience, 6*(11), 1141-1142.

Di Iorio, A., Cherubini, A., Volpato, S., Sparvieri, E., Lauretani, F., Franceschi, C., Senin, U., Abate, G., Paganelli, R., Martin, A., Andres-Lacueva, C., and Ferrucci, L. (2006). Markers of inflammation, vitamin E and peripheral nervous system function: The InCHIANTI study. *Neurobiology of Aging, 27*(9), 1280-1288.

Ding, J., Nieto, F.J., Beauchamp, N.J., Longstreth, W.T., Jr, Manolio, T.A., Hetmanski, J.B., and Fried, L.P. (2003). A prospective analysis of risk factors for white matter disease in the brain stem: The Cardiovascular Health Study. *Neuroepidemiology, 22*(5), 275-282.

Elosua, R., Bartali, B., Ordovas, J.M., Corsi, A.M., Lauretani, F., Ferrucci, L., and InCHIANTI Investigators. (2005). Association between physical activity, physical performance, and inflammatory biomarkers in an elderly population: The InCHIANTI study. *Journals of Gerontology Series A: Biological Sciences and Medical Sciences, 60*(6), 760-767.

Eley, T.C., Sugden, K., Corsico, A., Gregary, A.M., Sham, P., McGuffin, P., Plomin, R., and Craig, I.W. (2004). Gene-environment interaction analysis of serotonin system markers with adolescent depression. *Molecular Psychiatry, 9*, 908-915.

Epel, E.S., Blackburn, E.H., Lin, J., Dhabhar, F.S., Adler, N.E., Morrow, J.D., and Cawthon, R.M. (2004). Accelerated telomere shortening in response to life stress. *Proceedings of the National Academy of Sciences, USA, 101*(49), 17312-17315. Epub 2004 Dec 1.

Ewbank, D.C. (2002). Mortality differences by APOE genotype estimated from demographic synthesis. *Genetic Epidemiology*, (22), 146-155.

Ferrucci, L., Penninx, B.W., Volpato, S., Harris, T.B., Bandeen-Roche, K., Balfour, J., Leveille, S.G., Fried, L.P., and Guralnik, J.M. (2002). Change in muscle strength explains accelerated decline of physical function in older women with high interleukin-6 serum levels. *Journal of the American Geriatrics Society, 50*(12), 1947-1954.

Fratiglioni, L., Wang, H.X., Ericsson, K., Maytan, M., and Winblad, B. (2000). Influence of social network on occurrence of dementia: A community-based longitudinal study. *Lancet, 355*, 1315-1319.

Fried, L.P., Kronmal, R.A., Newman, A.B., Bild, D.E., Mittelmark, M.B., Polak, J.F., Robbins, J.A., and Gardin, J.M. (1998). Risk factors for 5-year mortality in older adults: The Cardiovascular Health Study. *Journal of the American Medical Association, 279*(8), 585-592.

Fried, L.P., Bandeen-Roche, K., Chaves, P.H.M., and Johnson, B.A. (2000). Preclinical mobility disability predicts incident mobility disability in older women. *Journals of Gerontology Series A: Biological Sciences and Medical Sciences, 55A*(1), M43-M52.

Fried, L.P., Tangen, C.M., Walston, J., Newman, A.B., Hirsch, C., Gottdiener, J., Seeman, T., Tracy, R., Kop, W.J., Burke, G., McBurnie, M.A., and Cardiovascular Health Study Collaborative Research Group. (2001). Frailty in older adults: Evidence for a phenotype. *Journals of Gerontology Series A: Biological Sciences and Medical Sciences, 56*(3), M146-M156.

Friedman, E.M., Hayney, M.S., Love, G.D., Urry, H.L., Rosenkranz, M.A., Davidson, R.J., Singer, B.H., and Ryff, C.D. (2005). Social relationships, sleep quality, and interleukin-6 in aging women. *Proceedings of the National Academy of Sciences, USA, 102*(51), 18757-18762.

Gatz, M., Pedersen, N.L., Berg, S., Johansson, B., et al. (1997). Heritability for Alzheimer's disease: The study of dementia in Swedish twins. *Journals of Gerontology Series A: Biological Sciences and Medical Sciences*, (52A), M117-M125.

Gillespie, N.A., Whitfield, J.B., Williams, B., Heath, A.C., and Martin, N. (2005). The relationship between stressful life events, the serotonin transporter (5-HTTLPR) genotype, and major depression. *Psychological Medicine, 35*, 101-111.

Gottesman, I.I., and Gould, T.D. (2003). The endophenotype concept in psychiatry: Etymology and strategic intentions. *American Journal of Psychiatry, 160*(4), 636-645.

Grabe, H.J., Lange, M., Wolff, B., Volzke, H., Lucht, M., Freyberger, H.J., John, U., and Cascorbi, I. (2005). Mental and physical distress is modulated by a polymorphism in the 5-HT transporter gene interacting with social stressors and chronic disease burden. *Molecular Psychiatry, 10*, 220-224.

Gruenewald, T.L., Seeman, T.E., Ryff, C.D., Karlamangla, A.S., and Singer, B.H. (2006). Early warning biomarkers: Combinations predictive of later life mortality. *Proceedings of the National Academy of Sciences, USA, 103*(3i8), 14158-14163.

Harris, J.R. (2005). Research on environmental effects in genetic studies of aging: Introduction. *The Journal of Gerontology B: Psychological and Social Sciences, 60*, Spec No. 1:5-6.

Harris, J.R. (2007). Genetics, social behaviors, social environments, and aging. *Twin Research and Human Genetics, 10*(2), 235-240.

Hawkley, L.C., and Cacioppo, J.T. (2003). Loneliness and pathways to disease. *Brain, Behavior, and Immunity, 17*, S98-S105.

Hawkley, L.C., Masi, C.M., Berry, J.D., and Cacioppo, J.T. (2006). Loneliness is a unique predictor of age-related differences in systolic blood pressure. *Psychology and Aging, 21*, 152-164.

Heller, D.A., de Faire, U., Pedersen, N.L., Dahlen, G., and McClearn, G.E. (1993). Genetic and environmental influences on serum lipid levels in twins. *New England Journal of Medicine, 22:328*(16), 1150-1156.

Heller, D.A., Pedersen, N.L., de Faire, U., and McClearn, G.E. (1994). Genetic and environmental correlations among serum lipids and apolipoproteins in elderly twins reared together and apart. *American Journal of Human Genetics*, (6), 1255-1267.

Hong, Y., Dahlen, G., Pedersen, N., Heller, D., McClearn, G.E., and deFaire, U. (1995). Potential environmental effects on adult lipoprotein(a) levels: Results from Swedish twins. *Atherosclerosis, 17*, 295-304.

Hong, Y., Pedersen, N.L., Brismar, K., and de Faire, U. (1997). Genetic and environmental architecture of the features of the insulin-resistance syndrome. *American Journal of Human Genetics, 60*(1), 143-152.

Hong, Y., Pedersen, N.L., Egberg, N., and de Faire, U. (1999). Genetic effects for plasma factor VII levels independent of and in common with triglycerides. *Journal of Thrombosis and Haemostasis, 81*(3), 382-386.

House, J.S., Landis, K.R., and Umberson, D. (1988). Social relationships and health. *Science, 241*, 540-545.

House, J.S., Lepkowski, J.M., Kinney, A.M., Mero, R.P., Kessler, R.C., and Herzog, A.R. (1994). The social stratification of aging and health. *Journal of Health and Social Behavior, 35*, 213-234.

Hu, P., Reuben, D.B., Crimmins, E., Harris, T.B., Huang, M.H., and Seeman, T.E. (2005). The effects of serum beta-carotene level and burden of inflammation on all-cause mortality in high-functioning older persons: MacArthur Studies of Successful Aging. *Journals of Gerontology Series A: Biological Sciences and Medical Sciences, 59*, 849-854.

Hu, P., Wagle, N., Goldman, N., Weinstein, M., and Seeman, T.E. (2006). The associations between socioeconomic status, allostatic load and measures of health in older Taiwanese persons: Taiwan social environment and biomarkers of aging study. *Journal of Biosocial Science, 20*, 1-12.

Huang, T.L., Zandi, P.P., Tucker, K.L., Fitzpatrick, A.L., Kuller, L.H., Fried, L.P., Burke, G.L., and Carlson, M.C. (2005). Benefits of fatty fish on dementia risk are stronger for those without APOE epsilon4. *Neurology, 65*(9), 1409-1414.

Iliadou, A., Lichtenstein, P., de Faire, U., and Pedersen, N.L. (2001). Variation in genetic and environmental influences in serum lipid and apolipoprotein levels across the lifespan in Swedish male and female twins. *American Journal of Medical Genetics, 22:102*(1), 48-58.

Iliadou, A., and Sneider, H. (2004). Genetic epidemiological approaches in the study of risk factors for cardiovascular disease. *European Journal of Epidemiology,* (19), 209-217.

Institute of Medicine. (2006). *Genes, behavior, and the social environment*. Washington, DC: The National Academies Press.

Jansson, M., Gatz, M., Berg, S., Johansson, B., Malmberg, B., McClearn, G.E., Schalling, M., and Pedersen, N.L. (2003). Association between depressed mood in the elderly and a 5-HTR2A gene variant. *American Journal of Medical Genetics Part B: Neuropsychiatric Genetics, 120*(1), 79-84.

Jenny, N.S., Arnold, A.M., Kuller, L.H., Sharrett, A.R., Fried, L.P., Psaty, B.M., and Tracy, R.P. (2006). Soluble intracellular adhesion molecule-1 is associated with cardiovascular disease risk and mortality in older adults. *Journal of Thrombosis and Haemostasis, 4*(1), 107-113.

Kado, D.M., Karlamangla, A.S., Huang, M.H., Troen, A., Rowe, J.W., Selhub, J., and Seeman, T.E. (2005). Homocysteine versus the vitamins folate, B_6 and B_{12} as predictors of cognitive function and decline in older high functioning adults: MacArthur Studies of Successful Aging. *American Journal of Medicine, 118*, 161-167.

Karlamangla, A.S., Singer, B.H., McEwen, B.S., Rowe, J.W., and Seeman, T.E. (2002). Allostatic lead as a predictor of functional decline: MacArthur Studies of Successful Aging. *Journal of Clinical Epidemiology, 55*(7), 696-710.

Karlamangla, A.S., Singer, B.S., Reuben, D., and Seeman, T.E. (2004). Increase in serum non-high-density lipoprotein cholesterol may be beneficial in some high-functioning older adults: MacArthur Studies of Successful Aging. *Journal of the American Geriatrics Society, 52,* 487-494.

Karlamangla, A.S., Singer, B., Greendale, G., and Seeman, T. (2005a). Increase in epinephrine excretion is associated with cognitive decline in elderly men: MacArthur Studies of Successful Aging. *Psychoneuroendocrinology, 30*(5), 453-460.

Karlamangla, A.S., Singer, B.H., Chodosh, J., McEwen, B.S., and Seeman, T.E. (2005b). Urinary cortisol excretion as a predictor of incident cognitive impairment. *Neurobiology of Aging, 26S,* S80-S84.

Karlamangla, A.S., Singer, B.H., and Seeman, T.E. (2006). Reduction in allostatic load in older adults is associated with lower all-cause mortality risk: MacArthur Studies of Successful Aging. *Psychosomatic Medicine, 68,* 500-507.

Kaufman, J., Yang, B.Z., Douglas-Palumberi, H., Houshyar, S., Lipshitz, D., Krystal, J., and Gelernter, J. (2004). Social supports and serotonin transporter gene moderate depression in maltreated children. *Proceedings of the National Academy of Sciences, USA, 101,* 17316-17321.

Kendler, K.S., Kuhn, J.W., Vittum, J., Prescott, C.A., and Riley, B. (2005). The interaction of stressful life events and a serotonin transporter polymorphism in the prediction of episodes of major depression: A replication. *Archives of General Psychiatry, 62,* 529-535.

Kiecolt-Glaser, J.K., Loving, T.J., Stowell, J.R., Malarkey, W.B., Lemeshow, S., Dickinson, S.L., and Glaser, R. (2005). Hostile marital interactions, proinflammatory cytokine production, and wound healing. *Archives of General Psychiatry, 62,* 1377-1384.

Koster, A., Penninx, B.W., Bosma, H., Kempen, G.I., Harris, T.B., Newman, A.B., Rooks, R.N., Rubin, S.M., Simonsick, E.M., van Eijk, J.T., and Kritchevsky, S.B. (2005a). Is there a biomedical explanation for socioeconomic differences in incident mobility limitation? *Journals of Gerontology Series A: Biological Sciences and Medical Sciences, 60*(8), 1022-1027.

Koster, A., Penninx, B.W., Bosma, H., Kempen, G.I., Newman, A.B., Rubin, S.M., Satterfield, S., Atkinson, H.H., Ayonayon, H.N., Rosano, C., Yaffe, K., Harris, T.B., Rooks, R.N., Van Eijk, J.T., and Kritchevsky, S.B. (2005b). Socioeconomic differences in cognitive decline and the role of biomedical factors. *Annals of Epidemiology, 15*(8), 564-571.

Koster, A., Bosma, H., Penninx, B.W., Newman, A.B., Harris, T.B., van Eijk, J.T., Kempen, G.I., Simonsick, E.M., Johnson, K.C., Rooks, R.N., Ayonayon, H.N., Rubin, S.M., Kritchevsky, S.B., and Health ABC Study. (2006). Association of inflammatory markers with socioeconomic status. *Journals of Gerontology Series A: Biological Sciences and Medical Sciences, 61*(3), 284-290.

Kubzansky, L.D., Kawachi, I., and Sparrow, D. (1999). Socioeconomic status, hostility, and risk factor clustering in the Normative Aging Study: Any help from the concept of allostatic load? *Annals of Behavioral Medicine, 21*(4), 330-338.

Lichtenstein, P., de Faire, U., Floderus, B., Svartenren, M., Svedberg, P., and Pedersen, N.L. (2002). The Swedish Twin Registry: A unique resource for clinical, epidemiological, and genetic studies. *Journal of International Medical Research,* (252), 184-205.

Loucks, E.B., Berkman, L.F., Gruenewald, T.L., and Seeman, T.E. (2006). Relation of social integration to inflammatory marker concentrations in men and women 70-79 years. *American Journal of Cardiology, 97*(7), 1010-1016.

McClearn, G.E., Johansson, B., Berg, S., Pedersen, N.L., Ahern, F., Petrill, S.A., and Plomin, R. (1997). Substantial genetic influence on cognitive abilities in twins 80 or more years old. *Science, 276*(5318), 1560-1563.

McDermott, M.M., Guralnik, J.M., Corsi, A., Albay, M., Macchi, C., Bandinelli, S., and Ferrucci, L. (2005). Patterns of inflammation associated with peripheral arterial disease: The InCHIANTI study. *American Heart Journal, 150*(2), 276-281.

McGuire, S., and Clifford, J. (2000). Genetic and environmental contributions to loneliness in children. *Psychological Sciences, 11,* 487-491.

Moceri, V.M., Kukull, W.A., Emanual, I., van Belle, G., Starr, J.R., Schellenberg, G.D., McCormick, W.C., Bowen, J.D., Teri, L., and Larson, E.B. (2001). Using census data and birth certificates to reconstruct the early-life socioeconomic environment and the relation to the development of Alzheimer's disease. *Epidemiology,* (12), 383-389.

National Research Council (2001a). *Cells and surveys: Should biological measures be included in social science research?* Committee on Population, C.E. Finch, J.W. Vaupel, and K. Kinsella, Eds. Commission on Behavioral and Social Sciences and Education. Washington, DC: National Academy Press.

National Research Council. (1997). *Between Zeus and the salmon: The biodemography of longevity.* Committee on Population. K.W. Wachter and C.E. Finch, Eds. Commission on Behavioral and Social Sciences and Education. Washington, DC: National Academy Press.

National Research Council (2001b). *New horizons in health: An integrative approach.* Committee on Future Directions for Behavioral and Social Sciences Research at the National Institutes of Health, B.H. Singer and C.D. Ryff, Eds. Board on Behavioral, Cognitive, and Sensory Sciences. Commission on Behavioral and Social Sciences and Education. Washington, DC: National Academy Press.

Niaura, R., Banks, S.M., Ward, K.D., Stoney, C.M, Spiro, A., III, Aldwin, C.M., Landsberg, L., and Weiss, S.T. (2000). Hostility and the metabolic syndrome in older males: The normative aging study. *Psychosomatic Medicine, 62*(1), 7-16.

Nilsson, S.E., Evrin, P.E., Tryding, N., Berg, S., McClearn, G., and Johansson, B. (2003a). Biochemical values in persons older than 82 years of age: Report from a population-based study of twins. *Scandinavian Journal of Clinical and Laboratory Investigation, 63*(1), 1-13.

Nilsson, S.E., Takkinen, S., Tryding, N., Evrin, P.E., Berg, S., McClearn, G., and Johansson, B. (2003b). Association of biochemical values with morbidity in the elderly: A population-based Swedish study of persons aged 82 or more years. *Scandinavian Journal of Clinical and Laboratory Investigation, 63*(7-8), 457-466.

Onder, G., Penninx, B.W., Cesari, M., Bandinelli, S., Lauretani, F., Bartali, B., Gori, A.M., Pahor, M., and Ferrucci, L. (2005). Anemia is associated with depression in older adults: Results from the InCHIANTI study. *Journals of Gerontology Series A: Biological Sciences and Medical Sciences, 60*(9), 1168-1172.

Payton, M., Riggs, K.M., Spiro, A., III, Weiss, S.T., and Hu, H. (1998). Relations of bone and blood lead to cognitive function: The VA Normative Aging Study. *Neurotoxicology and Teratology, 20*(1), 19-27.

Pedersen, N.L., McClearn, G.E., Plomin, R., Nesselroade, J.R., Berg, S., DeFaire, U., Mortimer, J.A., et al. (1991). The Swedish Adoption Twin Study of Aging: An update. *Acta Geneticae Medicae et Gemellologiae* (40), 7-20.

Penninx, B.W., Guralnik, J.M., Ferrucci, L., Fried, L.P., Allen, R.H., and Stabler, S.P. (2000). Vitamin B(12) deficiency and depression in physically disabled older women: Epidemiologic evidence from the Women's Health and Aging Study. *American Journal of Psychiatry, 157*(5), 715-721.

Penninx, B.W., Pahor, M., Cesari, M., Corsi, A.M., Woodman, R.C., Bandinelli, S., Guralnik, J.M., and Ferrucci, L. (2005). Anemia is associated with disability and decreased physical performance and muscle strength in the elderly. *Journal of the American Geriatric Society, 52*(5), 719-724.

Psaty, B.M., Anderson, M., Kronmal, R.A., Tracy, R.P., Orchard, T., Fried, L.P., Lumley, T., Robbins, J., Burke, G., Newman, A.B., and Furberg, C.D. (2004). The association between lipid levels and the risk of incident myocardial infarction, stroke, and total mortality: The Cardiovascular Health Study. *Journal of the American Geriatric Society*, 52(10), 1639-1647.

Rantanen, T., Guralnik, J.M., Sakari-Rantala, R., Leveille, S., Simonsick, E.M., Ling, S., and Fried, L.P. (1999). Disability, physical activity, and muscle strength in older women: The Women's Health and Aging Study. *Archives of Physical Medicine and Rehabilitation*, 80(2), 130-135.

Ray, A.L., Semba, R.D., Walston, J., Ferrucci, L., Cappola, A.R., Ricks, M.O., Xue, Q.L., and Fried, L.P. (2006). Low serum selenium and total carotenoids predict mortality among older women living in the community: The Women's Health and Aging Studies. *Journal of Nutrition*, 136(1), 172-176.

Reuben, D.B., Cheh, A.I., Harris, T.B., Ferrucci, L., Rowe, J.W., Tracy, R.P., and Seeman, T.E. (2002). Peripheral blood markers of inflammation predict mortality and functional decline among high functioning community-dwelling older persons. *Journal of the American Geriatric Society*, 50(4), 638-644.

Reuben, D.B., Judd-Hamilton, L., Harris, T.B., and Seeman, T.E. (2003, August). The associations between physical activity and inflammatory markers in high functioning older persons: MacArthur Studies of Successful Aging. *Journal of the American Geriatrics Society*, 51(8), 1125-1130.

Reynolds, C.A., Jansson, M., Gatz, M., and Pedersen, N.L. (2006a). Longitudinal change in memory performance associated with HTR2A polymorphism. *Neurobiology of Aging*, 27(1), 150-154.

Reynolds, C.A., Prince, J.A., Feuk, L., Brookes, A.J., Gatz, M., and Pedersen, N.L. (2006b). Longitudinal memory performance during normal aging: Twin association models of APOE and other Alzheimer candidate genes. *Behavioral Genetics*, 36(2), 185-194. Epub 2006 Jan 10.

Reynolds, C.A., Gatz, M., Berg, S., and Pedersen, N.L. (2007). Genotype-environment interactions: Cognitive aging and social factors. *Twin Research and Human Genetics*, 10(2), 241-254.

Rutter, M., Moffitt, T.E., and Caspi, A. (2006). Gene-environment interplay and psychopathology: Multiple varieties but real effects. *Journal of Child Psychology and Psychiatry*, 47(3-4), 226-261.

Ryff, C.D., and Singer, B.H. (Eds.) (2001). *Emotion, social relationships, and health*. New York: Oxford University Press.

Ryff, C., and Singer, B.H. (2005). Social environments and the genetics of aging: Advancing knowledge of protective health mechanisms. *The Journal of Gerontology B: Psychological and Social Sciences*, 60, 12-23.

Ryff, C.D., Singer, B.H., and Dienberg Love, G. (2004). Positive health: Connecting well-being with biology. *Philosophical Transactions of the Royal Society B: Biological Sciences*, 359, 1383-1394.

Ryff, C.D., Dienberg Love, G., Urry, H.L., Muller, D., Rosenkranz, M.A., Friedman, E.M., Davidson, R.J., and Singer, B. (2006). Psychological well-being and ill-being: Do they have distinct or mirrored biological correlates? *Psychotherapy and Psychosomatics*, 75(2), 85-95.

Saxton, J., Lopez, O.L., Ratcliff, G., Dulberg, C., Fried, L.P., Carlson, M.C., Newman, A.B., and Kuller, L. (2004). Preclinical Alzheimer disease: Neuropsychological test performance 1.5 to 8 years prior to onset. *Neurology*, 63(12), 2341-2347.

Seeman, T.E. (1996). Social ties and health. *Annals of Epidemiology*, 6, 442-451.

Seeman, T.E., Berkman, L.F., Blazer, D., and Rowe, J. (1994). Social ties and support and neuroendocrine function: MacArthur Studies of Successful Aging. *Annals of Behavioral Medicine, 16,* 95-106.

Seeman, T.E., Berkman, L.F., Kohout, F., LaCroix, A., Glynn, R., and Blazer, D. (1993). Intercommunity variations in the association between social ties and mortality in the elderly: A comparative analysis of three communities. *Annals of Epidemiology, 3,* 325-335.

Seeman, T.E., Bruce, M.L., and McAvay, G. (1996). Social network characteristics and onset of ADL disability: MacArthur Studies of Successful Aging. *Journal of Gerontology: Psychological Sciences, 51B,* S191-S200.

Seeman, T.E., Crimmins, E., Bucur, A., Huang, M.H., Singer, B., Gruenewald, T., Berkman, L.F., and Reuben, D.B. (2004a). Cumulative biological risk and socioeconomic differences in mortality: MacArthur Studies of Successful Aging. *Social Science and Medicine, 58,* 1985-1997.

Seeman, T.E., Glei, D., Goldman, N., Weinstein, M., Singer, B., and Lin, Y-H. (2004b). Social relationships and allostatic load in Taiwanese elderly and near elderly. *Social Science and Medicine, 59,* 2245-2257.

Seeman, T.E., Huang, M.H., Bretsky, P., Crimmins, E., Launer, L., and Guralnik, J.M. (2005). Education and APOE-ε4 in longitudinal cognitive decline: MacArthur Studies of Successful Aging. *Journal of Gerontology: Psychological Sciences, 60*(2), P74-P83.

Seeman, T.E., Kaplan, G.A., Knudsen, L., Cohen, R., and Guralnik, J. (1987). Social ties and mortality in the elderly: A comparative analysis of age-dependent patterns of association. *American Journal of Epidemiology, 126,* 714-723.

Seeman, T.E., Lusignolo, T., Berkman, L., and Albert, M. (2001a). Social environment characteristics and patterns of cognitive aging: MacArthur Studies of Successful Aging. *Health Psychology, 20,* 243-255.

Seeman, T.E., and McEwen, B.S. (1996). Impact of social environment characteristics on neuroendocrine regulation. *Psychosomatic Medicine, 58*(5), 459-471.

Seeman, T.E., McEwen, B.S., Singer, B., Albert, M., and Rowe, J.W. (1997a). Increase in urinary cortisol excretion and declines in memory: MacArthur Studies of Successful Aging. *Journal of Clinical Endocrinology & Metabolism, 82,* 2458-2465.

Seeman, T.E, Merkin, S., Crimmins, E., Koretz, B., and Karlamangla, A. (no date). Socioeconomic status, ethnicity, and cumulative biological risk profiles in a national sample of U.S. adults: NHANES III. (manuscript in preparation).

Seeman, T.E., Singer, B., Horwitz, R., and McEwen, B.S. (1997b). The price of adaptation—allostatic load and its health consequences: MacArthur Studies of Successful Aging. *Archives of Internal Medicine, 157,* 2259-2268.

Seeman, T.E., Singer, B., Rowe, J., and McEwen, B. (2001b). Exploring a new concept of cumulative biological risk—allostatic load and its health consequences: MacArthur Studies of Successful Aging. *Proceedings of the National Academy of Sciences, USA, 98*(8), 4770-4775.

Seeman, T.E., Singer, B., Ryff, C., and Levy-Storms, L. (2002, May/June). Psychosocial factors and the development of allostatic load. *Psychosomatic Medicine, 64,* 395-406.

Shadlen, M.F., Larson, E.B., Wang, L., Phelan, E.A., McCormick, W.C., Jolley, L., Teri, L., and van Belle, G. (2005). Education modifies the effect of apolipoprotein epsilon 4 on cognitive decline. *Neurobiology of Aging, 26*(1), 17-24.

Simonsick, E.M., Maffeo, C.E., Rogers, S.K., Skinner, E.A., Davis, D., Guralnik, J.M., and Fried, L.P. (1997). Methodology and feasibility of a home-based examination in disabled older women: The Women's Health and Aging Study. *Journals of Gerontology, 52A*(5), M264-M274.

Singer, B., and Ryff, C. (1999). Hierarchies of life histories and associated health risks. *Annals of New York Academy of Sciences, 896,* 96-115.

Singer, B., Ryff, C., and Seeman, T.E. (2004). Allostasis, homeostasis, and the cost of physiological adaptation. In J. Schulkin (Ed.), *Operationalizing allostatic load* (pp. 113-149). Cambridge, England: Cambridge University Press.

Sneider, H., van Doornen, L.J., and Boomsma, D.I. (1997). The age dependency of gene expression for plasma lipids, lipoproteins, and apolipoproteins. *American Journal of Human Genetics,* (60), 638-650.

Snowdon, D.A., Kemper, S.J., Mortimer, J.A, Wekstein, D.R., and Markesbery, W.R. (1996). Linguistic ability in early life and cognitive function and Alzheimer's disease in late life: Findings from the Nun study. *Journal of the American Medical Association, 275*(7), 528-532.

Surtees, P.G., Wainwright, N.W., Willis-Owen, S.A., Luben, R., Day, N.E., and Flint, J. (2006). Social adversity, the serotonin transporter (5-HTTLPR) polymorphism, and major depressive disorder. *Biological Psychology, 59,* 224-229.

Tang, M.X., Stern, Y., Marder, K., Bell, K., Gurland, B., Lantigua, R., Andrews, H., Feng, L., Tycko, B., and Mayeux, R. (1998). The APOE-epsilon4 allele and the risk of Alzheimer's disease among African Americans, Whites, and Hispanics. *Journal of the American Medical Association, 11,* 751-575.

Taylor, S.E., Repetti, R.L., and Seeman, T. (1997). Health psychology: What is an unhealthy environment and how does it get under the skin? *Annual Review of Psychology, 48,* 411-447.

Todaro, J.F., Con, A., Niaura, R., Spiro, A., III, Ward, K.D., and Roytberg, A. (2005). Combined effect of the metabolic syndrome and hostility on the incidence of myocardial infarction (the Normative Aging Study). *American Journal of Cardiology, 96*(2), 221-226.

Tomaka, J., Thompson, S., and Palacios, R.J. (2005). High homocysteine and low B vitamins predict cognitive decline in aging men: The Veterans Affairs Normative Aging Study. *American Journal of Clinical Nutrition, 82*(3), 627-635.

Tomaka, J., Thompson, S., and Palacios, R.J. (2006, June). The relation of social isolation, loneliness, and social support to disease outcomes among the elderly. *Aging Health, 18*(3), 359-384.

Tucker, K.L., Qiao, N., Scott, T., Rosenbert, I., and Spiro, A., III (2005). High homocysteine and low B vitamins predict cognitive decline in aging men: The Veterans Affairs normative aging study. *American Journal of Clinical Nutrition, 82*(3), 627-635.

Uchino, B.N., Cacioppo, J.T., and Kiecolt-Glase, J.K. (1996). The relationship between social support and physiological processes: A review with emphasis on underlying mechanisms and implications for health. *Psychological Bulletin, 119,* 488-531.

Visser, M., Pahor, M., Taaffe, D.R., Goodpaster, B.H., Simonsick, E.M., Newman, A.B., Nevitt, M., and Harris, T.B. (2002). Relationship of interlukin-6 and tumor necrosis factor-alpha with muscle mass and muscle strength in elderly men and women: The Health ABC Study. *Journals of Gerontology Series A: Biological Sciences and Medical Sciences, 57*(5), M326-M332.

Volpato, S., Guralnik, J.M., Fried, L.P., Remaley, A.T., Cappola, A.R., and Launer, L.J. (2002). Serum thyroxine level and cognitive decline in euthyroid older women. *Neurology, 58*(7), 1055-1061.

von Kanel, R., Dimsdale, J.E., Ancoli-Israel, S., Mills, P.J., Patterson, T.L., McKibbin, C.L., Archuleta, C., and Grant, I. (2006). Poor sleep is associated with higher plasma proinflammatory cytokine interleukin-6 and procoagulant marker fibrin D-dimer in older caregivers of people with Alzheimer's disease. *Journal of the American Geriatrics Society, 54*(3), 431-437.

Walston, J., McBurnie, M.A., Newman, A., Tracy, R.P., Kop, W.J., Hirsch, C.H., Gottdiener, J., and Fried, L.P. (2002). Frailty and activation of the inflammation and coagulation systems with and without clinical comorbidities: Results from the Cardiovascular Health Study. *Archives of Internal Medicine, 162*(20), 2333-2341.

Walston, J., Arking, D.E., Fallin, D., Li, T., Beamer, B., Xue, Q., Ferrucci, L., Fried, L.P., and Chakravarti, A. (2005). IL-6 gene variation is not associated with increased serum levels of IL-6, muscle, weakness, or frailty in older women. *Experimental Gerontology, 40*(4), 344-352.

Walston, J., Xue, Q., Semba, R.D., Ferrucci, L., Cappola, A.R., Ricks, M., Guralnik, J., and Fried, L.P. (2006). Serum antioxidants, inflammation, and total mortality in older women. *American Journal of Epidemiology, 163*(1), 18-26.

Weaver, J., Huang, M.H., Albert, M., Harris, T., Rowe, J., and Seeman, T.E. (2002). Interleukin-6 as a predictor of cognitive function and cognitive decline: MacArthur Studies of Successful Aging. *Neurology, 59*, 371-378.

Wen, M., Hawkley, L.C., and Cacioppo, J.T. (2006). Objective and perceived neighborhood environment, individual SES and psychosocial factors, and self-rated health: An analysis of older adults in Cook County, Illinois. *Social Science Medicine, 63*, 2575-2590.

Wilson, R.S., Bienias, J.L., Evans, D.A., et al. (2004). The Religious Orders Study: Overview and relation between change in cognitive and motor speed. *Aging Neuropsychology and Cognition, 11*, 280-303.

Yende, S., Waterer, G.W., Tolley, E.A., Newman, A.B., Bauer, D.C., Taaffe, D.R., Jensen, R., Crapo, R., Rubin, S., Nevitt, M., Simonsick, E.M., Satterfield, S., Harris, T., and Kritchevsky, S.B. (2006). Inflammatory markers are associated with ventilatory limitation and muscle dysfunction in obstructive lung disease in well functioning elderly subjects. *Thorax, 61*(1), 10-16.

Zhang, H.P., and Singer, B. (1999). *Recursive partitioning in the health sciences.* New York: Springer-Verlag.

6

The Women's Health Initiative: Lessons for the Population Study of Biomarkers

Robert B. Wallace

Much of the discussion in this workshop has been directed to the study of biomarkers in cohort studies. My commentary addresses some lessons learned from the Women's Health Initiative (WHI), a research program comprising a set of large, randomized clinical trials as well as a cohort study. These lessons derive from the intricate and complex management issues relevant to multisite collaborative studies, the challenges of establishing and maintaining a large biospecimen repository, the difficulties of providing support for many investigative groups, the response to the changing bioethical environment, and the potential for future biomarker applications in the health and social sciences.

WHI began in 1992 and is ongoing. Its main goal has been to evaluate a series of interventions with the potential to prevent the onset of important chronic conditions in postmenopausal women, but, as discussed below, once the study data and biological specimens were in place, the opportunities for additional studies, some not previously planned, became substantial. In this report I briefly review the structure of WHI, describe the biomarker experience to date, discuss some of the strengths, weaknesses, and issues that arose in this very large, multicenter study, and offer general suggestions for further exploitation and investigation of stored specimens. In addition, I address some of the social and organizational aspects of managing biomarker studies and study investigators, as well as the large data sets, specimen collections, and scientific applications.

STRUCTURE OF THE WHI

WHI is comprised of a program office at the National Heart, Lung, and Blood Institute, National Institutes of Health (NIH), 40 clinical sites at which the participants are seen, a clinical coordinating center at the Fred Hutchinson Cancer Center in Seattle, and other supporting academic and laboratory units. The study design has been described in detail (Women's Health Initiative Study Group, 1998). Basically, from 1993 to 1997, postmenopausal women ages 50-79 were recruited into a set of three large, preventive, randomized, placebo-controlled trials, listed as follows with the primary outcomes: (1) estrogen alone or estrogen plus progestin therapy to assess the effect on coronary heart disease; (2) low-fat diet to prevent incident breast cancer; and (3) calcium and vitamin D dietary supplements to prevent clinical fractures. The last trial, the calcium/vitamin D intervention, was a superimposed randomization on the participants of the other two trials and did not include separate participants. Of note, that makes the interpretation of the outcomes a bit more complex, despite the randomized experimental design, because the impact of the other intervention(s) must be accounted for in all analyses of outcomes.

Recruitment to WHI was conducted by the 40 participating clinical sites, using such methods as media advertising, direct population mailing, community lectures and public service announcements, and celebrity endorsements. This yielded about 353,000 women who were evaluated for participation in WHI. About 68,000 enrolled in the trials. Of the remainder, those who were ineligible, unwilling, or unable to participate in the clinical trials were invited into a cohort study, the Observational Study (OS), in which no interventions took place; about 93,000 women began full participation in the OS. There were many specific reasons why potential trial participants opted for OS participation. Three of the most important were not being able to spend the time undergoing the trials' rigors; unwillingness to be randomized with respect to the use of hormonal therapy; and having too low a percentage of dietary fat calories, making them ineligible for the dietary modification trial.

Participants in the clinical trials as well as the OS had the same, extensive baseline data collection, including medical history, risk factors for important chronic illnesses, social and behavioral characteristics, mental health characteristics, and diet and dietary supplement use. In addition, and important to this discussion, virtually all participants had blood obtained through venipuncture, which was aliquoted and stored, and much of it is available today for additional studies, in addition to the many ancillary biomarker studies that have already taken place. Additional blood samples were obtained from some participants at selected intervals after the study onset. Of interest, aside from some planned safety studies related to the preventive trials, there were no specific plans on how the specimens might be applied or what determinations would take place. Yet many important issues have been addressed, related to modern and evolving scientific hypotheses, and some specimens have been used to better understand certain adverse effects of the trials that were not anticipated when they began.

Although the major results of these trials have been published (Writing Group for the Women's Health Initiative, 2002; Women's Health Initiative Steering Committee, 2004; Prentice, 2006; Jackson et al., 2006), much more work is occurring, and WHI has resulted in over 100 publications to date. Of import to the application of biomarker studies, there are many secondary and ancillary scientific hypotheses being evaluated with WHI specimens. These studies have consumed some of the blood specimens acquired in WHI, an important issue in itself. The specimens collected in WHI continue to be used for many scientific purposes, under the heading of "ancillary studies," which are described below.

SPECIMEN ACQUISITION AND THE INSTITUTIONAL REVIEW BOARD

Although many issues regarding the ethics of specimen collection are beyond the scope of this discussion, some issues and situations concerning the WHI experience are worthy of note. A general problem of multicenter studies is the requirement that all institutional review boards (IRBs) approve study protocols. At times that was true of WHI, in which there were 40 clinical site IRBs, the coordinating center IRB, and ethical review at NIH and perhaps other sites, such as at institutions performing biomarker determinations. Because WHI took place over many years, scientific advances occurring after the study was designed offered the possibility of new biomarker determinations on existing specimens. However, the initial informed consent for blood or urine or other specimen collection may not have covered some of the new, unexpected, and unspecified determinations that had not yet been conceived or invented. This can be solved in part by general consent form language that allows future, unspecified measures to be studied.

Yet this may not cover all situations. An example is when a biomarker is developed that turns out to have the capacity to screen for a specific medical condition. For example, early studies of prostate-specific antigen (PSA) may have had biological import only, but as it became a candidate for clinical prostate cancer screening, studies conducting PSA determinations would need to have medical expertise available to dispose of clinically elevated levels. The study of biomarkers that turn out to have clinical import raises several important bioethical issues, requiring considerable thought and medical consideration. Because of changing scientific interests and biomarker applications, WHI made a decision to reconsent surviving participants in 2005 for further, unspecified use of stored blood specimens, in order to modernize consent form language, address new tests being performed, and remind participants that the specimens were still in use. Both in collaborative and individual studies that maintain bio-specimen repositories, oversight by IRBs is increasing and becoming more complex. For example, some universities require that all biorepositories have separate annual review, regardless of use.

MANAGING BIOMARKER STUDIES

The number of WHI investigators is large and may well become larger as other investigators take advantage of the biomarkers collected and stored by the study. In this section I discuss some of the logistical issues surrounding the management and disposition of blood biomarkers in WHI, with lessons for other long-term clinical trials as well as observational cohorts.

Some context is needed when describing the WHI experience. WHI investigators early on developed priorities for specimen use. The first priority was for quality control and management of the study. Other biomarker applications, such as explaining WHI findings, pursuing industrial applications, and joining collaborative biospecimen groups, are discussed below. In addition, a major disposition for collected specimens is WHI investigator-initiated ancillary studies, those not part of the original WHI mission but that are deemed to be high-quality science and to have relevance to women's health issues. Individual investigative groups in WHI may have collected additional specimens in local studies, but these were not provided to the parent WHI repository. The main WHI repository is gradually being made available to collaborating, non-WHI investigators.

While it is not the intent here to make a detailed account of the techniques or logistics of specimen collection, the following is an overview of selected procedures and activities in WHI. In specimen collection, all participating staff were trained in the methods of specimen acquisition and initial processing, not only to standardize and preserve the accuracy of the determinations, but also to protect them from untoward exposure to biohazards. This was true even for staff who had substantial experience in prior studies or such techniques as venipuncture. Equal attention was paid to the timing of specimen collection, initial processing, shipping methods, and shipping duration in order to ensure that distortion of biomarker findings was minimized (Landi and Caporaso, 1997). Although not performed in WHI, if specimens for cell and molecular biological determinations were being performed, such as using cell cryopreservation, additional steps would be needed (Holland, Smith, Eskenazi, and Bastaki, 2003). The same principles would be true for other types of biomarkers in population studies, such as trace elements in bodily fluids, microbiological specimens, or micronutrients (Wild, Andersson, O'Brien, Wilson, and Woods, 2001), each of which involves some separate processing issues.[1] Whatever the scientific goals for specimen collection, there is a need to ensure laboratory quality control, both at the start of the study and over time for longitudinal studies (Gompertz, 1997; Tworoger and Hankinson, 2006).[2] A corollary but critical issue is that the methods of specimen collection, handling, and storage should be well documented for use by future investigators. At the very least, such information should either be published in the indexed literature or made publicly available through protocol documents.

There are several issues related to the burdens of managing specimen distribution after they are collected and stored. With a large number of specimens, there is a problem of registering and tracking the specimens—where they are, how much has been extracted, and how much is left for subsequent determinations. Not surprisingly, all of this requires a computerized database that is frequently updated and provides information on the type and quantity of specimens, the number of times the specimens have been thawed, if at all, and, ideally, specific reference to databases in which previous determination results of those specimens can be found. Creating and maintaining a repository database can be a costly and demanding activity and must be confronted early in study design.

The issue of specimen stability and integrity over time is critical for biorepositories and their long-term applications. Scrupulous records must be kept on storage conditions, possible power outages, specimen retrievals, and other lapses in specimen maintenance. Even under optimal storage conditions, some potential and future biomarkers may not remain well preserved. In some cases, certain molecules are known to survive long periods, such as immune globulins and steroid hormones. However, other moieties, such as peptide hormones, may not. When proposing to use specimens that have been stored for long periods, the investigators should either cite the evidence for specimen stability or propose studies to demonstrate it.

There are complex logistics with regard to distributing and retrieving stored specimens. Can they all be identified and retrieved in a timely

---

[1]Additional useful information can be found at the website of the International Society for Biological and Environmental Repositories: http://www.isber.org.

[2]The September 15, 2006, issue of *Cancer Epidemiology Biomarkers and Prevention* has several useful articles related to biomarker acquisition, processing, and storing in population studies.

manner? How are the fates of the remnant specimens monitored, and will these specimens be sent back for storage and forwarded to another investigator for other determinations? Is there a need for the parent study to monitor the quality of laboratory determinations that are done as part of ancillary research studies? Addressing this latter point is particularly important, because some research determinations may have direct implications for participant health status. If biomarker determinations are to be used for clinical alerts or those results are returned to participants for possible clinical use, then there may need to be appropriate certification of the laboratories providing the determinations. As is true of all biorepositories, someone has to pay for the management of specimens used for clinical alerts, and this should be considered when such studies are designed and budgeted. In this case, all storage and retrieval activity was sponsored for WHI by the National Heart, Lung, and Blood Institute. However, as new cohorts and other studies contemplating the acquisition of biomarkers are founded, budgets must include provision for these activities.

A related but important concern is unapproved uses of distributed specimens to study investigators and their collaborators for ancillary research projects. If one gives an investigator specimens to do a certain number of tasks that have been scientifically endorsed, what if they use the specimens for another purpose, even if closely related to the approved application? For example, if an investigator is approved for determining certain candidate genes, is determining nearby regulatory genes or other, previously unspecified genes appropriate? In my view, this must be tightly controlled, because the specimens are commonly owned by all the investigators, both for their intellectual value and for their potential future applications. One must maintain the rules about how the specimens are used and about returning them when the assays are completed. Of course, it is possible to request additional applications of the specimens to the appropriate committee; in the case of WHI, it is the Ancillary Studies Committee.

Another question is whether ancillary study findings related to biomarkers should at some point be released to the scientific community or the public. That is, most WHI investigators think it is important to provide the findings from local ancillary studies to the WHI data bank back for further study after the primary scientific reports have become public, allowing other investigators to use the biomarker determinations for other hypotheses and avoiding duplication of cost or accelerated specimen depletion. In the longer term, NIH policies are likely to demand that the data become available from large studies such as WHI, but rapid archiving of information will accelerate this scientific process.

Box 6-1 is a partial list of selected biomarkers that have been proposed for or performed in ancillary studies in WHI, to give a sense of the scope

BOX 6-1
Examples of Biomarkers from Women's Health Initiative
Safety and Ancillary Study Protocols

Markers of Inflammation: IL-6, ultrasensitive C-reactive protein; tumor necrosis
    factor alpha
Micronutrients: alpha and beta carotene; tocopherol; vitamin C; vitamin D
Obesity-Related Hormones: adiponectin; ghrelin; leptin
Gonadal Hormones: estrogens, progestins; hormone-binding globulins; markers
    of testosterone metabolism
Pituitary Axis Hormones (free and bound): corticosteroids; growth hormone;
    IGF-1; thyroid and parathyroid hormones
Clotting Factors: Factor VIII; Factor IX; elements of the complement cascade;
    thrombin; prothrombin; fibrinogen
Lipids and Lipoproteins: LDL, HDL, and VLDL cholesterol and subfractions;
    lipoprotein particle size
Tissue and Vascular Matrix Factors: E-selectin; matrix metalloproteinase factors

of the biomarkers derived from investigator-initiated hypotheses that could be determined from frozen plasma, serum, and buffy coat/DNA. It is clear that many biomarkers of potential interest cannot be determined on these frozen blood materials. However, a potentially large number of determinations are available; an extensive discussion of organ- and disease-specific biomarkers is summarized in a recent compendium (Trull et al., 2002).

As of this writing, all large NIH-funded projects require a data-sharing plan. WHI has received requests for data from persons outside the study, and the usual issues of maintaining confidentiality as well as allowing investigators to complete and publish their original studies are important here. Biomarker studies in general may have special issues of confidentiality, particularly for unique genetic coding sequences, and this must be considered when contemplating the release of such information.

Whether distributing specimens to WHI investigators or to outside collaborators, the issue of developing procedures and priorities for specimen use is paramount. To begin, criteria are needed as to who is a suitable investigator to apply for the specimens and which investigators have priority. In the case of WHI, nonstudy investigators may apply only if they are affiliated with an existing WHI investigator or approved by the steering committee. Of course, the scientific promise and feasibility of a proposed ancillary study, and the relevance to the overall mission and themes of WHI, should be strong determinants of who receives specimens. Also, as an extra precaution with regard to retaining some control over specimen disposition, it is probably important to release specimens only to institutions that have IRB approval mechanisms, institutional willingness to accept responsibility for the study, and possibly also an ongoing relation to NIH, so that some level of administrative communication and quality assurance remains.

Whether a proposed biomarker study has funding in place at the time of application is a difficult conundrum. If repository specimens are promised and held for yet unfunded studies, the delay in acquiring funds may occupy specimens that otherwise could be put to an equally important use. However, applicants must demonstrate to a funding agency that the specimens will be adequate for their proposed goals (i.e., conduct pilot work) and also demonstrate ensured specimen access. The specimen distribution process would be simplified if the applicants had funding already in hand, but there is no easy solution to this dilemma.

Another issue with respect to specimen allocation priorities is that, as ancillary studies progress and specimens are consumed, certain types of specimens may be in relatively short supply. Often, these are specimens from participants who had the outcomes of primary interest, conditions such as myocardial infarction, breast cancer, or bone fracture. In WHI, as in some other studies, planning for specimen use priorities among those with the greatest call may require special planning and political negotiation. Even when agreed-upon rules are developed, they may have to be revisited for important specimens that are near exhaustion.

Despite the large amount of work and the committee process to create priorities for specimen allocation, there may be superimposed additional priorities that must be rationalized within the existing system. In the case of WHI, two major situations arose. One that commanded a special priority was the decision on the part of the investigators to explore some possible biological mechanisms related to the adverse events that occurred with the use of estrogen plus progestin. Substantial research using biomarkers is being conducted in WHI to determine possible causes of the adverse events as well as potentially to determine if specific high-risk groups can be defined. Laboratory investigations are being considered to explore other adverse effects that occurred with WHI interventions, such as an increased risk of dementia associated with hormone therapy (Shumaker et al., 2003) and a 17 percent increased risk of kidney stones associated among women taking calcium and vitamin D supplements.

A second general priority was invoked when WHI was contacted by certain private companies, which inquired as to whether specimens and health outcome data might be used to search for genetic causes of the outcomes using high-throughput methods that otherwise would not be available to WHI. After considerable consultation with investigators having suitable expertise and several discussions by WHI investigators, it was decided to allow a portion of the specimens to be used for this activity. In some instances, complex negotiations with regard to ownership of the findings and their publication were necessary. Work on this partnership activity is continuing, and the possibility of defining genetic and other molecular risk factors for the hormone-related adverse effects may be important in the future.

ADDITIONAL APPLICATIONS FOR
BIOMARKER STUDIES IN WHI

Participation in Large Scientific Consortia Emphasizing Biomarkers

In addition to the priority dispositions of biomarker specimens generated by WHI leadership and investigators, larger population studies and clinical trials should consider partnering with collaborative studies that pool individual study data and findings, and thus have the capacity to explore important scientific questions that no one study can attain. As an example, WHI has agreed to participate in CGEMS, the Cancer Genetic Markers of Susceptibility project (find more information at http://cgems. cancer.gov). This is a three-year project sponsored by the National Cancer Institute that explores genetic alterations that lead people to be at greater risk of breast and prostate cancer. Genetic case-control studies are performed with large sample sizes, evaluating over 500,000 tag single-nucleotide polymorphisms. Although such participation may dilute individual scientific and authorship credit, it may allow scientific discovery that otherwise would not be possible. In addition, negotiations are under way to utilize WHI biospecimens for proteomic studies.

Applications of WHI to the Social and Behavioral Sciences

It is perhaps not adequately appreciated that WHI has many measures of social and behavioral function that might be of use to investigators interested in those domains, in addition to more disease-oriented themes. There are measures of general health status and health-related quality of life. Within the latter measures are subscales of physical functioning, role limitations in general and due to emotional problems, bodily pain, energy and fatigue, social functioning, and mental health. There are also cognitive measures, such as the Modified Mini Mental State Examination, a depression scale, measures of sleep disturbance, and items related to sexual satisfaction and preference. There are some basic assessments of physical functional status (activities of daily living) and, in a subsample, tests of physical performance. The WHI database can be a very rich source on the health, behavior, and aging of postmenopausal women, including the sociology and psychology of aging.

Other Potential Research Directions

One area of biomarker application that can be exploited in such studies as WHI, in which both the sample size and geographic coverage are large, is to address the association of disease occurrence or biomarker levels with variation in the ambient environment. For example, if a new biomarker is becoming available that reflects an exposure to the physio-chemical environment, such as an airborne pollutant, studies like WHI can test whether such pollutant levels are perturbed by a general set of population characteristics, such as smoking, alcohol consumption, or variation in diet, as well as age, race, ethnicity, and many other potential confounders of interest. Once the properties of these emerging biomarkers are understood, it is possible to apply them to scientific hypotheses that will lead to more specific associations. Variation in geographic distributions of certain environmental pollutants measured though biomarker determinations can then also be explored, as well as relating these markers to physiological and disease outcomes.

Another use of the biomarkers has been to validate questionnaire responses. In the past, there have been substantial work on the validation of self-reports of cigarette smoking by determining nicotine metabolite levels. An additional example is the validation of dietary questionnaires. Self-report of diet has always been problematic, and while most dietary instruments used in population studies have some demonstrated validity, there may be problems with accuracy and precision. For example, accurate measures have been critical to understanding the outcomes of the dietary modification trial conducted by WHI. Measuring of micro-nutrient biomarkers has been one way to address this problem. WHI is addressing the accuracy of dietary intake instruments in other ways as well: a doubly labeled water study has been performed to obtain an objective biomarker of energy intake, and verification of protein intake was conducted by comparing dietary reports with urinary creatine levels. Biomarkers can thus be used in selected situations to validate other kinds of data collection.

It may be possible to use populations such as in WHI to explore health disparities, particularly those that may be manifest in biomarker determinations. With the range of ethnic and racial groups represented, the substantial representation of both rural and urban women, and the variation in health and functional status, this study in longitudinal context may provide substantial insight into health outcomes related to bio-logical measurement. Available outcomes, such as institutionalization and mortality, are likely to be of interest in disparities research. It should be noted, however, that WHI did not provide general medical care and required that medical insurance for certain clinical outcomes be provided by the participants. Thus, there were fewer uninsured persons in WHI than in the general population.

Finally, in addition to research applications, it is possible that population study biomarker repositories may provide a set of services for the participants, if that is desired, related to the clinical uses of the specimens. Archived blood components, for example, may have several applications, such as providing preinfection antibody levels if there is a question of an incident infectious process, in which diagnostic assistance is needed. Furthermore, there may be serological evidence of prior environmental toxicant exposures that could help in diagnosis. If a participant is deceased, stored DNA may be helpful to living family members if there is an issue of genetic transmission leading to a risk of a chronic illness. The use of such services may be infrequent, but their availability may be of public relations value to establishing and maintaining cohorts.

STRENGTHS AND WEAKNESSES OF BIOMARKERS STUDIES USING CLINCAL TRIAL POPULATIONS

It is important to note the issues that population scientists face when attempting to study clinical trial populations such as WHI and its companion observational cohort. There are many strengths to such an approach, including the very large sample size, the broad age distribution, the large number of minority participants, the broad range of social and behavioral measures available for study, the large and multidisciplinary group of investigators who can provide collaboration and consultation, the exquisitely documented nature of many of the health outcomes, including those that are less common and not part of the initial study design or primary outcomes, and the advanced planning for specimen storage and retrieval.

Yet some weaknesses of this approach to secondary data analysis associated with primary exploration of the blood specimens are apparent. Most importantly, all of the cohorts in WHI are comprised of volunteers meeting several inclusion and exclusion criteria; these individuals are not likely to be referent to any particular geographically defined population. Among the most important criteria is the absence of disabling chronic illness and known preterminal or institutionalizable conditions. The OS is even further selected, since these are participants who had originally sought participation in the clinical trials. This lack of generalizability can be difficult to overcome if directly relevant to a particular set of scientific hypotheses, but nonetheless there may be suitable variation in study measures to allow detailed exploration of many important scientific hypotheses. Other potential impediments include differing propensities of the 42 IRBs to allow ancillary studies and the potential IRB demand for reconsenting of participants for new, previously unplanned determinations on archived biological specimens. In addition, the blood or blood components for specific participants may be dwindling or not available, and that may thwart certain types of studies of participants with specific characteristics. The specimens most likely to be exhausted are those belonging to participants with diseases and conditions comprising the primary outcomes of the study.

CONCLUSION

Although WHI is perhaps a larger study than most investigators will ever encounter, there are important lessons for all population investigators about the acquisition, management, and disposition of biomarker collections. It is hoped that clinical trial populations will be considered as worthy populations for study. In many instances, the distinct strengths of this approach to large group scientific inquiry should more than counterbalance the lack of population reference.

ACKNOWLEDGMENTS

The author wishes to thank the many investigators and staff at the study sites, the Clinical Coordinating Center, the participating centers that provided additional study management and laboratory determinations, and the National Institutes of Health, where staff scientists were so instrumental in the genesis and stewardship of the Women's Health Initiative. In addition, the author is grateful to the over 161,000 participants who gave their time, energy, and dedication to exploring the frontiers of women's health and elder health.

REFERENCES

Gompertz, D. (1997). Quality control of biomarker measurement in epidemiology. *IARC Scientific Publications, 142*, 215-222.

Holland, N.T., Smith, M., Eskenazi, B., and Bastaki, M. (2003). Biological sample collection and processing for molecular epidemiological studies. *Mutation Research, 543*, 217-234.

Jackson, R., LaCroix, A., Gass, M., Wallace, R., et al. (2006). Calcium plus vitamin D supplementation, and the risk of fractures. *New England Journal of Medicine, 354*, 669-683.

Landi, M.T., and Caporaso, N. (1997). Sample collection, processing, and storage. *IARC Scientific Publication, 142*, 223-236.

Prentice, RL. (2006). Research opportunities and needs in the study of dietary modification and cancer risk reduction: The role of biomarkers. *Journal of Nutrition, 136,* 2668S-2670S.

Shumaker, S., Legault, C., Rapp, S., Thal, L., Wallace, R., Ockene, J., Hendrix, S., Jones, B., III, Assaf, A., Jackson, N., Kotchen, J., Wassertheil-Smoller, S., and Wactawski-Wende, J. (2003). Estrogen plus progestin and the incidence of dementia and mild cognitive impairment in post-menopausal women. The Women's Health Initiative Memory Study: A Randomized Controlled Trial. *Journal of the American Medical Association, 289,* 2652-2662.

Trull, A., Demers, L., Holt, D., Johnston, A., Tredger, M., and Price, C.P. (2002). *Biomarkers of disease: An evidence-based approach.* Cambridge, England: Cambridge University Press.

Tworoger, S., and Hankinson, S. (2006). Use of biomarkers in epidemiological studies: Minimizing the influence of measurement error in the study design and analysis. *Cancer Causes and Control, 17,* 889-899.

Wild, C., Andersson, C., O'Brien, N., Wilson, L., and Woods, J. (2001). A critical evaluation of the application of biomarkers in epidemiological studies on diet and health. *British Journal of Nutrition, 86,* (Suppl.)1, S37-S53.

Women's Health Initiative Steering Committee. (2004). Effects of conjugated equine estrogen in post-menopausal women with hysterectomy. *Journal of the American Medical Association, 291,* 1701-1712.

Women's Health Initiative Study Group. (1998). Design of the Women's Health Initiative clinical trial and observational study. *Controlled Clinical Trials, 19,* 61-109.

Writing Group for the Women's Health Initiative. (2002). Risks and benefits of estrogen plus progestin in healthy post-menopausal women: Principal results from the Women's Health Initiative. *Journal of the American Medical Association, 288,* 321-333.

7

Comments on Collecting and Utilizing Biological Indicators in Social Science Surveys

Duncan Thomas and *Elizabeth Frankenberg*

These are exciting times for population scientists. This volume describes several innovative population surveys that include biological and genetic information and have shed new light on important questions in the social and health sciences. With the rapid rate of technological change in the collection and measurement of biological information and associated reduction in costs, it is increasingly common for biomarkers to be included in large-scale population-based socioeconomic and demographic surveys. It is unlikely that the benefits to science of these innovations will not be realized until study designs more fully integrate the theoretical insights from the biological, medical, and social sciences. Such integration is likely to yield broad new vistas of inquiry and has the potential to herald a new era of scientific discovery at the interface of the health and population sciences.

The chapters in Part I provide a rich and textured description of some pioneering projects from across the globe that integrate social surveys with biomarker measurement. Each chapter tells a tale of exemplary creativity in the design and implementation of studies that have yielded insights regarding a broad array of indicators of health status, their prevalence in specific populations, and, in some cases, their associations with individual characteristics and the social environment. These studies provide a tantalizing array of intriguing facts and promise extraordinary opportunities to better understand the complexities underlying the interplay between social behavior and biology. The studies also highlight the contributions of research that span different social and ecological environments and provide sources of variation that are necessary to successfully discriminate among hypotheses.

CHALLENGES FOR THE FUTURE

The research in this volume reflects the current state of knowledge and the chapters in Part I also suggest several challenges that are likely to be faced in the future. Studies have clearly demonstrated the feasibility of including biomarkers in large-scale population surveys. Important questions remain about how to select biomarkers for inclusion in particular studies as well as questions about the trade-offs between resources allocated to biomarker measurement relative to other dimensions of a survey. The latter includes resources for the collection of other survey items, sample composition and sample size, including the age range of the sample, whether to include other family members in the study, whether to purposively sample particular subpopulations, and whether to follow subjects over time in order to collect longitudinal information on health and social status. These are not easy choices, and one size does not fit all.

Questions about the selection of biomarkers arise, in part, because there is little clarity on which biological, nutritional, and genetic markers should have the highest priority for inclusion in broad-purpose demographic and socioeconomic surveys. The typical population survey is large in scale and multipurpose in scope and designed to support the testing of an array of different hypotheses. Because health is multidimensional, it is tempting to collect a wide array of different biometric indicators to parallel both the broad-purpose nature of population surveys and the broad-based nature of most self-assessed indicators of health status reported in these surveys. Succumbing to this temptation quickly becomes prohibitively expensive. It is difficult to overstate the contributions of recent developments of simple, low-cost methods for the collection, storage, and analysis of biological samples, and future innovations promise to revolutionize the field. However, progress will be limited if we rely on innovation driven by measurement alone. It is important that theory-driven hypotheses also influence the development of technologies for measurement of biomarkers so they can be included in population surveys.

Some of the most influential social science surveys in the last 40 years have been guided by the integration of theory with measurement across multiple disciplines. For example, in economics, ideas that were pioneered by T.W. Schultz, Gary Becker, and their colleagues have influenced the design of many surveys across the globe. These include highlighting that decisions in any single domain of an individual's life (such as health) affect decisions in other domains (such as work); that choices today affect options tomorrow and are thus affected by choices yesterday; that choices are constrained by the opportunities available to an individual; and that the household and the family play a key role in decision making by individuals.

To fully harness the benefits of integrating biomarker collection with social science surveys, it will be profitable to invest in integrating theory from the biological, biochemical, genetic, and behavioral science fields. This integration will provide direction for data collection and analyses that are hypothesis driven, and the evidence from those studies will serve as guideposts for further inquiries.

An Example: The Work and Iron Status Evaluation

An ongoing nutrition intervention that we have been conducting in Indonesia provides an illustration. The study seeks to test the hypothesis that health has a causal impact on economic and social prosperity. Its design was influenced by one of our earlier studies, the 1997 wave of the Indonesia Family Life Survey (IFLS), in which we collected a series of biomarkers. These included anthropometry, lung capacity (with a puff test), blood pressure, physical mobility, and, key for this discussion, hemoglobin (Hb). Hb is assessed using an in-home Hemocue photometer, which is safe, inexpensive, and simple to field, calling for no more than blood from a finger prick.

Data from IFLS indicate that around a third of reproductive-age women (ages 15 to 40) in Indonesia have low levels of iron, as indicated by Hb levels below 120g/L (the cutoff recommended by the World Health Organization). High rates of iron deficiency among women in this age group have been documented in many countries across the globe, and these women, along with infant children, are the focus of most iron supplementation programs. However, there are reasons to suppose that iron deficiency rates may be high among other demographic groups in Indonesia. The typical diet contains little by way of iron-rich foods (such as red meat), and rice, the primary staple, retards iron absorption (particularly from nonanimal sources). Because there was no evidence on prevalence rates at the population level in Indonesia, in IFLS we collected Hb from over 27,000 male and female respondents across the entire age distribution.

Among women, Hb levels *decline* with age after menopause and almost half of all women ages 65 and older in Indonesia are iron deficient (as indicated by Hb < 120g/L). The recommended cutoff for men is higher (Hb < 130g/L). The incidence of low iron is lower among prime-age men, with about one in six men ages 20-29 having Hb below the cutoff. However, as with women, Hb levels decline with age beginning around age

30. Slightly over half of all men ages 65 and older are iron deficient (as indicated by Hb < 130g/L) (Thomas and Frankenberg, 2002).

This is important. Iron deficiency is associated with elevated susceptibility to disease and fatigue. Moreover, there is a substantial body of literature that demonstrates that iron deficient anemia—the combination of low Hb and inadequate iron stores—results in reduced work capacity, as measured by, for example, VO_2max. This has been demonstrated in rigorous treatment-control studies with both humans and animals. Because iron plays a critical role in transporting oxygen through the blood, the mechanisms underlying the relationship between iron and work capacity are well understood right down to the cell level. What is not known is whether increased work capacity translates into higher hourly earnings, more income, and improved well-being. It is possible that as health improves and work capacity increases, hourly earnings increase and, with no change in work effort, income will increase proportionately. However, it is also possible that as hourly earnings increase, individuals will allocate less time to work, so that there is no change in income (or income could even decline). Economic theory suggests that the typical behavioral response will depend on the subject's expectations about the longevity of the change in hourly earnings and is likely to lie between these two extremes.

With this background, we designed the Work and Iron Status Evaluation (WISE) to address these questions. The study follows approximately 5,000 older people (ages 30 through 70) and some 12,000 other household members for a period of six years. The study design and some results are described in Thomas et al. (2006, 2007).

Subjects were randomly assigned to receive a weekly iron supplement or an identical placebo for slightly over a year. During the first three years, subjects were interviewed every four months and administered a very detailed survey instrument that collects information on type of work, hours of work, own farm and nonfarm businesses, earnings, wealth and savings, consumption and spending, transfers to noncoresident family, participation in community activities, time allocation, risk and intertemporal preferences, along with self-assessed physical and psychosocial health.

Considerable time and effort were invested in ensuring that subjects took the tablets every week. Supplements and placebos were provided in blister packs of four tablets and, during the first four weeks, subjects were visited twice a week to ensure compliance with the protocols. The frequency of visits was reduced if subjects appeared to be following the protocols. Blister packs were resupplied every four weeks. At every visit to a household, the number of tablets remaining in the blister packs was recorded. Counting the number of tablets removed from the blister packs can provide only an upper bound on the number of tablets ingested. An advantage of focusing on the impact of iron supplementation is that we can use biomarker assessments as a more direct indicator of compliance. Specifically, at each of the four-monthly interviews in WISE, several biomarkers were assessed, including Hb. Changes in Hb for each subject across waves of the study provide biological evidence on compliance with the treatment protocols, at least among those who were iron deficient and received the treatment. In addition, by monitoring changes in Hb every four months, we are able to trace the time path of the effect of the iron supplementation on iron in the blood. We did this during the year that tablets were distributed and also after the supplements ended, in order to assess the longevity of any effect on Hb levels.

Because low Hb alone does not indicate iron deficient anemia, we also measured iron stores with transferrin receptors using dried blood spots and an Elisa assay (McDade and Shell-Duncan, 2002) that were collected at the same time that Hb levels were measured. These data demonstrate that subjects who received iron and had low levels of iron at baseline, had higher levels of Hb and iron stores (as measured by lower transferrin receptors) relative to baseline levels when supplementation ended. For a subsample of respondents, we also measured work capacity by having each subject pedal to exhaustion (or a maximum of 18 minutes) on a stationary ergocycle with weights, while carefully monitoring the subject's heart. The subjects who received iron supplements and who had low Hb at baseline were able to bike longer after the supplementation than similar controls.

The evidence from WISE thus far essentially replicates the results in the nutrition literature. The biomarkers provide direct evidence that among iron deficient subjects who received iron supplements, iron levels have increased. We have not ruled out that part of the low Hb is driven by a hemoglobinopathy. A potentially important candidate is inflammation. Because Hb tends to be depressed by inflammation and, at least in the United States, inflammation rates rise with age, it may be that the age profile of Hb reflects elevated levels of inflammation among older Indonesians. To investigate that issue, we draw on the dried blood spots that were collected as part of the survey and measured a marker for inflammation, C-reactive protein (Cordon, Elborn, Hiller, and Shale, 1991). While inflammation is probably a factor in the observed low levels of Hb in the Indonesian population, the evidence from our data clearly demonstrates that the age profile of C-reactive protein cannot possibly explain the age profile of Hb (Crimmins, Frankenberg, McDade, Seeman, and Thomas, 2007).

Armed with the evidence from biomarker measures, which indicate that the treatment-control intervention has been implemented successfully, we turn to measurement of the impact of a randomly assigned health shock on social and economic prosperity. Consider, first, productivity at work, which we measure by hourly earnings. It is unlikely that someone whose hourly wage is not directly related to productivity would see any increase in wages if their iron levels were improved by the treatment. However, people whose hourly earnings are a reflection of productivity are likely to have higher hourly earnings if they were iron deficient at baseline and received the treatment. We therefore examined self-employed men—most of whom are farmers or construction workers or run bicycle taxis—whose earnings are likely to be closely related to their own productivity. After six months of supplementation, relative to a year before, the hourly earnings of men who received the treatment and who had low iron at baseline were significantly higher than comparable controls. There were no differences among those working in the wage sector. Among men working in the wage sector at baseline, there were no productivity differences between treatments and controls but there were differences in how they allocated their time.

CONCLUSION

This example provides an illustration of the potential value of integrating theoretical insights from the nutrition and biochemistry literatures with those from the social sciences. The credibility of this study relies crucially on the combination of direction from the health sciences on the likely impact of iron on health and well-being, knowledge of the appropriate markers to assess changes in iron (and thus health) status, and the insights suggested by economic theory on likely responses to changes in productivity associated with improved health. It is hard to imagine designing and implementing a study like this without guidance from theory and practice in all of these disciplines. It is also hard to imagine successfully completing a project like this without an interdisciplinary team of scientists involved in all aspects of the study.

REFERENCES

Cordon, S., Elborn, J., Hiller, E., and Shale, D. (1991). C-reactive protein measured in dried blood spots from patients with cystic fibrosis. *Journal of Immunological Methods*, 143, 69-72.

Crimmins, E., Frankenberg, E., McDade, T., Seeman, T., and Thomas, D. (2007). Inflammation and socio-economic status in a low income population. Unpublished manuscript, University of California, Los Angeles.

McDade, T., and Shell-Duncan, B. (2002). Whole blood collected on filter paper provides a minimally invasive method for assessing human Transferrin Receptor level. *Journal of Nutrition, 132,* 3760-3763.

Thomas, D., and Frankenberg, E. (2002). The measurement and interpretation of health in social surveys. In C. Murray, J. Salomon, C. Mathers, and A. Lopez (Eds.), *Summary measures of population health: Concepts, ethics, measurement, and applications* (Chapter 8.2, pp. 387-420). Geneva, Switzerland: World Health Organization.

Thomas, D., Frankenberg, E., Friedman, J., Habicht, J-P., Hakimi, M., Ingwersen, N., Jaswadi, Jones, N., McKelvey, C., Pelto, G., Seeman, T., Sikoki, B., Smith, J.P., Sumantri, C., Suriastini, W., and Wilopo, S. (2006). *Causal effect of health on labor market outcomes: Experimental evidence.* (CCPR working paper #2006-70.) Los Angeles: California Center for Population Research, University of California.

Thomas, D., Frankenberg, E., Friedman, J., Habicht, J.-P., McKelvey, C., Jones, N., Pelto, G., Sikoki, B., Sumantri, C., and Suriastini, W. (2007). Iron supplementation, iron deficient anemia, and hemoglobinopathies: Evidence from older adults in rural Indonesia. Unpublished manuscript, University of California, Los Angeles.

8

Biomarkers in Social Science Research on Health and Aging: A Review of Theory and Practice

Douglas C. Ewbank

Five years ago, *Cells and Surveys: Should Biological Measures Be Included in Social Science Research?* (National Research Council, 2001a) identified biomarkers as an important tool for understanding the association of socioeconomic status with health and mortality. Since then, biomarkers have rapidly become a standard feature in large-scale social surveys, and there has been growth of a new, rich literature on this topic. This is an appropriate time to pause and review the theories behind the use of biomarkers and to assess how far we have progressed in applying them in social science research.

There are many reasons to collect biomarkers in population-based longitudinal studies. For example, Harris, Gruenewald, and Seeman (Chapter 5 in this volume) are primarily interested in understanding the role of various biological systems with a secondary interest in how they interact with social factors. Similarly, Straub, Cutolo, Zietz, and Scholmerich (2001) call for the inclusion of biomarkers in longitudinal epidemiological studies to document the progression of the "vicious cycle of immunosenescence, endocrinosenescence, and neurosenescence." However, from the perspective of social scientists studying aging and health, the research priorities are generally more limited. Biomarkers can be useful for studying a variety of social behaviors and environments. They can operate as underlying risk factors (e.g., genetics or birth weight) or as intermediate variables. They can also provide an alternative to self-reports. For example, serum cotinine concentration has been used to evaluate the accuracy of self-reports of cigarette smoking (Perez-Stable, Marin, Marin,

Brody, and Benowitz, 1990). However, much of the recent research on aging has focused on the effects of chronic social and psychological stress and measures associated with the biological concepts of "robustness" and "allostatic load."

Social scientists studying health and aging face a great challenge. The health outcomes we study involve interactions among a wide variety of physical and cognitive conditions. The social and behavioral environments we want to relate to health problems are equally complex and difficult to measure. We therefore seek associations between very heterogeneous causes and equally heterogeneous outcomes. Biomarkers offer more narrowly defined, more proximal intermediate outcomes or, as discussed below, more specific characterization of social and behavioral environments.

Before trying to evaluate recent progress in using biomarkers to study health and aging, it is useful to review a few of the social and biological concepts that are central to much of the research. This review therefore begins with discussions of chronic social–psychological stress and biological robustness. The next section considers strategies for identifying which biomarkers are potentially most valuable for social science research on stress and aging. This is followed by a discussion of life-cycle approaches to understanding the aging process from conception to senescence. The chapter concludes with a few words of caution for future research and an overview of the progress in using biomarkers in social science research on aging.

SOCIAL AND BEHAVIORAL STRESS

There are both short-term and long-term perspectives for studying the health impact of stress. In the short term, stress generally refers to events or circumstances that elicit an emotional (usually negative) response. Episodes of stress can affect the immune system and therefore susceptibility to disease (Segerstrom and Miller, 2004). Recently, social scientists have begun to study the cumulative effects of repeated episodes of a wider range of social-psychological, physical, and environmental stressors. This literature emphasizes links between stress and such chronic diseases as cardiovascular disease and diabetes.

We have a lot to learn about the sources of stress. For example, the Whitehall study quickly showed that the concept of "executive stress" was not consistent with the gradient they observed; the executives at the top of the hierarchy had the lowest risk of several major causes of death. This raises the question of whether we know real stress when we see it or whether we can identify sources of chronic stress.

Ironically, biomarkers can help us recognize stressful circumstances.

They provide one type of evidence about which individuals in an organization or society are most stressed, what kinds of social situations are most stressful, what kinds of social organizations relieve stress, and how individuals differ in their ability to handle stress. Much of the early research on this topic relied on measures associated with cardiovascular activation, which generally reflects reactions to recent events (up to 24 hours). Few of these studies were population-based. For example, catecholamines (epinephrine and norepinephrine) excretion rates (James and Brown, 1997) were used to identify which individuals in Otmoor villages in Oxfordshire were most stressed (Harrison, 1981) and which occupations were most stressful (Jenner, Harrison, Day, Huizinga, and Salzano, 1982). However, as James and Brown have noted, these measures reflect "the complexity of the circumstance, including the subject's perceptions and cognitive state, the nature of the social situation, potential food, stimulant and exercise effects, and the ambient environmental conditions, such as temperature or altitude" (James and Brown, 1997, p. 329).

Recent research has emphasized biomarkers that reflect longer term damage to the body's regulatory systems (see below). For example, chronic elevation of C-reactive protein (CRP) is a good marker for systemic infection and is closely associated with coronary heart disease (Danesh et al., 2004). CRP levels have been used to examine the long-term effect of stress associated with socioeconomic status in Whitehall II (Owen, Poulton, Hay, Mohamed-Ali, and Steptoe, 2003), in Chicago (McDade, Hawkley, and Cacioppo, 2006), and in the United States as a whole (Alley et al., 2006).

Psychologists studying the effects of stress have proposed five categories of stress (Segerstrom and Miller, 2004). Most of their research has involved episodes of acute stress. *Acute time-limited stressors* involve artificial experiences produced in laboratory experiments, such as tasks involving mental arithmetic. *Brief naturalistic stressors* involve such events as students taking SAT exams. Although the effect of these brief stressors on immune function may cumulate to create serious health problems in the long run, they will be hard to document in large longitudinal surveys (Graham, Christian, and Kiecolt-Glaser, 2006).

Social scientists are more apt to be interested in sources of stress that have longer effects. *Stressful event sequences* involve a single event that has a number of consequences for one's life. Examples include natural disasters and the loss of a spouse or child. Although the effects may last for months or years, they can be expected to subside over time. *Chronic stressors* usually require restructuring social roles or personal identity. These include becoming a caregiver for someone with dementia or a traumatic injury that causes permanent disability (Graham et al., 2006). The final category is *distant stressors,* traumatic experiences, such as sexual assaults in the past, that can have long-term psychological effects.

Studies of the long-term effect of stress on health, such as those related to occupational or race/ethnic relations, may involve frequent (perhaps daily) exposure to a variety of stressors, some of which don't fit neatly into these categories. For example, we might want to include unreasonable expectations, distrust, or lack of respect, which may reflect the general social environment, rather than simply the effects of single incidents.

There is no single biomarker that will identify all of the social and psychological circumstances that we consider stressful. In particular, it appears that different types of psychological stress can have very different effects on biomarkers. For example, within the category of stressful event sequences, the loss of a spouse is generally associated with a decline in natural inflammatory responses, which is not apparent following a breast biopsy (Segerstrom and Miller, 2004).[1]

The complex relationship between social environments and stress is apparent from animal models. Sapolsky (2005) has reviewed the literature on the physiological effects of social hierarchies in primates. He concludes that "no consensus exists as to whether dominant or subordinate animals are more physiologically 'stressed.' . . . Rank means different things in different societies and populations." He offers a number of generalizations.

1. In social groups with a relatively fixed hierarchy, subordinate individuals feel stress from subjugation. However, when structures are more fluid, dominant individuals tend to be more stressed, since they must struggle to maintain their dominance. This suggests that results based on the Whitehall study that high-status jobs are associated with less stress may not be generalizable across a wide range of social organizations. In particular, a study of men in the Otmoor villages in Oxfordshire found higher levels of adrenaline associated with high social class (professional and manual versus nonmanual labor) (Jenner, Reynolds, and Harrison, 1980; Harrison, 1981).

2. Subordination is more stressful in species with greatest inequality in access to resources.

3. Coping strategies and social support including the presence of kin can ameliorate the effects of stressful circumstances.

4. Subordinates can develop social strategies for avoiding interactions with dominants.

---

[1]Segerstrom and Miller provide an excellent meta-analysis of the effects of various types of stressors on a wide range of immune indicators, including 24 articles on the effects of chronic stress and 44 articles on the effects of important life events (Segerstrom and Miller, 2004).

These observations about primate societies are quite consistent with hypotheses that social scientists are beginning to study. For example, some discussions of the effects of social inequality have been framed in terms of stress (Geronimus, Bound, Waidmann, Colen, and Steffick, 2001), and researchers have used biomarkers to examine the importance of social support (Seeman et al., 2002) and social ties (Seeman et al., 2004).

BIOLOGICAL ROBUSTNESS

The challenge for social scientists is to identify a manageable number of biomarkers that capture the most important features of the aging process and the development of chronic diseases. Two related concepts—biological robustness and allostatic load—provide a theoretical framework for selecting biomarkers for social science research.

Systems analysis provides a useful way of thinking about how an organism manages the numerous changes in chemistry required to deal with regular and irregular challenges that result from the interactions with the environment. The challenges come from the physical environment, stressful situations, persistent anxiety, and such behaviors as diet and smoking. This approach involves consideration of feedback loops, redundancy, and diversity of function. A central theme is the concept of robustness. Kitano states: "[r]obustness is a property that allows a system to maintain its functions despite external and internal perturbations. It is one of the fundamental and ubiquitously observed systems-level phenomena that cannot be understood by looking at the individual components" (2004, p. 826).

Over time, the functioning of biological systems leads to the accumulation of damage that reduces robustness. Initially this can result in a delayed return to equilibrium following a challenge. Longer periods spent in disequilibrium cause additional damage. The rate of accumulation is determined, in part, by the frequency and severity of challenges as well as genetic and developmental factors that determine initial robustness. Eventually this process can lead to a state of chronic, severe disequilibrium. The concept of robust response to challenge is central to many ways of thinking about the aging process.

Proper functioning of biological systems does not rely on robust maintenance of a single equilibrium point (homeostasis). Instead, the appropriate balance point changes to minimize energy requirements (and expected requirements) in given circumstances. For example, when there is a perception of a need for defensiveness and vigilance, the mobilization of energy requirements is dominated by the sympathetic nervous system. At other times, the parasympathetic system imposes cognitive control that can better balance resources. Therefore, the appropriate balance point depends, in part, on perceptions about the environment.

These perceptions are managed by the prefrontal cortex of the brain, which provides the likely link between social and psychological stress and health. Thayer and Friedman note that "in modern society . . . inhibition [of the sympathetic system], delayed response, and cognitive flexibility are vital for successful adjustment and self-regulation." However, such emotional states as worry and anxiety, reinforced by feedback loops, can lead to "prolonged prefrontal inactivity [that] can lead to hypervigilance, defensiveness, and perseveration" (Thayer and Friedman, 2004, p. 577). They postulate that the resulting imbalance between the sympathetic and parasympathetic systems may be the link between persistent feelings of stress and the development of chronic diseases.

Social science research on the relationship between stress and health need not involve complex modeling of the biological linkages between cognition and other biological systems. Instead, we can use biomarkers that reflect important aspects of these systems as intermediate variables. Most research using biomarkers has examined associations with single biomarkers. However, McEwen and colleagues have described a more holistic approach that links social and psychological events to long-term health outcomes (McEwen and Stellar, 1993; Seeman, Singer, Rowe, Horwitz, and McEwen, 1997; McEwen, 1998). They use the term "allostasis" much like robustness to refer to a system's maintenance of stability.[2] The term "allostatic load" (AL) then refers to the cumulated damage resulting from repeated cycles of allostasis. McEwen and Stellar defined it as "the strain on the body produced by repeated ups and downs of physiologic response, as well as by the elevated activity of physiologic systems under challenge, and changes in metabolism . . . that can predispose the organism to disease" (McEwen et al., 1993, p. 2094).

Insulin resistance and type II diabetes offer a simple example. Insulin resistance involves reduced sensitivity of cells to insulin-stimulated glucose absorption. This leads to an increased secretion of insulin and a slower response to increases in serum glucose levels (glucose intolerance).[3] As the condition worsens over a period of years, the body loses its

---

[2]McEwen points out that the terms "homeostasis" and "allostasis" are sometimes used in the narrow sense of stable levels of those functions immediately necessary to sustain life (e.g., temperature, pH, oxygen tension) and at other times they are expanded to include systems that are more variable (e.g., hormone levels, heart rate, blood pressure). He prefers to limit homeostasis to mean stability in those systems that are most essential to life. He uses the term allostasis in the broader sense to include those systems that maintain homeostasis by responding to changing environment (McEwen, 2000).

[3]An alternative view is that this progression begins with inadequate production of insulin that only becomes a problem when there is decreased sensitivity to insulin. In either case, the

ability to produce enough insulin to stabilize serum glucose levels, which results in diabetes, a chronic disequilibrium in glucose levels. Excess serum glucose leads to the glycosylation of proteins that causes damage to numerous other biological systems (especially the circulatory system). One result is a reduction in the ability of the pancreas to produce insulin.[4] This simple model of insulin response to changes in serum glucose levels illustrates the progression from poor short-term responsiveness to chronic instability that develops in many systems in response to challenges

This "glucose-centric" view of insulin resistance is grossly oversimplified. For example, the insulin resistance (or metabolic) syndrome is generally defined as including elevated insulin levels, glucose intolerance, obesity (especially a concentration of fat around the waist), elevated blood pressure, elevated serum triglycerides, and decreased high-density lipoproteins (HDL).[5] The last four are also commonly recognized risk factors for another major health problem: heart disease. The symptoms of the metabolic syndrome are related through complex feedback loops involving adiposity, triglycerides, insulin, leptin, interleukin-6 (IL-6), and tumor-necrosis factor α (TNFα) (Kitano et al., 2004). IL-6 and TNFα involve stimulation of the immune system, which plays an important role in the feedback loops that lead to chronic insulin resistance and which may contribute to the development of vascular disease. Their release is attenuated by the parasympathetic nervous system, thus providing an important link to emotional stress (Thayer and Friedman, 2004). This might explain, in part, the association of insulin resistance with affective disorders (including depression) and Alzheimer's disease in addition to diabetes and heart disease (Rasgon and Jarvik, 2004).

SELECTING THE MOST USEFUL BIOMARKERS

The list of biomarkers that could be included in large social science surveys is expanding rapidly (McDade et al., 2006). Singer, Ryff, and Seeman state that the operationalization of allostatic load is still in its infancy, and they provide guidelines for selecting biomarkers that provide insights into the concept. They propose that "any operationalization of the concept of AL should signal pending onset of diverse kinds of dis-

---

initial result is a decreased robustness in the body's ability to deal with a sudden increase in serum glucose.

[4]For a complete description of the metabolic syndrome from the perspective of system analysis and robustness, see Kitano (2004).

[5]There is extensive evidence of statistical associations among the components of this syndrome, but the biological and clinical significance of these associations and the exact list of elements of the syndrome are still hotly debated (Kahn, Buse, Ferrannini, and Stern, 2005; Kahn, 2006).

ability and disease . . . [and] should represent the biological signature of cumulative antecedent challenges" (Singer, Ryff, and Seeman, 2005, pp. 113-114). Therefore, we want both leading indicators as well as measures of cumulative damage. To limit the number of indicators, we need measures associated with multiple health conditions.

We can see aspects of these criteria in the 10 measures included in the original operationalization of allostatic load (Seeman et al., 1997). The list included six items that are generally part of the metabolic syndrome: systolic and diastolic blood pressure, HDL, the ratio of total cholesterol to HDL, glycosylated hemoglobin, and waist-hip ratio (a measure of abdominal obesity). Three are chemical messengers that are primary mediators of allostatic load: cortisol (a major stress hormone) and its antagonist, DHEAS, as well as two measures of sympathetic nervous system activity, epinephrine, and norepinephrine.

Different measures are used to measure different stages of the accumulation of allostatic load. We have already seen that the metabolic syndrome is associated with multiple important diseases, including diabetes and heart disease. Among the elements of the metabolic syndrome, the waist-hip ratio is a risk factor or precursor of disease, whereas glycosylated hemoglobin measures the cumulative effect of serum glucose levels over the past few months. The Whitehall II study included a fasting glucose tolerance test that directly measures the body's response to challenge.

Singer, Ryff, and Seeman note that the complex dynamics of the immune system "suggest[s] that a large battery of the immune measures be incorporated in an allostatic load scoring scheme. On the other hand, because our larger objective is to use biological indicators that reflect possible malfunction across multiple systems . . . it is important to have a limited set of biomarkers in place for each system, but use those that are *sensitive to broadly based dysregulation*" (Singer et al., 2005, pp. 130-131, emphasis in original).

In this context, they discuss the value of adding IL-6 to the list of biomarkers. Trauma, infection, fever, and stressful experiences are all associated with elevated levels of IL-6 (Singer et al., 2005). Elevated levels are also associated with a wide range of health conditions and with several forms of disability and poor health, including depression.[6] The immune system is tightly integrated with the nervous system and the endocrine system (Cacioppo, Berntson, Sheridan, and McClintock, 2000; Glaser and Kiecolt-Glaser, 2005), which means that immune markers are associated with broader dysregulation. Measurements of Il-6 were included in the Social Environment and Biomarkers of Aging Study in Taiwan (Goldman,

---

[6]Note that these characteristics diminish its usefulness in a clinical context.

Glei, Seplaki, Liu, and Weinstein, 2005), a nested case-control study in Whitehall II (Brunner et al., 2002) and the Health ABC Study (Koster et al., 2006).

Similar criteria can be used to select genes for social science research on health.[7] For example, a relatively common allele of the ApoE gene has been shown to be associated with increased risk of ischemic heart disease and Alzheimer disease (Farrer et al., 1997; Eichner et al., 2002). In addition, both ApoE genotype and serum concentration of ApoE are involved in regulating the acquired immune response (Colton, Brown, and Vitek, 2005). Similarly, the central role of IL-6 in the immune system has led to a number of studies of mortality associated with a common allele of the gene for IL-6 (Christensen, Johnson, and Vaupel, 2006).

Several studies have examined the association of various causes of morbidity or mortality with individual biomarkers or subindexes containing some of the elements of allostatic load (Seeman, McEwen, Rowe, and Singer, 2001; Goldman et al., 2005; Chapter 3 in this volume). This type of analysis is important for testing the value of including specific biomarkers, and it may help to understand the specific mechanisms at work in a population. However, the potential value of an index of allostatic load is that it doesn't simply measure a number of important biological systems. Rather it should recognize that these systems are not independent. Although the early signs of chronic dysregulation may first appear in one system or another, examining biomarkers of separate systems risks missing the forest for the trees. Although early reports are encouraging, the value of indexes of biomarkers based on the theory of allostatic load is still uncertain.

BIOMARKERS IN A LIFE-CYCLE CONTEXT

Aging is a process. Much recent research is based on the hypothesis that many common chronic health problems, such as cardiovascular disease, diabetes, and dementia, can result from the cumulated effects of repeated or sustained social and physical challenges (Garruto, Little, James, and Brown, 1999). Yet socioeconomic characteristics can change over time and environmental or social change (e.g., moving to a new location, becoming unemployed or widowed) is itself associated with increased psychological stress. There is evidence that the negative effects

---

[7]Genetic material collected in social science surveys can be used to perform a large number of genetic tests. They can therefore contribute important information on genotype frequencies for a large number of genes and may be useful for studying genetic risk factors for disease. However, only a few genotypes might be important enough to include in analyses of the associations between social factors and health.

of stress begin in utero (Graham et al., 2006). Documenting a cohort's history of early life conditions, social and psychological stress, and the appearance of dysfunction of several systems would require a longitudinal study of extraordinary duration. For this reason, it will always be necessary to piece together results from different studies to create synthetic cohorts. A good example of this approach is work by Seeman and colleagues that compares results from the Wisconsin Longitudinal Study (ages 58 and 59) and the MacArthur Study of Successful Aging (initially 70-79) (Seeman et al., 2002).

Figure 8-1 suggests how a life-cycle approach could piece together results from a number of longitudinal surveys. The list of studies could be expanded to include research at younger ages on the short-term effects of stress at different parts of the life cycle (Lundberg, 2005). Among adolescents and young adults, research should focus on the prevalence of risk factors and on measures of short-term responses to stress. For example, some studies on this topic suggest that responsiveness to stress at younger ages may interact with genetic predisposition to cause problems of hypertension at later ages (Light et al., 1999). The Whitehall II study covers

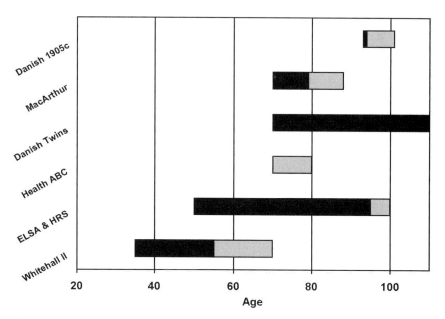

FIGURE 8-1 A life-cycle depiction of the ages included in six studies that employed biomarkers. The black rectangles show the ages included in the baseline and the grey indicate ages included in the follow-up surveys.

a middle-aged group and provides evidence about both the response to acute stress and the effects of chronic stress. It included nested case-control studies to examine differences in heart rate, blood pressure, and cortisol levels through the work day that can be compared with simpler case-control studies (Matthews et al., 2000; Steptoe et al., 2003). They have also examined heart rate variability (Hemingway et al., 2005), which Thayer and Friedman emphasize as an important marker for parasympathetic nervous system (Thayer and Friedman, 2004). At the older ages, research can focus on chronic stress and chronic dysregulation.

The life-cycle approaches used in the social sciences parallel the more biological "life history" approach to studying development and aging. This approach is based on the fact that available resources must be allocated among three processes: growth, maintenance, and reproduction. It has been used to examine the development of the immune system at early ages that can have serious implications for mortality at the older ages (McDade, 2005). It is also the basis for the "disposable soma" theory of the evolution of aging (Kirkwood and Austad, 2000).

Life-cycle models have been used extensively to study a wide range of topics in sociology and demography. However, the simplicity in diagrams of such models is deceiving. Issues of causal ordering and feedback loops raise some of the most serious modeling and statistical issues facing social scientists (National Research Council, 2001b). In addition, there are risks in combining results from different types of samples in different populations. At this point, we may not know enough about what is important to be able to tell when results from studies in one population can reliably be extrapolated to other populations.

A FEW WORDS OF CAUTION

The extensive literature on the relationship between the ApoE genotype and mortality suggests two important lessons that apply to all research on biomarkers. First, type 2 errors may be more frequent than type 1 errors in some types of research and are potentially more serious. The search for genes associated with longevity has led to many (apparently) false positives. However, there have also been false negatives in longitudinal studies of single genes (Bader, Zuliani, Kostner, and Fellin, 1998; Juva et al., 2000). Social scientists tend to obsess about avoiding false positives (p-values) and we are rarely concerned with the risk of false negatives (power). This attitude seems to be heightened in biomedical research. One reason is that clinicians have to be cautious in applying new research findings since the cost of type 1 errors can have serious effects. A second factor is that in genetic epidemiology scans of the genome generally involve a large number of multiple comparisons, which increases the risk of type 2 errors. However, in certain circumstances, false negatives may outnumber false positives. For example, standard power calculations set the risk of a type 1 error at 0.05 and the risk of a type 2 error at 0.20. False negatives may also be more damaging than false positives because negative results tend to discourage further research. This emphasizes the need for replication and multiple tests of good hypotheses even when they don't pan out immediately.[8]

Second, at this stage of research using biomarkers, there is a serious risk of unwarranted generalization. The use of biomarkers in social science research is still quite new. We are eager to draw conclusions and to base the design of new studies on previous results. However, a true negative in one population might be a true positive in another population. This can happen for both biological and sociocultural reasons. For example, the ApoE e34 genotype is not associated with excess risk of death in African Americans but it is associated with excess risk in Europeans (Ewbank, 2007). If research on the topic had started with studies of African Americans, this line of research might have ended after only one or two studies. Similar problems can arise with social variables. As noted above, similar social positions (e.g., manager) or events (e.g., widowhood) may be associated with very different levels of stress in different social settings. As scientists, we are trained to detect patterns and draw conclusions. However, we have to be cautious when results are not confirmed in subsequent studies—especially when the studies are of populations that are biologically and socially as different as Whitehall and Taiwan.

CONCLUDING COMMENTS

The use of biomarkers in social science research holds great promise. Biomarkers are quantifiable intermediate variables that can act between the complex and heterogeneous social factors and the equally diverse and complex health outcomes. Chang and colleagues note that "it is too early for us to answer the really big question: Does the inclusion of biomarkers in household surveys help us to understand SES differences in health, particularly with regard to the role of stressful experience?" (Chapter 3 in this volume). It is still early, but I am convinced that biomarkers will enable us to greatly expand our understanding of the effects of social factors on health. We are still testing new, promising biomarkers, and we are just beginning to get the data needed to sort out the effects of various kinds

---

[8]Many biomarkers measure normal levels of hormones that fluctuate over time. In many cases, it may be that the easiest way to increase the power of a test is to carry out multiple measurements on each individual (Jenner et al., 1980; Segerstrom, Lubach, and Coe, 2006).

of stress at different ages. Work is still needed to better operationalize and test the concept of allostatic load.

We have learned a lot in the past 10 years. First, a number of researchers have gained experience collecting samples and working with medical labs that can analyze them. This is a fundamental step and one that required a significant amount of work. Second, there are now data on several biomarkers in very different populations. We are just beginning to see comparable analyses of similar data in different settings. We have expanded the list of biomarkers that have been tested, and we have moved beyond the list of 10 initially used to measure allostatic load. In addition to these lessons, the new data from Taiwan should teach us to be cautious about drawing conclusions from cross-sectional data (see Chapter 3). Similarly, the Whitehall II study has demonstrated the advantages of working in a defined, structured population in which it is possible to collect a wider range of biomarkers and the social structure is easier to define. It is not necessary for all studies to be nationally representative.

The experience thus far also offers lessons to guide future research. First, research at the older ages needs to be combined with studies of the early signs of reduced robustness in younger adults. The research on younger ages will require using biological challenges, like two-hour glucose tolerance tests, that are difficult to carry out in large population studies. Second, we need to be aware that some biomarkers might be associated with stress in different ways at different ages. There may also be different associations with different types of stress. This will complicate comparisons among studies until we have a sufficient amount of data to begin disentangling these relationships.

Finally, I think the most encouraging development in the past 10 years is the emergence of a cohort of young researchers who feel that biomarkers are central to what we do and how we do it. The pioneering research makes it easy to imagine what this approach has to offer and provides valuable models for future research.

ACKNOWLEDGMENTS

This research was supported by grant number R01-AG-016683 from the National Institute on Aging.

REFERENCES

Alley, D.E., Seeman, T.E., Kim, J.K., Karlamangla, A., Hu, P., and Crimmins, E.M. (2006). Socioeconomic status and C-reactive protein levels in the U.S. population: NHANES IV. *Brain, Behavior, and Immunity, 20*(5), 498-504.

Bader G., Zuliani, G., Kostner, G.M., and Fellin, R. (1998). Apolipoprotein E polymorphism is not associated with longevity or disability in a sample of Italian octo- and nonagenarians. *Gerontology, 44*(5), 293-299.

Brunner, E.J., Hemingway, H., Walker, B.R., Page, M., Clarke, P., Juneja, M., Shipley, M.J., Kumari, M., Andrew, R., Seckl, J.R., Papadopoulos, A., Checkley, S., Rumley, A., Lowe, G.D.O., Stansfeld, S.A., and Marmot, M.G. (2002). Adrenocortical, autonomic, and inflammatory causes of the metabolic syndrome: Nested case-control study. *Circulation, 106*(21), 2659-2665.

Cacioppo, J.T., Berntson, G.G., Sheridan, J.F., and McClintock, M.K. (2000). Multilevel integrative analyses of human behavior: Social neuroscience and the complementing nature of social and biological approaches. *Psychological Bulletin, 126*(6), 829-843.

Christensen, K., Johnson, T.E., and Vaupel, J.W. (2006). The quest for genetic determinants of human longevity: Challenges and insights. *Nature Reviews Genetics, 7*(6), 436-448.

Colton, C.A., Brown, C.M., and Vitek, M.P. (2005). Sex steroids, APOE genotype, and the innate immune system. *Neurobiology of Aging, 26*(3), 363-372.

Danesh, J., Wheeler, J.G., Hirschfield, G.M., Eda, S., Eiriksdottir, G., Rumley, A., Lowe, G.D.O., Pepys, M.B., and Gudnason, V. (2004). C-reactive protein and other circulating markers of inflammation in the prediction of coronary heart disease. *New England Journal of Medicine, 350*(14), 1387-1397.

Eichner, J.E., Dunn, S.T., Perveen, G., Thompson, D.M., Stewart, K.E., and Stoehla, B.C. (2002). Apolipoprotein E polymorphism and cardiovascular disease: A HuGE review. *American Journal of Epidemiology, 155*(6), 487-495.

Ewbank, D.C. (2007). Differences in the association between APOE genotype and mortality across populations. *Journals of Gerontology Series A: Biological Sciences and Medical Sciences, 62A*(8), 899-907.

Farrer, L.A., Cupples, L.A., Haines, J.L., Hyman, B., Kukull, W.A., Mayeux, R., Myers, R.H., Pericak-Vance, M.A., Risch, N., and Van Duijn, C.M. (1997). Effects of age, sex, and ethnicity on the association between apolipoprotein E genotype and Alzheimer disease. *Journal of the American Medical Association, 278*(16), 1349-1356.

Garruto, R.M., Little, M.A., James, G.D., and Brown, D.E. (1999). Natural experimental models: The global search for biomedical paradigms among traditional, modernizing, and modern populations. *Proceedings of the National Academy of Sciences of the United States of America, 96*(18), 10536-10543.

Geronimus, A.T., Bound, J., Waidmann, T.A., Colen, C., and Steffick, D. (2001). Inequality in life expectancy, functional status, and active life expectancy across selected black and white populations in the United States. *Demography, 38*(2), 227-251.

Glaser, R., and Kiecolt-Glaser, J.K. (2005). Science and society: Stress-induced immune dysfunction: Implications for health. *Nature Reviews Immunology, 5*(3), 243-251.

Goldman, N., Glei, D.A., Seplaki, C., Liu, I., and Weinstein, M. (2005). Perceived stress and physiological dysregulation in older adults. *Stress, 8*(2), 95-105.

Graham, J.E., Christian, L.M., and Kiecolt-Glaser, J.K. (2006). Stress, age, and immune function: Toward a lifespan approach. *Journal of Behavioral Medicine, 29*(4), 389-400.

Harrison, G.A. (1981). Catecholamine excretion rates in relation to life-styles in the male population of Otmoor, Oxfordshire. *Annals of Human Biology, 8*(3), 197-209.

Hemingway, H., Shipley, M., Brunner, E., Britton, A., Malik, M., and Marmot, M. (2005). Does autonomic function link social position to coronary risk?: The Whitehall II Study. *Circulation, 111*(23), 3071-3077.

James, G.D., and Brown, D.E. (1997). The biological stress response and lifestyle: Catecholamines and blood pressure. *Annual Review of Anthropology, 26*, 313-335.

Jenner, D.A., Harrison, G.A., Day, J.A., Huizinga, J., and Salzano, F.M. (1982). Inter-population comparisons of urinary catecholamines: A pilot study. *Annals of Human Biology, 9*(6), 579-582.

Jenner, D.A., Reynolds, V., and Harrison, G.A. (1980). Catecholamine excretion rates and occupation. *Ergonomics, 23*(3), 237-246.

Juva, K., Verkkoniemi, A., Viramo, P., Polvikoski, T., Kainulainen, K., Kontula, K., and Sulkava, R. (2000). APOE epsilon4 does not predict mortality, cognitive decline, or dementia in the oldest old. *Neurology, 54*(2), 412-415.

Kahn, R. (2006). The metabolic syndrome (emperor) wears no clothes. *Diabetes Care, 29*(7), 1693-1696.

Kahn, R., Buse, J., Ferrannini, E., and Stern, M. (2005). The metabolic syndrome: Time for a critical appraisal: Joint statement from the American Diabetes Association and the European Association for the Study of Diabetes. *Diabetes Care, 28*(9), 2289-2304.

Kirkwood, T.B.L., and Austad, S.N. (2000). Why do we age? *Nature, 408*(6809), 233-238.

Kitano, H. (2004). Biological robustness. *Nature Reviews Genetics, 5*(11), 826-837.

Kitano, H , Oda, K., Kimura, T., Matsuoka, Y., Csete, M., Doyle, J., and Muramatsu, M. (2004). Metabolic syndrome and robustness tradeoffs. *Diabetes, 53*(Suppl. 3): S6-S15.

Koster, A., Bosma, H., Penninx, B.W.J.H., Newman, A.B., Harris, T.B., van Eijk, J.Th.M., Kempen, G.I.J.M., Simonsick, E.M., Johnson, K.C., Rooks, R.N., Ayonayon, H.N., Rubin, S.M., and Kritchevsky, S.B., for the Health ABC Study. (2006). Association of inflammatory markers with socioeconomic status. *Journals of Gerontology Series A: Biological Sciences and Medical Sciences, 61*(3), 284-290.

Light, K.C., Girdler, S.S., and Sherwood, A., Bragdon, E.E., Brownley, K.A., West, S.G., and Hinderliter, A.L. (1999). High stress responsivity predicts later blood pressure only in combination with positive family history and high life stress. *Hypertension, 33,* 1458-1464.

Lundberg, U. (2005). Stress hormones in health and illness: The roles of work and gender. *Psychoneuroendocrinology, 30*(10), 1017-1021.

Matthews, K.A., Raikkonen, K., Everson, S.A., Flory, J.D., Marco, C.A., Owens, J.F., and Lloyd, C.F. (2000). Do the daily experiences of healthy men and women vary according to occupational prestige and work strain? *Psychosomatic Medicine, 62*(3), 346-353.

McDade, T.W. (2005). Life history, maintenance, and the early origins of immune function. *American Journal of Human Biology, 17*(1), 81-94.

McDade, T.W., Hawkley, L.C., and Cacioppo, J.T. (2006). Psychosocial and behavioral predictors of inflammation in middle-aged and older adults: The Chicago health, aging, and social relations study. *Psychosomatic Medicine, 68*(3), 376-381.

McEwen, B.S. (1998). Protective and damaging effects of stress mediators. *New England Journal of Medicine, 338*(3), 171-179.

McEwen, B.S. (2000). The neurobiology of stress: From serendipity to clinical relevance. *Brain Research, 886*(1-2), 172-189.

McEwen, B.S., and Stellar, E. (1993). Stress and the individual: Mechanisms leading to disease. *Archives of Internal Medicine, 153*(18), 2093-2101.

National Research Council (2001a). *Cells and surveys: Should biological measures be included in social science research?* Committee on Population, C.E. Finch, J.W. Vaupel, and K. Kinsella, Eds. Commission on Behavioral and Social Sciences and Education. Washington, DC: National Academy Press.

National Research Council (2001b). *New horizons in health: An integrative approach.* Committee on Future Directions for Behavioral and Social Sciences Research at the National Institutes of Health, B.H. Singer and C.D. Ryff, Eds. Board on Behavioral, Cognitive, and Sensory Sciences. Commission on Behavioral and Social Sciences and Education. Washington, DC: National Academy Press.

Owen, N., Poulton, T., Hay, F.C., Mohamed-Ali, V., and Steptoe, A. (2003). Socioeconomic status, C-reactive protein, immune factors, and responses to acute mental stress: Special issue on psychological risk factors and immune system involvement in cardiovascular disease. *Brain, Behavior, and Immunity, 17*(4), 286-295.

Perez-Stable, E.J., Marin, B.V., Marin, G., Brody, D.J., and Benowitz, N.L. (1990). Apparent underreporting of cigarette consumption among Mexican American smokers. *American Journal of Public Health, 80*(9), 1057-1061.

Rasgon, N., and Jarvik, L. (2004). Insulin resistance, affective disorders, and Alzheimer's disease: Review and hypothesis. *Journals of Gerontology Series A: Biological Sciences and Medical Sciences, 59*(2), M178-M183.

Sapolsky, R.M. (2005). The influence of social hierarchy on primate health. *Science, 308*(5722), 648-652.

Seeman, T.E., Glei, D.A., Goldman, N., Weinstein, M., Singer, B., and Lin, H.-Y. (2004). Social relationships and allostatic load in Taiwanese elderly and near elderly. *Social Science and Medicine, 59*(11), 2245-2257.

Seeman, T.E., McEwen, B.S., Rowe, J.W., and Singer, B.H. (2001). Allostatic load as a marker of cumulative biological risk: MacArthur studies of successful aging. *Proceedings of the National Academy of Sciences, USA, 98*(8), 4770-4775.

Seeman, T.E., Singer, B.H., Rowe, J.W., Horwitz, R.I., and McEwen, B.S. (1997). Price of adaptation: Allostatic load and its health consequences. MacArthur studies of successful aging. *Archives of Internal Medicine, 157*(19), 2259-2268.

Seeman, T.E., Singer, B.H., Ryff, C.D., Dienberg Love, G., and Levy-Storms, L. (2002). Social relationships, gender, and allostatic load across two age cohorts. *Psychosomatic Medicine, 64*(3), 395-406.

Segerstrom, S.C., Lubach, G.R., and Coe, C.L. (2006). Identifying immune traits and biobehavioral correlates: Generalizability and reliability of immune responses in rhesus macaques. *Brain, Behavior, and Immunity, 20*(4), 349-358.

Segerstrom, S.C., and Miller, G.E. (2004). Psychological stress and the human immune system: A meta-analytic study of 30 years of inquiry. *Psychological Bulletin, 130*(4), 601-630.

Singer, B., Ryff, C.D., and Seeman, T.E. (2005). Operationalizing allostatic load. In J. Shulkin (Ed.), *Allostasis, homeostasis, and the costs of physiological adaptation* (pp. 113-149). Cambridge, England: Cambridge University Press.

Steptoe, A., Kunz-Ebrecht, S., Owen, N., Feldman, P.J., Willemsen, G., Kirschbaum, C., and Marmot, M. (2003). Socioeconomic status and stress-related biological responses over the working day. *Psychosomatic Medicine, 65*(3), 461-470.

Straub, R.H., Cutolo, M., Zietz, B., and Scholmerich, J. (2001). The process of aging changes the interplay of the immune, endocrine, and nervous systems. *Mechanisms of Ageing and Development, 122*(14), 1591-1611.

Thayer, J.F., and Friedman, B.H. (2004). A neurovisceral integration model of health disparities in aging. In National Research Council, *Critical perspectives on racial and ethnic differences in health in late life* (pp. 567-603). N.B. Anderson, R. Bulatao, and B. Cohen, Eds. Panel on Race, Ethnicity and Health in Later Life. Committee on Population, Division of Behavioral and Social Sciences and Education. Washington, DC: The National Academies Press.

Part II:
The Potential and Pitfalls of Genetic Information

9

Are Genes Good Markers
of Biological Traits?

Mary Jane West-Eberhard

T he 20th century has been called "the century of the gene" (Keller, 2000), an era of unprecedented progress in the understanding of inheritance. The 20th century began with the rediscovery of Mendel's laws, witnessed the discovery of the structure and nature of DNA, and concluded with the launching of the human genome project. The result has been an accumulation of genetic information far beyond the comprehension of any single scientist. Nongeneticists are confronted almost daily with new facts about genes that seem to be of enormous relevance to their lives. What is one to make of these discoveries? How should one think about them?

I recently had to deal with this problem as a biologist writing about development in relation to evolution. Genetics is the foundation of modern evolutionary biology, so I had no doubt about the importance of genes. But I wanted to explain the evolution of phenotypes—-the observable behavioral, physiological, and morphological characteristics of organisms. My main expertise is in the social behavior of insects, which made me keenly aware of the adaptive flexibility of behavior and the hormonally mediated links between gene expression and morphological diversification. I was especially aware that the dramatic differences between sterile social-insect "workers" and egg-laying "queens" do not depend on genetic differences between individuals, but instead spring from differences in their social environments, in particular the dominance relations among adults, and their diets as larvae. From such a background it is clear that phenotypic traits are not determined by genes alone. How then should one think about the formative role of genes? In this essay I discuss some results of my own struggles with this problem that may help other nongeneticists think about genes. The companion volume to this one, *Cells and Surveys* (National Research Council, 2001) contains a chapter (Wallace, 2001) on genetic markers in population surveys of human traits whose language could serve as a model of meticulous accuracy in the discussion of genetic data. This chapter is intended as a kind of reader's guide for how to relate genes to phenotypic traits in general, in order to better interpret research results and public discussions that attempt to relate genes to particular human characteristics. For a more thorough discussion, see West-Eberhard (2003, Part I).

VALUE OF A DEVELOPMENTAL-EVOLUTIONARY PERSPECTIVE FOR GENETIC ANALYSES OF TRAITS

To understand how any complex apparatus works, it helps to know how it was put together. For traits of organisms, this means understanding how they develop and how they have evolved. Sometimes evolution is depicted as a process of random genetic mutation and selection. Advances in the molecular genetics of gene expression and development, as well as in phylogenetic methods that permit more accurate histories of organismic change, support a different view: novel traits originate via the developmental reorganization of ancestral phenotypes, not just by a series of random mutations and their cumulative new effects. That is, the traits one observes have been assembled via the reorganization of older traits, with old genes used in new combinations. Furthermore, developmental reorganization can be initiated by environmental factors, as well as by mutations. In keeping with the universally acknowledged importance of environment in development, environmental induction can play an important role in the reorganizational origins of novel traits (for a summary and extensive documentation see West-Eberhard, 2003, Chapters 9-18, on evolution by developmental reorganization; Chapters 6, 20, 26 on the role of environmental factors).

These findings are relevant to the search for genetic markers—genetic loci whose different alleles correlate strongly with, and can therefore be used to predict, variation in human traits. First, due to change by reorganization of gene expression, related organisms or populations can have markedly distinctive characteristics, or "phenotypes," without having a large number, or any, distinctive new genes or genetic alleles (alternative DNA sequences at the same chromosomal locus). This is illustrated by the small genetic distance between humans and chimpanzees despite considerable differences in their behavioral and morphological phenotypes (King and Wilson, 1975). Second, the reuse of the same genes in different contexts means that a gene found to be crucial for variation in a trait of interest—a disease phenotype or a demographic property like longevity or fertility—may prove to be expressed commonly in other contexts or may have different effects at different life stages (see, e.g., Ewbank, 2000, on the ApoE gene). Research on markers needs to take into account life stage and other contextual and environmental variables in the search for reliable predictors of particular traits. The role of the environment in the induction of genetically complex, reorganized phenotypes, as when fetal undernutrition affects the expression of obesity, diabetes, coronary heart disease, and hypertension in large numbers of adults (Osmond and Barker, 2000), is a reminder that human populations can contain appreciable frequencies of complex, well-defined phenotypic variants that do not correspond to genetic variants.

An evolutionary perspective also adds to the arsenal of techniques that can be brought to bear in the search for genetic causes. Evolutionary biologists routinely compare different species and populations to illuminate the functions and causes of particular traits by ascertaining their correlations with other traits and with the conditions of life. Comparative study can facilitate the selection of organisms that are particularly likely to throw light on particular questions of interest to demographers and social scientists, and can pinpoint traits of unsuspected value for testing certain ideas.

For example, a recent survey of a very wide range of species revealed a correlation between slow embryonic development and rapid aging among different species of birds and mammals (Ricklefs, 2006). And a phylogenetic study of parrots revealed peculiarities of the mitochondrial genome that may be associated with the unusual longevity of parrots among birds (Eberhard, Wright, and Bermingham, 2001; Wright and Eberhard, in press). In addition, evolutionary biologists are practiced in the analysis of the population genetics of genetically complex (polygenic) traits, like most of those of interest to epidemiologists and demographers (Sing, Haviland, Templeton, Zerba, and Reilly, 1992; Sing, Haviland, Templeton, and Reilly, 1995). And the disciplined practice of wondering about the functional, adaptive significance of particular kinds of genes can lead to new insights with practical results. For example, an evolutionary hypothesis regarding the significance of parentally imprinted genes—genes whose expression depends on the parent of origin—has led to a focus on these particular genes as possibly being involved in the development of autism and related disorders (Badcock and Crespi, 2006).

The past belief in the importance of mutation for the origin of novel traits has helped to perpetuate misunderstanding of the role of single genes in the evolution and development of complex traits. The notion of one-gene, one-phenotype is now widely acknowledged by biologists to be mistaken, but it is still an intuitively attractive idea that is reinforced by textbook exercises on Mendelian genetics and by research on bacterial genomes and their molecular phenotypes (e.g., Ptashne, 1992). Modern research on development and evolution in multicellular organisms helps pave the way for a more realistic view of the role of genes in the production of complex phenotypes in humans and other organisms. The question of single-gene control is further explored in the next two sections.

GENETIC STRUCTURE OF A TRAIT

A "trait" is simply a somewhat discrete characteristic of an organism. It could be an aspect of morphology, a physiological state, a behavior, a molecule, or a disease, but the implication is that it is a product of development that is qualitatively distinct relative to other aspects of the organism. Some authors use the term "module" to describe a discrete trait. In operational terms, a discrete or modular trait can be defined as a product of a separate developmental pathway. But it is more accurate to say that a trait is "somewhat discrete" rather than "discrete," or that it is "modular" rather than "a module" because no trait is completely independent of all other traits in an integrated individual organism. In addition to the discrete on-off qualitative traits of organisms, there are other traits, such as body size or longevity, that are "quantitative traits"—features that are described in terms of their numerically measurable (quantifiable) values (e.g., weight, mass, or life span). Discrete, qualitative traits have dimensions (for example, the length of a bone, the duration of a behavior) that can be measured as quantitatively variable traits.

Examples of discrete traits are differentiated tissues like skin, bone, or blood; a differentiated sex (male or female); a behavior like courtship, laughter, or aggressive attack; or a disease like schizophrenia or the flu. Each of these complex traits involves the expression of a specific set of genes or the use of a specific set of gene products. In the development of an individual organism, a discrete trait is manifested or "expressed" when a threshold for its production is passed. Particular genes and their products are not always "on" or in use. An outburst of laughter, for example, presumably has some threshold of perception or sensitivity that is passed when it is triggered, and then there is another threshold for bringing it to an end. The same is true for physiological states and for the growth and differentiation of morphological traits. The timing of the "on" and the "off" determines the value of the quantitative dimension of a discrete trait, so one can measure the duration of laughter, the strength of a physiological response like a muscle contraction, the length of a bone, or the time of onset and duration of a disease.

To illustrate the genetic structure of a complex trait, consider a relatively simple and distinctive behavioral trait like laughter. How can one characterize the genetic underpinnings of such a trait? A laughter gene—or, more accurately, a gene that influences the ability of an individual to laugh—could be a gene that affects the structure of the vocal chords, the form of the facial muscles that participate in the expression of mirth, and the muscles of the diaphragm that enable bursts of air to produce the characteristic sounds of laughter. These sets of genes—those that are expressed or whose products are used when a trait is manifested by an individual organism—are *modifiers of form*. Another set of genes—*modifiers of regulation*—influence whether or not, and when, the trait is expressed, or turned on and off. The modifiers of regulation might include genes that affect the sensory acuity (of vision, audition, and touch) that enable perception of laughter-provoking stimuli and the genes that affect the central processing of those stimuli in the brain.

Sometimes the modifiers of regulation are described as acting "upstream" of the threshold point that turns the trait on; the modifiers of form are described as being expressed "downstream" of the switch (Figure 9-1). The threshold, or switch point, is a decision point, the point at which the expression of a trait is said to be determined. Clearly, both the modification of regulation and the modification of form can be highly polygenic, since all of these contributing systems are themselves genetically complex.

The genes themselves do not *determine* the expression of a trait like laughter—or any other trait. The genetically influenced laughter apparatus requires some environmental factor—a joke, a comical event, or a siege of tickling—to be turned on. The expression of the trait is jointly modified by genes (e.g., those influencing the level of the threshold) and the environment. The environment further influences the structure that responds, for no gene can act alone to produce a structure (e.g., muscle, nerve, and bone) without stimuli and materials that are environmental in origin. Environmental factors that influence regulation and form can be as specific and precise in their properties as are the genomic instructions themselves, as illustrated by the precision of day-length cues that trigger hibernation and diapause in the winter physiologies of organisms, and the dietary elements, such as specific vitamins and amino acids, that are required for normal human development.

The intertwined genomic and environmental influences on trait expression are further complicated by the fact that previous episodes of gene expression themselves add to the effective environment: genes act within cells, and gene products become part of the internal environment for subsequent gene expression. Previous gene expression and interaction with the environment also contribute to the physical and social environments of any subsequent episode of gene expression. In spite of this

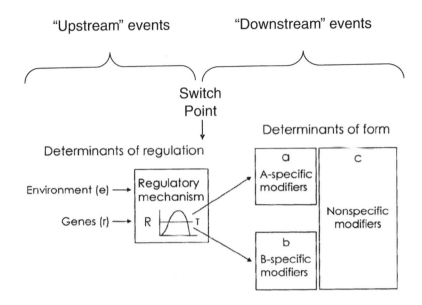

FIGURE 9-1 The development of a phenotypic trait. Development can be visualized as a series of branching pathways, each modular trait being the end point of a developmental pathway or branch. Trait determination occurs when a new branch is formed at a decision or switch point, depicted here as governed by a threshold that characterizes some environmentally sensitive, genetically influenced developmental mechanism. Alternative pathways lead to alternative phenotypic forms, which involve the expression of alternative sets of genetic modifiers or use of alternative sets of gene products. Environmental factors may also be determinants (building blocks) of phenotypic form. The events upstream and downstream of trait determination are influenced by different sets of genes.

intricate history of genetic influence, it would be inaccurate to think that the genome can run everything, for no gene can act to produce a developmental signal or a protein product without materials, energy, and often cues ultimately derived from outside the organism.

The complexity of trait structure is sometimes revealed by partial expression of the trait. In the full expression of laughter, for example, a single input (such as a tickle) affects a threshold of trait determination (the initiation of laughter) that coordinates a compound response, with various, otherwise somewhat independent, aspects of the phenotype, such as respiration, facial expression, and vocal output brought into play in synchrony. At a low level of stimulation, some of these elements of the compound response might be omitted, if the threshold for all is not passed.

In such a compound or mosaic trait, several different semi-independent components, each influenced by a separate set of underlying genes, are brought into play. Similarly, an infection may induce a "typical" full set of disease symptoms, but some may be absent at sub- or near-threshold expressions of the disease.

Variation in quantitative traits likewise typically depends on a multitude of genes and on environmental factors. Body size, for example, may be influenced by metabolic efficiency, appetite, foraging ability, disease resistance, and the ability to learn how to obtain food—all genetically complex aspects of the individual phenotype. And obviously, body size is affected by such environmental factors as food availability, the energy demands of fluctuating physical conditions, and the incidence of predation or disease.

This is a grossly simplified characterization of the genetic underpinnings of trait development and variation,[1] but it is a useful guide for relating genes to observable phenotypic traits. Complex off-on traits, as well as quantitative traits, are universally polygenic in both regulation and form, as well as universally environmentally sensitive in their expression. In addition, discrete trait expression involves two major sets of genes: those upstream of the trait, whose expression or products influence whether or not and when the trait is produced; and those downstream of the trait, whose expression or whose products modify the form of the trait itself, rendering it distinctive relative to other traits of the same organism.

SINGLE-GENE TRAIT MARKERS: THE MEANING OF A GENE "FOR" A TRAIT

In spite of the genetic complexity of most traits of interest to biologists, medical researchers, and social scientists, it has long been common for theoretical geneticists to refer to genes "for" particular traits, like coat color or altruism. It is also common for biologists to think in terms of a gene for a trait in classroom explanations of how Mendelian genetics and trait evolution work. So language that refers to a gene "for" a complex trait is by no means new or unusual among biologists. But this linguistic convention has taken on a new significance in the present age, when genes associated with particular traits can be biochemically characterized, assigned names, and precisely located on particular chromosomes. What

---

[1]For a discussion of developmental processes actually affected by modifiers of regulation and form, of alternative terms like "regulatory genes" and "structural genes," and a more thorough discussion of modularity and of genetics in relation to the development and evolution of novel traits, see West-Eberhard (2003). Gerhart and Kirschner (1997) and Kirschner and Gerhart (2005) discuss the nonrandom nature of novel traits derived from preexisting responses of cells.

was formerly an abstract manner of speaking has become a real assertion about the genetic determination of traits.

For some years I have been making a collection of genes "for" human traits announced in scientific journals and the popular press. The collection began in the early 1990s with the discovery of "the obesity gene." It now includes, among others, announcements of genes for language, intelligence, alcoholism, bipolar illness, deafness, schizophrenia, asthma, longevity, and maternal solicitude. There is also a warrior gene, a "gay" (sexual orientation) gene, and even a graceful jaw gene responsible for a human facial trait not present in other primates.

What is the empirical basis for these announcements? Usually the gene-for-trait claim is based on the discovery of a mutation or a genetic knockout that dramatically affects the probability of expressing the trait. What this means in developmental terms is that one of the many modifiers of regulation of the trait, for example, one of the polygenes that affects its threshold of expression, has been altered and acquires a large, "major gene" effect on trait expression. That is, the mutation's effect makes one of the polygenes predominate in trait determination so that it alone can control the switch for expression of the trait. This is a useful research tool because it renders one of the modifiers of regulation identifiable, whereas in its normal unmutated state its effect might have been so small as to be indistinguishable from those of the many other modifiers of regulation. The mutation's large effect gives the impression of single-gene control of the trait. In fact, the expression or nonexpression of the phenotype in individuals of the population at large is subject to polygenic and environmental influence.

Mutations of large effect are commonly responsible for familial, inherited diseases, giving the false impression of a genetically simple phenotype devoid of environmental effects. Bipolar (manic-depressive) illness is an example of a "genetic" disease inherited within families. Patterns of inheritance within affected populations suggest that the genome of some ancestor suffered a mutation, whose bearers among descendents have an increased probability of manifesting the disease.

If the model of polygenic genetic architecture I have just outlined is correct, several predictions are possible. First, mutations at different genetic loci should be associated with the disease in different families or populations. Second, there should be evidence of environmental influence on incidence of the disease. And third, there should be phenotypic (physiological, morphological, behavioral) evidence of the complexity of mechanisms underlying the disease.

All of these predictions hold for bipolar illness. The disease has been linked to regions on at least eight different chromosomes (Blackwood, Visscher, and Muir, 2001; Kelsoe et al., 2001), and mutations at different chromosomal locations are important in different populations. For example, genetic markers are located on the X chromosome in a Finnish population (Pekkarinen, Terwilliger, Bredhacka, Lannqvist, and Peltonen, 1995); chromosome 18 in Costa Rican families (Freimer et al., 1996); chromosome 22 in a general North American population (Kelsoe et al., 2001) and (at a different locus) in northern European families of Caucasian ancestry (Barrett et al., 2003); and chromosome 12 in a Danish population and (with some chromosomal overlap) in an isolated population on the North Atlantic Faroe Islands, a region colonized by Scandinavians (Degn et al., 2001).

Environmental influence on the expression of bipolar illness is suggested by the fact that occurrence of the depressive phase correlates with day length (incidence is higher in winter); and from the fact that 20 percent of identical twins of affected individuals do not show the disease (Gershon, 1990). So the genetic mutation can be present without being expressed.

The mosaic or compound nature of the bipolar disease phenotype is indicated by clinical variation in disease symptoms of this and related disorders, which has been referred to as the "bipolar spectrum" (Gershon, 1990, p. 380). In bipolar type I disorder, patients have both manic and depressive phases; in bipolar type II disorder, they show hypomania but not a full manic phase; and twin studies have indicated that there is some overlap in vulnerability for unipolar (depressive) and bipolar (manic-depressive) illness. There is also some evidence from studies of genetic relatives for an association between bipolar disease and other illnesses, such as schizoaffective disorder and cyclothymic personality disorder (Gershon, 1990).

Finally, the complexity of mechanisms underlying bipolar disease is evident from various studies. For example, research on signal transduction mechanisms in the brain suggest that bipolar disease may be due to the interaction of many kinds of abnormalities in these systems (Bezchlibnyk and Young, 2002), a conclusion supported by the fact that lithium, a substance long known empirically to bring relief to bipolar patients, appears to act by stabilizing neuronal activities, including signaling activities of several kinds and at multiple levels of influence on neural plasticity (Jope, 1999).

Mutation studies can aid in the genetic dissection of the causes of complex phenotypes. Study of the mutations and brain tissues in bipolar disease, for example, have identified altered function and levels of particular effector molecules, such as protein kinase Lambda (PKΛ) and protein kinase C (PKC) (Bezchlibnyk and Young, 2002), and of G protein receptor kinase 3 (GRK3) (Barrett et al., 2003), as involved with the signal transduction abnormalities associated with the disease.

In sum, single-gene-locus studies are clearly valuable because they permit the identification of genes and pathways that influence the development of complex traits. But they do not tell us that a particular genetic locus "controls" development of the trait. With the exception of a very few, rare diseases that appear to be truly under single-locus control (e.g., cystic fibrosis, Huntington's disease (Huntington's chorea), phenylketonuria, and Smith-Lendi-Opitz syndrome), mutant loci associated with inherited disease are unreliable global markers because disease genotype may vary with the population of origin.

Inaccurate statements about the "genetic determination" or the "environmental determination" of traits are easily fixed by simply using the word "influenced" rather that "determined." Similarly, a gene "for" a particular trait is more accurately described as a gene that "influences" a particular trait. The single word "influence" turns a misleading headline into an accurate one. This is a linguistic mutation whose spread should be encouraged under strong selective pressure from biologists and social scientists alike.

QUANTITATIVE TRAIT MARKERS: LONGEVITY GENES

Three of the genes in my collection of genes "for" traits are genes for quantitatively variable traits—obesity, longevity, and intelligence. The polygenic nature and environmental sensitivity of variation in these traits is so obvious as to need no evidence beyond common experience. The control of the development of a polygenic quantitative trait can be overwhelmed by a mutation of major effect, just as the determination of a discrete trait can be. So the same reservations regarding single-gene markers outlined in the previous section apply to single-gene markers of quantitative traits as well. Here I focus on longevity genes because they are "demo-genes" (Ewbank, 2000)—genes that affect a parameter of interest to demographers.

Evolutionary theory has a branch that deals with the evolution of longevity and senescence. Although this is not my specialty, one idea strikes me as particularly relevant to the search for a developmental-genetic basis of longevity. Williams (1957) has suggested that a postreproductive acceleration of senescence is expected due to a kind of antagonistic pleiotropy—the accumulation of *negative* pleiotropic postreproductive effects of genes that have *positive* effects on survival and reproduction at younger (reproductive and prereproductive) stages. Natural selection can only affect traits (and genes) expressed prior to the end of reproduction; postreproductive expression does not usually affect relative reproductive success (selection) unless it somehow affects the reproductive success of cocarriers of the same genetic alleles (e.g., relatives). This is one argument used to explain the "grandmother effect"—the long postmenopausal survival and activity of human females, who often make a substantial postreproductive contribution to rearing their grandchildren (Hawkes, O'Connell, Jones, Alvarez, and Charnov, 1998). Antagonistic pleiotropy predicts that the diseases of aging are especially likely to be associated with early benefits (Williams and Nesse, 1991) and that some genes that contribute to senescence in the elderly may show little allelic variation (polymorphism) within populations, being under strong positive selection during earlier life stages.

Reading the literature on the search for longevity genes, one senses a conviction that there *must* be some underlying general basis for a long life span if only one could find it. It is a search older than science itself, dating back to the medieval alchemists' search for the elixir that would prolong life. Longevity genes remain the quantitative-trait genes of widest public interest. Not everyone has severely limited intelligence or an obesity problem, but everyone has a limited life span.

Notwithstanding our deep-seated desires, commonsense reasoning and findings to date provide little hope for success in the discovery of major-effect longevity genes. Any of the enormous number of genes that contribute to survival could be considered longevity genes. Among the candidates that have been proposed are several heat-shock protein loci, which are given special attention because they are widespread among organisms, including humans, and are general-purpose responses to several kinds of systemic stress, such as heat stress, oxidants, and starvation. Some heat-shock proteins have been shown to prolong life in transgenic lines of fruit flies subjected to these stresses (Wang, Kazemi-Esfarjani, and Benzer, 2004). I can see no reason to grant higher status to these genes as longevity genes than genes that affect, say, wing development, which likewise influences fruit fly survival in multiple contexts (e.g., location of food sources under starvation, movement into the shade under heat stress, flight to escape predators). Still, the Wang et al. (2004) study represents an advance in molecular research on longevity because, instead of ignoring the environment, it experimentally manipulated the environment to investigate the longevity effects of particular genes.

Ewbank (2000) lists four criteria that a longevity gene must have to be considered demographically useful markers able to predict variation in lifespan: (1) association with the most common causes of death; (2) multiple alleles (genetic polymorphism) of the gene, associated with substantial variation in mortality; (3) large variation in frequencies of these alleles across populations; and (4) correlations of alleles with environmental or behavioral characteristics considered to be associated with mortality rates. Ewbank reports that only one human gene is known to satisfy all of these criteria, the apolipoprotein E (ApoE) gene. The ApoE gene (1) is a major risk factor for ischemic heart disease and Alzheimer disease; (2) has three common alleles (ApoE-2, ApoE-3, ApoE-4), two combinations of them (ApoE-3/ApoE-4 and ApoE-4/ApoE-4) associated with increased risk of both diseases, and two others (ApoE-2/ApoE-2 and ApoE-2/ApoE-3) associated with decreased risk of both diseases—these risks vary with both age and genetic background (Ewbank, 2000, pp. 73-74)—and with environmental circumstances (fat content of diet; Ewbank, 2000, pp. 78-79); (3) frequencies of the three alleles vary geographically (for example, the ApoE-4 allele is unusually common in Africa, where its frequency is .20); and (4) mortality rate associated with genotype is affected by diet, an environmental and behavioral trait of interest. Nonetheless, cautions Ewbank, "it is not likely that any single genotype will explain much of the heterogeneity of mortality under age 80. . . . To put this in perspective, I estimate that the APOE e4/4 genotype is associated with a relative risk of death at age 80 of about 2 relative to the most common genotype, e3/3. Less than 5 percent of the population has the e4/4 genotype" (Ewbank, 2000, p. 83).

This very reserved conclusion regarding the one gene that fits minimal characteristics of demographic utility suggests that particular genes are not promising markers of longevity. Perhaps there will prove to be genes or patterns of gene expression that do affect general energy or stamina over the long term. If so, parrots may be better model organisms than nematodes or fruit flies in the search for genomic characteristics associated with longevity. For studying phenotypic rather than genetic correlates of longevity with the needed large samples, insects have proven more suitable than either birds or mice (Carey, 2003).

GENETIC MARKERS: CRITERIA AND TRAPS

Criteria

Following Ewbank's (2000) lead, we can list the criteria to be satisfied by a dependable single-locus genetic marker of a phenotypic trait:

1. The phenotypic trait whose future occurrence is to be predicted by the marker has to be uniformly and operationally well defined, for example, as a specific measurable value of a quantitative trait or a set of consistently associated distinctive characteristics of a qualitative, discrete trait. Intelligence, for example, is an insufficiently well-defined trait to serve as the basis for a biomarker search. Some particular aspect of what we call intelligence, as measured by some well-researched standardized test, would need to be used.

2. To be of use as a predictor, the marker has to be detectable before the trait appears during individual development. A particular DNA sequence might satisfy this criterion, for genotype does not change during development. A protein specified by a particular DNA sequence is a less dependable marker, for reasons discussed below.
3. There must be multiple alleles (genetic polymorphism) of the marker gene, highly correlated with measurable variation in the expression of the trait. For qualitative (discrete) traits, this means variation in whether or not the trait is expressed, not variation in some feature of the trait once it is present.
4. The confidence in the marker is increased if it can be shown that its expression is involved in developmental mechanisms or pathways known to be involved in the development of the trait versus the nondevelopment of a discrete trait or lesser development of a quantitative trait.

Potential Traps

Given the structure of development in relation to gene expression and the polygenic nature of complex traits, there are several kinds of errors likely to appear in discussions of genetic markers and their evaluation by researchers.

Type 1 error. The assumption that a gene (a genetic allele) commonly expressed in bearers of a trait and not expressed in nonbearers of the trait is a major cause of the trait. This is erroneous for genes expressed downstream of discrete-trait determination, for modifiers of form, while they affect the characteristics of the trait, do not affect the likelihood of its expression during development. A type 1 error is akin to taking a symptom (such as fever) to be a cause of a disease.

Type 2 error. Unwarranted extrapolation between populations. A mutation that causes a highly heritable, familial trait may prove a useful genetic marker in the descendents of the mutant individual. But other populations with the same trait may have a different mutant cause, as in bipolar illness (see above). The same kind of variation in genetic underpinnings could occur in different populations for any polygenic trait.

Type 3 error. Use of a protein consistently expressed in individuals bearing the trait. Proteins are sometimes used as gene identifiers, because the specific form of a protein reflects the expression of a particular genetic allele. But the presence of a protein means that the gene has already been expressed, so to serve as a predictor of the future development of the trait, it would have to be the product of a modifier of *regulation* known to affect the probability of developing the trait, and upstream of trait determination, not a modifier of form (which would not affect probability of trait expression). Consider a disease phenotype, for example. A protein that is consistently a *symptom* of the disease could not serve as a predictive marker, even though it would be present in all carriers of the disease and might be absent in all noncarriers. Proteins that are expressed very early in trait development may sometimes be useful predictive markers.

Type 4 error. Use of a genetic allele that produces a protein associated with a trait. For example, having discovered that a particular protein is consistently symptomatic of a disease and not present in nonbearers of the disease, one might erroneously suppose that the presence of the allele responsible for that protein in a genotyped young individual would be a good predictor of the later development of the trait. It is important to realize, however, that genes can be present without ever being expressed, as in the unaffected identical twins of bipolar patients. The presence of the allele would be a necessary, but not a sufficient, condition for expression of the trait.

These kinds of errors are especially likely to appear in pharmaceutical frenzies—races to find genetic markers that enable prediction of disease and allow the identification of molecular processes to be attacked by drugs. Facile discussions of "drug biomarkers" based on proteins and genes belie the complexity of traits and the stringent criteria required of dependable markers. An article on cancer research in *Science* (Kaiser, 2004) cited an experienced cancer researcher as arguing that, although only a handful of biomarkers are widely used, the sequencing of the human genome and the debut of new, automated mass spectrometry machines for detecting proteins leaves the field ripe for new breakthroughs in the search for biomarkers. Although the map of the human genome can help locate identified genes, by itself it provides no information on genetic variation between individuals and populations of the kind crucial to biomarker research, for each segment of the map is (necessarily, for technical reasons) based on the genome of a single individual (Marshall, 1996); masses of protein data generated by automated machines would need to be accompanied by masses of correlated data on polymorphic genotypes, clinically well-defined phenotypes, life stages, and geographic locations of samples to be useful in biomarker research.

The pessimistic view presented here regarding single-gene markers need not apply to the search for genetic markers in general. But it does suggest that the search should be redirected, away from single genes and perhaps toward "collective markers"—sets of genetic alleles whose summed expression raises the probability of expression of a trait of interest by a specified amount. By first carefully focusing on upstream genetic modifiers of regulation, which are better predictive markers than are modifiers of form, it may be possible to construct collective markers that reflect the polygenic nature of complex traits (see, e.g., Sing et al., 1992, 1995).

MARKERS OR MARKETING?

Genes for traits have become great publicity devices for scientists anxious to claim applied significance and obtain increased funding for their research. Good advertising demands nice clean language, so "the obesity gene" or "the gene for alcoholism" readily displaces the more accurate "a gene that influences obesity" or "enhances the likelihood of addiction to alcohol." The public could easily be educated, in classrooms and in the press, to exercise common sense about genetic explanations and to realize that many genes must affect traits like obesity, intelligence, and alcoholism, whose causes are obviously complex. People seem to prefer simple explanations and the promise of simple solutions to complex problems. All too often wishful ignorance prevails, and "genes for traits" have become modern elixirs that can turn wishful ignorance to commercial gold.

My favorite marker-icon is still the obesity gene, a lucrative leader in the gene-age parade that would be delightful in its absurdity had it not been so effectively misleading. Not surprisingly, given the complexity of obesity as a quantitative trait, there are now at least 10 known "obesity genes," some of them with several allelic variants (Perusse et al., 2005). The first of the several obesity genes was hastily patented and brought millions of dollars for obesity research to the university where it was discovered. Even as I was finishing this chapter, a new gene for obesity was announced in *Science* (Herbert et al., 2006) and in the *Washington Post* (April 17, 2006, p. A6). The search for obesity genes is potentially virtually endless, given the great complexity of the trait and the variation in genetic composition of different populations for alleles that influence obesity (e.g., see Ewbank, 2000, and above, on the ApoE gene).

I have heard scientists defend obesity-gene language as "only a manner of speaking, for everyone knows that there is more to obesity than this one gene." It is quite common for scientists, especially the many who try to be accurate in interviews, to blame the gene-for-trait language on the press, but this seems unfair if eminent scientists engage the same brand of genetic hyperbole in the scientific and public media. Nobel laureate David Baltimore (2001), for example, during the euphoria of human genome announcements, wrote in *Nature* that "Analysis of Single Nucleotide Poly-morphisms will provide us with the power to uncover the genetic basis of our individual capabilities such as mathematical ability, memory, physical coordination, and even, perhaps, creativity"; and James Watson, another Nobel laureate, proclaimed (*Associated Press*, 2000) that "Now we have the instruction book for human life. . . ."

The human genome project was the most expensive genetic research project in history, a funding triumph as well as a scientific one. It is dif-ficult to escape the impression that one reason for genetic hyperbole is the desire of scientists to bring public attention to their research and thereby increase their access to funding, space in prestigious journals, and career success. But marketing is not the only explanation for the exaggerated status of genes, for there is also genuine ignorance and misunderstand-ing among scientists of the role of single genes in the development and evolution of complex traits.

In addition to neglect of the developmental role of the environment and of the complexity of the genetic architecture of traits that I have discussed in this article, neglected features of evolutionary genetics also illuminate some aspects of human genetics that otherwise seem mys-terious. For example, just as trait-specific genes are hard to find, so are human-specific genes, when the human genome is compared with that of other primates (Ridley, 1999). Both facts are explicable by the observation, already mentioned, that it is common to find that ancestral phenotypes, and the underlying genes, have been recombined during evolution in new coexpressed sets to produce novel phenotypic traits. Due to such "developmental" or phenotypic recombination (West-Eberhard, 2003) complexly distinctive human characteristics, such as many aspects of language, could have evolved, via reorganized gene expression, with few or even no new genes. Patterns like these require attention to whole organisms and to comparisons among related species that are increasingly rare in biology and in the training of modern biologists.

Although the social sciences naturally look to biology for guidance in understanding genes, in fact the social sciences may be able to inject some sense into the genetic interpretations of biologists. Social scientists are experts in a model organism—*Homo sapiens*—that is unquestionably the best studied vertebrate. They are fully aware of environmental influ-ence on phenotypes and of individuals as integrated wholes. The present century promises to be an exciting era for cross-disciplinary research in which the relatively holistic interests of social scientists and whole-organism biologists will converge with those of geneticists. Genetics is moving toward genomic studies that are increasingly concerned with gene expression and therefore with development and the phenotype. This will mean increased attention to the conditions and mechanisms of development, including hormone action, the nervous system, and behavior. Ultimately, and unavoidably, understanding gene action will have to address variation in environments, including the social milieu. This will have profound consequences for understanding the properties of human populations that are the subjects of this volume.

ACKNOWLEDGMENTS

I thank James R. Carey, William G. Eberhard, and an anonymous reviewer for helpful suggestions. Tim Wright permitted access to an unpublished manuscript, and Neal G. Smith and Lynne C. Hartshorn provided many essential references.

REFERENCES

Associated Press. (2000, June 26). Scientists announce DNA mapping.

Badcock, B., and Crespi, B. (2006). Imbalanced genomic imprinting in brain development: An evolutionary basis for the aetiology of autism. *Journal of Evolutionary Biology, 10,* 1420-9101.

Baltimore, D. (2001). Our genome unveiled. *Nature, 409,* 814-816.

Barrett, T.B., Hanger, R.I., Kennedy, J.L., Sadovnick, A.D., Remick, R.A., Keck, P.E., McElroy, S.L., Alexander, M., Shaw, S.H., and Kelsoe, J.R. (2003). Evidence that a single nucleotide polymorphism in the promoter of the G protein receptor kinase 3 gene is associated with bipolar disorder. *Molecular Psychiatry, 8*(5), 546-557.

Bezchlibnyk, Y., and Young, L.T. (2002). The neurobiology of bipolar disorder: Focus on signal transduction pathways and the regulation of gene expression. *Canadian Journal of Psychiatry, 47*(2), 135-148.

Blackwood, D.H.R., Visscher, P.M., and Muir, W.J. (2001). Genetic studies of bipolar affective disorder in large families. *British Journal of Psychiatry, 178,* 134-136.

Carey, J.R. (2003). *Longevity, the biology and demography of life span.* Princeton, NJ: Princeton University Press.

Degn, B., Lundorf, M.D., Wang, A., Vang, M., Mors, O., Kruse, T.A., and Ewald, H. (2001). Further evidence for a bipolar risk gene on chromosome 12q24 suggested by investigation of haplotype sharing and allelic association in patients from the Faroe Islands. *Nature, 6*(4), 450-455.

Eberhard, J.R., Wright, T.F., and Bermingham, E. (2001). Duplication and concerted evolution of the mitochondrial control region in the parrot genus *Amazona. Molecular Biology and Evolution, 18,* 1330-1342.

Ewbank, D. (2000). Demography in the age of genomics: A first look at the prospects. In National Research, *Cells and surveys: Should biological measures be included in social science research?* (pp. 64-109). Committee on Population. C.E. Finch, J.W. Vaupel, and K. Kinsella (Eds.) Commission on Behavioral and Social Sciences and Education. Washington, DC: National Academy Press.

Freimer, N.B., Reus, V.I., Escamilla, M., McInnes, A., Spesny, M., Leon, P., Service, S., Smith, L., Silva, S., Rojas, E., Gallegos, A., Meza, L., Fournier, E., Baharloo, S., Blankenship, K., Tyler, D., Batki, S., Vinogradov, S., Weissenbach, J., Barondes, S., Sandkuijl, L.A. (1996). Genetic mapping using haplotypes, association, and linkage methods suggests a locus for severe bipolar disorder (BPI) at 18q22-q23. *Nature Genetics, 12,* 436-441.

Gerhart, J., and Kirschner, M. (1997). *Cells, embryos, and evolution: Toward a cellular and developmental understanding of phenotypic variation and evolutionary adaptability*. Malden, MA: Blackwell.

Gershon, E.S. (1990). Genetics. In F.K. Goodwin and K.R. Redfield (Eds.), *Manic-depressive illness* (pp. 373-401). New York: Oxford University Press.

Hawkes, K., O'Connell, J.F., Jones, N.G.B., Alvarez, H., and Charnov, E.L. (1998). Grandmothering, menopause, and the evolution of life history traits. *Proceedings of the National Academy of Sciences, USA, 953*, 1336-1339.

Herbert, A., Gerry, N.P., McQueen, M.B., Heid, I.M., Pfeufer, A., Illig, T., Wichmann, H.-E., Meitinger, T., Hunter, D., Hu, F.B., Colditz, G., Hinney, A., Hebebrand, J., Koberwitz, K., Zhu, X., Cooper, R., Ardlie, K., Lyon, H., Hirschhorn, J.N., Laird, N.M., Lenburg, M.E., Lange, C., and Christman, M.F. (2006). A common genetic variant is associated with adult and childhood obesity. *Science, 312*(5771), 279-283.

Jope, R.S. (1999). Anti-bipolar therapy: Mechanism of action of lithium. *Molecular Psychiatry, 4*(2), 117-128.

Kaiser, J. (2004). NCI hears a pitch for biomarker studies. *Science, 306*, 1119.

Keller, H.F. (2000). *The century of the gene*. Cambridge, MA: Harvard University Press.

Kelsoe, J.R., Spence, M.A., Loetscher, E., Foguet, M., Sadovnick, A.D., Remick, R.A., Flodman, P., Khristich, J., Mroczkowski-Parker, Z., Brown, J.L., Masser, D., Ungerleider, S., Rapaport, M.H., Wishart, W.L., and Luebbert, H. (2001). A genome survey indicates a possible susceptibility locus for bipolar disorder on chromosome 22. *Proceedings National Academy of Sciences, USA, 98*(2), 585-590.

King, M.-C., and Wilson, A.C. (1975). Evolution at two levels: Molecular similarities and biological differences between humans and chimpanzees. *Science, 188*, 107-116.

Kirschner, M., and Gerhart, J. (2005). *The plausibility of life*. New Haven, CT: Yale University Press.

Marshall, E. (1996). Whose genome is it anyway? *Science, 273*, 1788-1789.

National Research Council (2001). *Cells and surveys: Should biological measures be included in social science research?* Committee on Population, C.E. Finch, J.W. Vaupel, and K. Kinsella, Eds. Commission on Behavioral and Social Sciences and Education. Washington, DC: National Academy Press.

Osmond, C., and Barker, D.J.P. (2000). Fetal infant and childhood growth are predictors of coronary heart disease, diabetes, and hypertension in adult men and women. *Environmental Health Perspectives Supplements, 108*(S3), 545-553.

Pekkarinen, P., Terwilliger, J., Bredhacka, P.E., Lannqvist, J. and Peltonen, L. (1995). Evidence of a predisposing locus to bipolar disorder on Xq24-q27.1 in an extended Finnish pedigree. *Genome Research, 5*, 105-115.

Perusse, L., Rankinen, T., Zuberi, A., Chagnon, Y., Weisnagel, S.J., Argyropoulos, G., Walts, B., Snyder, E.E., and Bouchard, C. (2005). The human obesity gene map: The 2004 update. *Obesity Research, 13*(3), 381-490.

Ptashne, M. (1992). *A genetic switch, second edition*. Cambridge, MA: Cell Press and Blackwell Scientific.

Ricklefs, R.E. (2006). Embryo development and ageing in birds and mammals. *Proceedings Royal Society London B*, doi:10.1098/rspb2006.3544, pp. 1-6.

Ridley, M. (1999). *Genome*. London, England: Fourth Estate.

Sing, C.F., Haviland, M.B., Templeton, A.R., Zerba, K.E., and Reilly, S.L. (1992). Biological complexity and strategies for finding DNA variations responsible for inter-individual variation in risk of a common chronic disease, coronary artery disease. *Annals of Medicine, 24*, 539-547.

Sing, C.F., Haviland, M.B., Templeton, A.R., and Reilly, S.L. (1995). Alternative genetic strategies for predicting risk of atherosclerosis. In F.P. Woodford, J. Davignon, and A. Sniderman (Eds.), *Proceedings of the 10th International Symposium on Atherosclerosis, Montreal* (pp. 638-644). New York: Elsevier.

Wallace, R.B. (2001). Applying genetic study designs to social and behavioral population surveys. In National Research Council, *Cells and surveys: Should biological measures be included in social science research?* (pp. 229-249). Committee on Population. C.E. Finch, J.W. Vaupel, and K. Kinsella, Eds. Commission on Behavioral and Social Sciences and Education. Washington, DC: National Academy Press.

Wang, H.-D., Kazemi-Esfarjani, P., and Benzer, S. (2004). Multiple-stress analysis for isolation of *Drosophila* longevity genes. *Proceedings National Academy of Sciences, USA, 101*(34), 12610-12615.

West-Eberhard, M.J. (2003). *Developmental plasticity and evolution.* New York: Oxford University Press.

Williams, G.C. (1957). Pleiotropy, natural selection, and the evolution of senescence. *Evolution, 11*, 398-411.

Williams, G.C., and Nesse, R.M. (1991). The dawn of Darwinian medicine. *Quarterly Review of Biology, 66*(1), 1-22.

Wright, T.F., Lackey, L.B., Schirtzinger, E.E., Gonzalez, L.A., and Eberhard, J.R. (in press). Mitochondrial control-region duplications and longevity in parrots.

10

Genetic Markers in Social Science Research: Opportunities and Pitfalls

George P. Vogler and *Gerald E. McClearn*

Social science survey research provides rich databases that are informative regarding factors that influence the health and well-being of target populations. These studies frequently are rich in behavioral assessments that can include measures as diverse as cognitive assessment and personality characteristics, social support patterns, health behaviors, and the availability and utilization of health care services. Such research frequently has major advantages, including large, carefully selected samples that are nationally representative or that target particular populations, frequently with long-term longitudinal data to monitor secular changes in risk and protective factors over time or over developmental periods. Examples of such research include such well-known studies as the Health and Retirement Study (HRS, Juster and Suzman, 1995), the National Long Term Care Survey (NLTCS, Research Triangle Institute, 2002), and the National Health and Nutrition Examination Survey (NHANES, Centers for Disease Control and Prevention, 2006).

Recently there have been efforts to incorporate collection of DNA and other biomarkers into these types of surveys through the collection of blood samples or cheek swabs. The incorporation of genetic markers into studies that have health as an emphasis has the potential to provide a more thorough understanding of the mechanisms of the complex factors that influence health outcomes, when considered with the other domains that are frequently assessed, including behavioral, psychological, social, economic, and environmental indicators. Studies such as the MacArthur Study of Successful Aging that have incorporated DNA and biological markers (Crimmins and Seeman, 2001) have demonstrated strong relationships between biological indicators and traditional demographic variables and health outcomes. In the following discussion, issues are considered regarding the incorporation of DNA collection into studies that were designed to address social science questions.

INCORPORATION OF DNA COLLECTION INTO SOCIAL SCIENCE SURVEYS

Incorporation of DNA collection into social science surveys has the potential to add considerable value to such studies. The utility of genetic information is based on the observation that there is variability in the allele form of individual genes that can be measured and that can contribute to variability in health-related outcomes. Potential uses include investigation of genes that have well-established influences, such as the Apolipoprotein E (ApoE) gene and its relationship to risk for Alzheimer disease and cardiovascular disease; identification of genes that had not been previously known to have relationships with the outcome variables; investigation of whether an individual's genetic status contributes to variability in the way in which other factors influence outcome variables (gene-environment interaction or gene-gene interaction); investigation of correlations between genetic and environmental factors; and other, more novel uses, such as controlling for genetic status to obtain a more accurate picture of how nongenetic factors are related to the outcome variables. For example, by controlling for genetic status at the ApoE gene, a clearer picture could emerge of the effect of other factors, such as cognitive activity, on the development of dementia.

COMPLEXITY OF THE COACTION OF GENES AND ENVIRONMENT

A central question is whether the complexity of the kinds of outcome variables that are considered in social science survey research is too great to be meaningfully considered in a genetic context. This argument might be considered to be valid if one is limited to the Mendelian perspective, which is the notion that genetic influences on a trait consist of the effect of a single major gene. Much early genetics research was concerned with determining the applicability of the Mendelian rules of transmission to an ever-broadening range of phenotypes. The phenotypes were typically dichotomous (presence or absence of a disease condition is a common type of Mendelian trait), and the principal concern was whether the relative numbers of organisms that were assignable to the different categories conformed to expectations derived from the Mendelian theory. Frequency statistics were adequate to assess the outcomes. If category assignment was reasonably unambiguous, variability in the phenotype within the categories was usually ignorable, along with any influence on that variability due to environmental factors or to genes other than the primary one under investigation. The Mendelian perspective dominated medical genetics through much of the middle of the 20th century, when major progress was made in identifying single major genes that influenced a variety of medical conditions, usually devastating but rare diseases (National Center for Biotechnology Information, 2006).

A parallel avenue of research into heredity was that of biometric or quantitative genetics, which was explicitly concerned with phenotypes that were continuously distributed. The analytical statistics in this approach concerned means, variances, and covariances, and the variance decomposition algorithms provided for a component assignable to heredity, another to environment, and others to correlations and interactions between these agencies. It is this perspective that is most likely to be the more appropriate one for considering the nature of genetic effects on complex traits that are of interest in social science survey studies. It is interesting to note that there has been a substantial emphasis on the analysis of complex traits in the recent genetic literature. This approach is applied widely in psychiatric genetics, behavioral genetics, and genetic epidemiological investigations of complex health-related factors, such as risk factors for cardiovascular disease.

For historical reasons that cannot be detailed here, in certain scientific circles, heredity, under the label of "nature," came to be regarded as antagonistic or oppositional to environment, labeled "nurture." The antagonism of these terms was already implied in the presumed source, Shakespeare's "The Tempest," wherein Caliban is described as "a devil, a born devil, on whose nature nurture can never stick." Galton (1883) brought the alliteration of nature-nurture into scientific discourse with the same intimation of hostility. Although his perspective was considerably more nuanced, his statement in respect to human faculty that "there is no escape from the conclusion that nature prevails enormously over nurture" was easily translated into a view of nature *versus* nurture. This perception had significant influence in the behavioral and social sciences, and led to a strong predilection to seek explanations from the environmental realm. Indeed, in some arenas, there was strong a priori rejection of any possibility of genetic involvement. In recent decades, however, a more balanced view has emerged, partaking increasingly of the quantitative genetics perspective attributing main effects to factors from both genetic and environmental domains, and with all of the subtle possibilities arising from the inevitable interaction terms. In its simplest version, the model can conceive of causal influences arising from these separate domains and combining additively to produce the phenotype. But it is increasingly clear that the causal relationships can be much more complex than this, with significant correlations and interactions between nature and nurture. These coactions have substantial implications for the design and interpretation of laboratory and field studies on health-related phenomena (see Moffitt, Caspi, and Rutter, 2005; Plomin and Rutter, 1998).

Plomin, DeFries, and Loehlin (1977) provided a now-classic description and terminology for various types of correlations and interactions in the context of behavioral genetics. Gene-environment correlations occur when the distribution of environmental influences on a particular phenotype is not independent of the distribution of genetic influences on that phenotype. Three classes are distinguished: passive, reactive, and active. A frequently cited example of the passive situation is the provision to children of an intellectually stimulating environment of books, education, and approbation of scholarly pursuits by parents who have also transmitted genes that positively affect cognitive abilities. Reactive correlation arises from situations in which the (genetically influenced) behavior of an individual evokes a particular response from the social environment that tends to support (or inhibit, as the case may be) that behavior. Active correlation ensues when an individual seeks out or alters her or his environment to be consistent with a genetically influenced predilection.

Various sorts of gene-by-environment (G×E) interaction can also be identified (see, for example, Kearsey and Pooni, 1996). In general terms, however, a compact definition can encompass those circumstances in which the effect of a difference in genotype (single gene or polygenic set) depends on features of the environment or, equivalently, in which the effect of differences in features of the environment are dependent on genotype.

The quantitative genetic perspective has been transformed by the wide availability of information about measured polymorphisms from DNA. This can be in the form of measurable polymorphisms in functional genes or as information on massive numbers of polymorphic markers distributed throughout the genome that are not necessarily functional polymorphisms in genes that affect the trait. The incorporation of measured genetic variability has the potential to increase greatly the ability to investigate complex traits, including a genetic perspective, particularly with respect to G×E interactions and other approaches that go beyond the idea of a main effect of an individual gene on a trait.

By virtue of the degree of control that can be exercised in experimental studies with animal models, G×E interactions can be displayed with striking clarity. Because these provide illustrations of the type of interactive phenomena that might also be expected in human studies, we present here some examples to illustrate the phenomena.

Animal Model Studies

Some interactions have emerged from studies in which the environmental differences were in effect for a sustained period of time. Blizard and Randt (1974), for example, reared two inbred strains of mice under one of three housing conditions—standard caging and relatively enriched or impoverished in terms of sensory stimulation. (The animals of an inbred strain are approximately uniform genetically, and different inbred strains differ genetically from each other). The measured phenotype, novel object-oriented activity, was an assessment of exploratory or anxiety behavior exhibited when the animals were placed in a novel environment. Briefly, animals of one strain (C57BL/6) were totally unaffected by differences in rearing environment; animals of the other strain (DBA/2) displayed a strong effect, with animals from the enriched condition displaying twice the level of activity as the impoverished group.

A possible toxic risk factor for development of Alzheimer disease—aluminum in the diet—was explored in a mouse study by Fosmire, Focht, and McClearn (1993). In each of five inbred strains, two groups were established, one being fed a normal control diet and the other a diet enriched in aluminum. The brain aluminum levels of three of the strains were unaltered by diet; one showed a trend toward increase, and one displayed a threefold increase.

Falconer (see Falconer and Mackay, 1996) selectively bred mice for high or low growth from three to six weeks of age on either a standard diet, or on one for which nutritional value had been degraded by dilution with nondigestible fiber. Response to selection occurred in both conditions. There was thus a line of animals that grew large on the good diet and another line that grew less on that diet; another line that grew large on the poor diet and one that grew less on that diet. Testing these groups under the alternate dietary condition made possible an evaluation of the extent to which the same genes influenced growth in the two environments. The genetic correlation was .66, indicating a large "overlap" of the genes involved, but indicating also a very substantial involvement of different genes in the two nutritional environments.

A further example may be drawn from the research literature on aging. Vieira and colleagues (2000) maintained Drosophila under different environmental circumstances throughout their lives. The five environments differed in the temperature of the incubator in which the animals lived—standard control, higher than control, lower than control, a control temperature but with a heat shock administered during pupal stage, and a reduced diet in a standard temperature. Seventeen quantitative trait loci (QTLs) affecting longevity were identified. (QTLs are genes associated with the phenotype but whose location on the chromosomes is known only approximately.) Not one of the QTLs was uniformly influential in all environments. In several cases, the effect was detectable in one environment only. For some, the allele of the QTL that was associated with longer life in one environment was associated with shorter life in another; for some the influence was shown in one sex only; for some, in one particular environment, the "increasing" allele in females was a "decreasing" allele in males. In short, all of the genetic variance was involved in genotype-environment interaction, genetic-sex interaction, or both.

Human Studies

A common feature of these examples is that a presumptive causal element, genetic or environmental, may have differing effect depending on the context of other environmental and genetic elements that are present. Paraphrasing in terms of the health sciences, it may be expected that the virulence of risk factors and the beneficence of remedial or preventive interventions will vary greatly from individual to individual, depending on the unique context presented by each individual. As in the case of the animal model studies, we provide a sampling of the relevant literature, which is large and burgeoning.

Bouchard and colleagues (1990) investigated a putative risk factor—the effects of long-term overfeeding—in young adult male monozygotic twins. For a variety of outcome variables, including gain in body weight, percentage of fat, fat mass, and estimated subcutaneous fat, the variance among pairs was about threefold the variance within pairs. This intrapair similarity is suggestive of involvement of genotype in response to this nutritional environmental intervention (shared environmental influence cannot be ruled out, however).

A polymorphism in the angiotensin-converting enzyme (ACE) that is characterized by the absence (deletion) of a 287-base pair marker has been shown to be related to the response to exercise. In general, the insertion allele (presence of the 287-base pair marker) is associated with greater endurance, whereas the deletion allele is associated with greater muscular strength, although the results have been somewhat conflicting (Folland et al., 2000; Gayagay et al., 1998; Montgomery et al., 1997; Myerson et al., 2001). On the basis of these findings, the relationship of ACE to efficacy of a health-promoting behavior in the elderly was explored by Kritchevsky and colleagues (2005). The risk to septuagenarians of developing mobility limitation was studied as a function of spontaneous activity level. Overall, those more active at baseline were less likely than those who did not exercise to incur limitations to their activity during a follow-up period. However, an interaction was present in that individuals with a particular genotype at this locus (designated II) benefited less than did the other two genotypes (DD or DI).

A well-studied gene with respect to gene-environment interactions related to cardiovascular health is the ApoE gene. This is one of the few genes identified to date that have common variants with well documented and consistent main effects on traditional social science outcomes. Numerous studies have investigated the effect of the ApoE isoform on response to dietary intervention, with inconsistent results. In a meta-analysis of a number of studies, Ordovas et al. (1995) found that the effect of the ApoE-4 showed a higher low-density lipoprotein (LDL) response than other ApoE alleles in diets that reduced total dietary fat, no difference in response in diets that reduced only dietary cholesterol, and a lower LDL response in diets in which the fat saturation (but not total amount of fat) was modified, although not all studies have found this effect (e.g., LeFevre et al., 1997). A differential response of LDL, high-density lipoprotein, and triglyceride levels to diet as a function of the ApoE-2 and ApoE-4 alleles was observed in a Finnish study of children and young adults who were in a free-living situation rather than a dietary intervention study (Lehtimäki et al., 1995). Cardiovascular responsivity to mental stress has also been shown to vary as a function of the ApoE polymorphism (Ravaja et al., 1997). Another ApoE polymorphism in the gene's promoter region (-219G→T) has been shown to affect LDL concentrations differentially (Moreno et al., 2004) and insulin sensitivity (Moreno et al., 2005) in response to a high-fat dietary modification.

Jaffee and colleagues (2005) examined conduct problems in young twins as a function of genetic risk (estimated from co-twin's behavior) and from physical maltreatment. Maltreatment increased probability of a conduct disorder diagnosis by 2 percent in those with the lowest genetic risk and by 24 percent among children with the highest genetic risk.

The intent of this discussion has been to illustrate the potential presence of interactions, not to imply that they are omnipresent. There are clearly variables in both the environmental and genetic domains that have powerful main effects. But interactions *may* be present, and they can be subtle and powerful. One implication is that identification of a gene as *the* gene for some attribute must be tentative; it might be without effect or with a different effect in some environment other than that in which it is first described. Similarly, environmental influences may have greater or lesser impact, depending on the genotype of the individuals on whom they impinge. Confidence in generality of the effect will accrue with subsequent observations in altered contexts. An interesting issue is whether an investigation of interaction effects should be undertaken in the absence of observation of main effects. There are methodological issues that arise related to the number of tests for interactions that are possible, resulting in the danger that a blind search for interaction effects in the absence of main effects could lead to a serious problem of false positive results. However, this should not preclude assessment of interaction effects in the absence of main effects under any circumstance. It is possible that subsequent work could provide a line of evidence in favor of a more careful search for interactions without main effects. The example from the meta-analysis of the differential effect of dietary intervention as a function of ApoE allele status noted above is a pertinent example, in which the differential effect of the alleles were in opposite directions, depending on the precise form of the dietary intervention.

In some respects and for some purposes, correlations and interactions such as we have described can be seen as vexing. A more positive view is perhaps warranted. It may well be that the interactions that appear are indications of key processes in the functioning of the complex systems mediating genetic and environmental influences on complex phenotypes. They may be signposts to particularly productive avenues of research.

In summary, a genetic perspective can be incorporated into research despite the likely complexity of genetic and environmental influences and how they interact in complex ways in social science traits. By discarding the nature versus nurture perspective, researchers studying the genetics of complex traits have embraced a more interactive model. The incorporation of measured genetic variability into social science survey research, which frequently has excellent environmental assessment, opens up rich new opportunities to investigate G×E interactions in large samples, with the potential of providing a more realistic description of factors that affect social science survey outcomes.

POTENTIAL ADVANTAGES AND DISADVANTAGES

The potential to investigate more complex, and more realistic, models that incorporate measured genetic information into social science survey studies is a major advantage of obtaining DNA. As noted above, the complexity of the outcomes makes utilizing DNA more challenging than simply including another predictor into a regression model. Researchers involved in genetic studies of complex traits, however, have recognized this complexity and have developed methodologies to deal appropriately with it. In the past, G×E interaction effects could be considered only in the context of latent variable models, which have extremely limited power to detect interactions. The recent explosion in the availability of measurable DNA variability through the use of DNA markers, coupled with advances in the assessment of environmental factors, opens up opportunities to make real progress in understanding these effects on complex traits. There is a danger of improperly dealing with the complexity, resulting in misleading or biased conclusions. It is therefore essential that the use of DNA is rigorously evaluated from a methodological perspective to limit the opportunities for erroneous conclusions to be drawn.

Potential Advantages

In addition to the broad potential advantages in making scientific progress in understanding more completely the nature of genetic and environmental factors on complex social science traits, there are several other practical advantages. One is the potential to have DNA in long-term availability. It is possible to create either immortalized cell lines that provide unlimited access to the DNA indefinitely into the future, or to extract a finite amount of DNA from a blood, cheek cell, or saliva sample for long-term storage. Consequently, there is the potential to have long-term access to DNA for research purposes not yet conceived or for use with technology not yet developed that will make more efficient use of limited samples. Timely collection of DNA can be important in survey research, particularly longitudinal research, in which attrition can result in substantial loss to follow-up of participants in later waves of assessment. For surveys that focus on aging-related issues, there is the more significant issue of loss due to mortality.

At first glance, the issue of G×E interactions seems to introduce an annoying degree of complexity. When an investigator is interested primarily in general characteristics of a population or specific subgroup, interactions do introduce an additional source of variability that can mask the primary effects that are of interest. However, if one takes a different perspective that more closely resembles that of a clinician, the ability to identify individual-level risk factors that are not constant across the group is significant. It has the potential to provide more effective individual-level prediction of outcomes based on risk factors that are unique to the individual.

A related issue is that the identification of individual-level risk factors can be used to develop prospective procedures for identifying individuals who are at high risk for developing adverse outcomes. At a minimum, such individuals can be informed of their elevated risk status so that they can be monitored more closely than standard. Of potentially greater interest is the possibility of targeting interventions specifically to at-risk individuals. The significance of this approach is that while interventions might clearly have an advantage at the overall population level, individuals could respond to an intervention differentially, with its being effective for some, neutral for others, and potentially even damaging to still others.

Potential Disadvantages

The incorporation of DNA collection into social science surveys has several potential disadvantages. At a practical level, there are logistical and financial constraints surrounding the process of sample collection. How severe these constraints are depends on the method of DNA collection and whether it is coupled with assessment of a more extensive battery of biomarker measures. If a survey is primarily conducted by mail or telephone, DNA samples of limited yield can be collected using cheek swabs that are returned through the mail. DNA can be extracted and stored for nominal cost. The drawback is that the amount of DNA is limited, and the failure rate for obtaining a useful DNA sample can be substantial. If blood is drawn, the yield and quality of the DNA improves substantially. However, this requires contact between the participant and a trained professional for the drawing of blood. If the study involves face-to-face contact, the logistical burden can be manageable, but otherwise logistical factors can be challenging. Although the costs of obtaining a DNA sample can be minimal, the cost of genotyping large numbers of individuals from large surveys can be prohibitive unless the genetic component of the study is very tightly focused (for example, limited to genotyping of the ApoE gene).

Another potential disadvantage is related to the participant burden and the purpose of the study. Participants might be recruited to a survey study on the basis of their willingness to be involved in social science research. It might not be clear to participants why they are being asked to provide DNA samples for a study that has been described to them as a social science survey. If participants have been involved longitudinally in a survey, they would not have consented to providing a DNA sample, so additional consent would be required. A request for DNA potentially imposes an additional response burden that participants had not anticipated. If the purpose and procedures for collecting DNA are not thoroughly and properly planned, there is the potential to introduce a bias into the sample that is related to willingness to provide DNA and not related to the primary purpose of the study.

An issue that is not so much a disadvantage but a challenge involves the difficulties that arise when there is a large amount of data generated, in terms of potential type 1 and type 2 error rates and the practical issues that arise in trying to limit the number of statistical tests that are done and account for multiple comparisons. If a study opts to do a full genome scan using single-nucleotide polymorphisms (SNPs) assayed using DNA microchips, there is the potential to generate a million or more DNA markers. Dealing with such massive amounts of data requires specialized techniques in data reduction and haplotype block identification.

Substantial methodological work in statistical genetics is currently being conducted to address this issue effectively.

ETHICAL ISSUES

In addition to the more practical issues discussed above, incorporation of DNA collection raises a number of ethical issues (reviewed by Durfy, 2001, in the context of aging-related surveys). Ownership of genetic information is an area that is potentially troublesome, and issues regarding ownership need to be worked out explicitly. Some studies have a policy that requires all information be made publicly available to any investigator who desires access. Other studies involving private biotechnology companies consider the genetic information to be proprietary. It is important for issues of ownership to be clearly planned, and information regarding ownership should be made clear to the participants in the informed consent process.

In addition to access issues, it is important to have a clear plan for the use that will be made of the genetic data, and these issues should be incorporated into the informed consent process. Questions regarding use include whether the use of DNA goes beyond the initial scientific goals of the survey study, whether the data will be made available to other investigators outside the scope of the original project, whether the data will be retained indefinitely so that use in the future cannot be anticipated, and whether results of DNA testing will be provided back to the participant. The last issue is important, since there are few genetic results that are clear-cut risk factors for individuals, particularly in the context of complex traits. In the context of other modifying environmental and genetic factors, individual-level information on even well-known risk factors can be difficult to convey accurately to participants, since risk is generally reported in probabilistic terms. If the decision is to provide information to participants, considerable attention must be paid to how the information is presented. Conversely, if there is clear information about a risk factor, the investigators must consider their obligations to provide this information to participants, either now or if future work clarifies the nature of a genetic risk factor.

While there are important issues regarding privacy and maintaining anonymity with public data sets for any potential identifying information, including demographic factors, these issues merit special attention for genetic data, which contain the risks associated with sensitive medical data in general but also have additional risks. Depending on the extent to which genotyping is undertaken, genetic data can contain sufficient information to identify an individual uniquely. There also is a component of risk related to the familial nature of genetic data in which there is the potential to identify individuals who have not consented to participate in the research but who are related to participants. Finally, misuse of genetic information following breach of anonymity can result in problems associated with stigmatization and discrimination on a variety of factors, such as the availability or cost of health insurance. Because of these concerns, attention is warranted to issues of risk of breach of anonymity in terms of making data publicly available to researchers outside the context of the original study (Annas, Glanz, and Roche, 1995; National Bioethics Advisory Commission, 1999).

WHAT CAN ONE LEARN FROM DNA IN SOCIAL SCIENCE SURVEYS?

The incorporation of DNA into social science surveys can contribute to a more accurate understanding of both genetic and environmental factors that contribute to complex outcomes. Particularly for surveys that focus on health, DNA has a potentially important role to play. Reproductive outcomes, health and diseases, life span and aging, and cognitive function and personality all have substantial genetic influences that are not fully understood. Examples of areas for which genes have been identified include cardiovascular risk factors, diabetes, hypertension, obesity, markers of inflammation, lung function, cognitive function, and addictive behaviors. The extensive environmental assessment information in social science surveys provides a unique opportunity to improve understanding of how these genes function in a broader context, as well as how they contribute to the context in which environmental factors influence social science outcomes.

REFERENCES

Annas, G.J., Glanz, L.H., and Roche P.A. (1995). Drafting the Genetic Privacy Act: Science, policy, and practical considerations. *Journal of Law, Medicine, and Ethics, 23,* 360-366.

Blizard, D.A., and Randt, C.T. (1974). Genotype interaction with undernutrition and external environment in early life. *Nature, 251,* 705-707.

Bouchard, C., Tremblay, A., Després, J.-P., Nadeau, A., Lupien, P.J., Thériault, G., Dussault, J., Moorjani, S., Pinault, S., and Fournier, G. (1990). The response to long-term overfeeding in identical twins. *The New England Journal of Medicine, 322,* 1477-1482.

Centers for Disease Control and Prevention (CDC). (2006). *National Health and Nutrition Examination Survey data.* Hyattsville, MD: U.S. Department of Health and Human Services, Centers for Disease Control and Prevention. Available: http://www.cdc.gov/nchs/data/nhanes/nhanes3/nh3gui.pdf [accessed January 15, 2007].

Crimmins, E., and Seeman, T. (2001). Integrating biology into demographic research on health and aging (with a focus on the MacArthur Study of Successful Aging). In National Research Council, *Cells and surveys: Should biological measures be included in social science research?* (pp. 9-41). Committee on Population. C.E. Finch, J.W. Vaupel, and K. Kinsella, Eds. Commission on Behavioral and Social Sciences and Education. Washington, DC: National Academy Press.

Durfy, S.J. (2001). Ethical and social issues in incorporating genetic research into survey studies. In National Research Council, *Cells and surveys: Should biological measures be included in social science research?* (pp. 303-328). Committee on Population, C.E. Finch, J.W. Vaupel, and K. Kinsella (Eds.), Commission on Behavioral and Social Sciences and Education. Washington, DC: National Academy Press.

Falconer, D.S., and Mackay, T.F.C. (1996). *Introduction to quantitative genetics, fourth edition.* Essex, England: Longman Group Ltd.

Folland, J., Leach, B., Little, T., Hawker, K., Myerson, S., Montgomery, H., and Jones, D. (2000). Angiotensin-converting enzyme genotype affects the response of human skeletal muscle to functional overload. *Experimental Physiology, 85,* 575-579.

Fosmire, G.J., Focht, S.J., and McClearn, G.E. (1993). Genetic influences on tissue deposition of aluminum in mice. *Biological Trace Element Research, 37,* 115-121.

Galton, F. (1883). *Inquiries into human faculty and its development* (p. 241). London, England: Macmillan.

Gayagay, G., Yu, B., Hambly, B., Boston, T., Hahn, A., Celermajer, D.S., and Trent, R.J. (1998). Elite endurance athletes and the ACE I allele: The role of genes in athletic performance. *Human Genetics, 103,* 48-50.

Jaffee, S.R., Caspi, A., Moffitt, T.E., Dodge, K.A., Rutter, M., Taylor, A., and Tully, L.A. (2005). Nature X nurture: Genetic vulnerabilities interact with physical maltreatment to promote conduct problems. *Development and Psychopathology, 17,* 67-84.

Juster, F.T., and Suzman, R. (1995). An overview of the health and retirement study. *The Journal of Human Resources, 30*(Suppl. 1995), S7-S56.

Kearsey, M.J., and Pooni, H.S. (1996). *The genetical analysis of quantitative traits.* London, England: Chapman and Hall.

Kritchevsky, S.B., Nicklas, B.J., Visser, M., Simonsick, E.M., Newman, A.B., Harris, T.B., Lange, E.M., Penninx, B.W., Goodpaster, B.H., Satterfield, S., Colbert, L.H., Rubin, S.M., and Pahor, M. (2005). Angiotensin-converting enzyme insertion/deletion genotype, exercise, and physical decline. *Journal of the American Medical Association, 294,* 691-698.

LeFevre, M., Ginsberg, H.N., Kris-Etherton, P.M., Elmer, P.J., Stewart, P.W., Ershow, A., Pearson, T.A., Roheim, P.S., Ramakrishnan, R., Derr, J., Gordon, D.J., and Reed, R. (1997). ApoE genotype does not predict lipid response to changes in dietary saturated fatty acids in a heterogeneous normolipidemic population. The DELTA Research Group. Dietary Effects on Lipoproteins and Thrombogenic Activity. *Arteriosclerosis, Thrombosis, and Vascular Biology, 17,* 2914-2923.

Lehtimäki, T., Moilanen, T., Porkka, K., Åkerblom, H.K., Rönnemaa, T., Räsänen, L., Viikari, J., Ehnholm, C., and Nikkari, T. (1995). Association between serum lipids and apolipoprotein E phenotype is influenced by diet in a population-based sample of free-living children and young adults: The Cardiovascular Risk in Young Finns Study. *Journal of Lipid Research, 36,* 653-661.

Moffitt, T.E., Caspi, A., and Rutter, M. (2005). Strategy for investigating interactions between measured genes and measured environments. *Archives of General Psychiatry, 62,* 473-481.

Montgomery, H.E., Clarkson, P., Dollery, C.M., Prasad, K., Losi, M.A., Hemingway, H., Statters, D., Jubb, M., Girvain, M., Varnava, A., World, M., Deanfield, J., Talmud, P., McEwan, J.R., McKenna, W.J., and Humphries, S. (1997). Association of angiotensin-converting enzyme gene I/D polymorphism with change in left ventricular mass in response to physical training. *Circulation, 96*, 741-747.

Moreno, J.A., Pérez-Jiminéz, F., Marín, C., Gómez, P., Pérez-Martínez, P., Moreno, R., Bellido, C., Fuentes, F., and López-Miranda, J. (2004). Apolipoprotein E gene promoter -219G→ T polymorphism increases LDL-cholesterol concentrations and susceptibility to oxidation in response to a diet rich in saturated fat. *American Journal of Clinical Nutrition, 80*, 1404-1409.

Moreno, J.A., Pérez-Jiminéz, F., Marín, C., Pérez-Martínez, P., Moreno, R., Gómez, P., Paniagua, J.A., Lairon, D., and López-Miranda, J. (2005). The apolipoprotein E gene promoter (-219G/T) polymorphism determines insulin sensitivity in response to dietary fat in healthy young adults. *Journal of Nutrition, 135*, 2535-2540.

Myerson, S.G., Montgomery, H.E., Whittingham, M., Jubb, M., World, M.J., Humphries, S.E., and Pennell, D.J. (2001). Left ventricular hypertrophy with exercise and ACE gene insertion/deletion polymorphism: A randomized controlled trial with losartan. *Circulation*, 226-230.

National Bioethics Advisory Commission. (1999). *Research involving human biological materials: Ethical issues and policy guidance.* Rockville, MD: National Bioethics Advisory Commission.

National Center for Biotechnology Information. (2006). *Online Mendelian Inheritance in Man, OMIM™.* Available: http://www.ncbi.nlm.nih.gov/omim/ [accessed Sept. 2007].

Ordovas, J.M., Lopez-Miranda, J., Mata, P., Perez-Jiminez, F., Lichtenstein, A.H., and Schaefer, E.J. (1995). Gene-diet interaction in determining plasma lipid response to dietary intervention. *Atherosclerosis, 118*, S11-S27.

Plomin, R., DeFries, J.C., and Loehlin, J.C. (1977). Genotype-environment interaction and correlation in the analysis of human behavior. *Psychological Bulletin, 84*, 302-322.

Plomin, R., and Rutter, M. (1998). Child development, molecular genetics, and what to do with genes once they are found. *Child Development, 69*, 1223-1242.

Ravaja, N., Räikkönen, K., Lyytinen, H., Lehtimäki, T., and Keltikangas-Järvinen, L. (1997). Apolipoprotein E phenotypes and cardiovascular responses to experimentally induced mental stress in adolescent boys. *Journal of Behavioral Medicine, 20*, 571-587.

Research Triangle Institute. (2002). *1999 National Long Term Care Survey: Functional and health changes of the elderly. Venipuncture, buccal cell, kin, and next-of-kin supplemental studies.* Final Report: RTI Project 7798. Research Triangle Park, NC: Research Triangle Institute. Available: http://www.cds.duke.edu/nltcs/doc_other.htm [accessed Nov. 2006].

Rhodes, J.S., and Crabbe, J.C. (2005). Gene expression induced by drugs of abuse. *Current Opinion in Pharmacology, 5*, 26-33.

Smith, M.A., Banerjee, S., Gold, P.W., and Glowa, J. (1992). Induction of c-fos in rat brain by conditioned and unconditioned stressors. *Brain Research, 578*, 135-141.

Vieira, C., Pasyukova, E.G., Zeng, Z.-B., Hackett, J.B., Lyman, R.F., and Mackay, T. (2000). Genotype-environment interaction for quantitative trait loci affecting lifespan in Drosophila melanogaster. *Genetics, 154*, 213-227.

11

Comments on the Utility of Social Science Surveys for the Discovery and Validation of Genes Influencing Complex Traits

Harald H.H. Göring

The debate about the relative importance of "nature" and "nurture" in determining behavior and emotional and physical well-being has persisted to this day and is conducted among scientists, educators, parents, politicians, philosophers, and many other members of the society. In this phrasing, "nature" is used to represent those forces shaping our existence over which we have virtually no control (as of yet), namely our genetic constitution. "Nurture" is used as a short form to suggest those influences on our life that we (think we) can shape to at least some degree, namely our environment. We are born with our genes, but we have some control over our exposure to certain aspects of the environment.

It is clear that virtually all traits are influenced by both genetic and environmental factors. This applies to the rare and often serious diseases for which a genetic defect is necessary and sufficient to bring about disease, but whose manifestation is nonetheless influenced by various other factors (that we often do not know) (Wexler et al., 2004). This likewise holds for infectious diseases, which are often viewed from the perspective of environmental exposure to the pathogen alone. As we now know, the genetic constitution of the human host greatly influences the risk of exposure, infection and disease, as well as its course and severity (Allison, 1954, 1961; Kulkarni et al., 2003).

The pendulum of prevailing social views keeps swinging back and forth between the two extremes. At the present time, the emphasis is clearly on the importance of genetic factors. The popular press contains daily reports of the discovery of yet another gene influencing yet another disorder, and predictions that scientists will unravel the genetic mysteries of most conditions in only a few more years abound, often coupled with enormous promises for the prevention and cure of disease in the near future. Under these circumstances, it is not surprising that many individuals have a very deterministic perception of the action of genes and think that there is a gene for every condition, with the condition being fully and accurately determined by this gene, independently of anything else. Geneticists are not blameless for this situation, as they often do not correct such views, unintentionally promote them by using sloppy terminology consistent with such opinions, or even intentionally further them by making exaggerated claims about the future impact of their area of research, perhaps in an effort to improve funding. It is in this environment that many researchers in other fields have begun thinking about whether they should and can incorporate gene discovery into their own studies.

In this chapter, I comment on the utility of large-scale social science surveys for the discovery and validation of genes influencing conditions of interest to social scientists. I start with a brief overview of the nature of so-called complex traits and highlight some of the concepts behind study designs that are being used for the identification of genes. I attempt to contrast the traits for which gene-mapping studies have succeeded and the designs of gene discovery experiments to social science surveys, with a focus of the suitability of the latter for gene identification. I close with a few remarks on how such surveys may be useful for gene discovery and validation from my perspective.

ETIOLOGICAL ARCHITECTURE OF COMPLEX TRAITS

There is no accepted definition of what constitutes a so-called complex or multifactorial trait. The term is generally used to denote the opposite of a so-called Mendelian trait, in which a defect in a gene by itself can cause a specific phenotype (the focus is often on a disease). In contrast, the relationship between genotype and phenotype is not as deterministic in complex traits, for which individual genetic variants merely modulate the probability of presenting a particular phenotype.

For many traits, we have absolutely no idea about the identity of environmental factors and genes whose variants account for some of the variability in the phenotype in the population, and the designation of a trait as complex simply acknowledges the belief—based on common sense, failed gene mapping attempts, analogies with similar traits about which we have a better understanding, or evolutionary considerations—that a multitude of genetic and environmental factors must influence the phenotype. It may well turn out that a trait is not as complicated as first assumed, such as when gene mapping studies readily succeed in pinpointing the location of genes of substantial effect. The distinction between Mendelian and complex traits is by no means black and white. The terms merely refer to the two extremes, with most traits falling somewhere in the middle.

Figure 11-1 provides a schematic view of the etiological architecture of a prototypical complex, multifactorial trait. The phenotype of an individual depends on a number of genetic and environmental factors that in concert determine the phenotypic outcome. The individual components of the etiological spectrum modulate the probability of manifesting a dichotomous phenotype, such as a disease, or they may alter the expected value of a quantitative trait. Not all components are equally important.

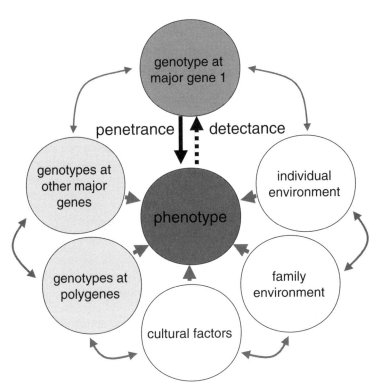

FIGURE 11-1 Schematic of the etiological architecture of a prototypical complex trait.
NOTE: The key goal when designing a gene-mapping study is to reduce the etiological complexity as much as possible. In the ideal case, all influences on the trait of interest are eliminated except for the gene that is to be localized and identified. In this situation, the phenotypes of study participants are good predictors of the genotypes at the underlying gene, facilitating its discovery.

Among the genetic factors, one generally distinguishes between major genes and polygenes. Major genes have a substantial influence on the phenotype and may be individually identifiable with gene-mapping approaches. Polygenes may have a sizable effect in the aggregate, but their individual influence is small, making them impossible to identify using statistical genetic methods. The various etiological components may act independently from one another, or there may be interaction, such as between different genetic factors (gene-gene interaction), between specific genes and aspects of the environment (gene-environment interaction), or among different environmental factors (environment-environment interaction).

The term "interaction" is widely used in scientific discourse, but the word is often misused and misunderstood because it has different meanings in everyday language and in statistics (Cox, 1984). The fact that each of two factors influences a trait does not imply the existence of interaction between them. A well-known and often replicated, though not uncontroversial, example of gene-environment interaction in complex trait etiology is the moderating effect of a variant of the 5-HTT gene on the impact of stressful life events on depression (Caspi et al., 2003).

While genetic factors are shared among related individuals in predictable ways, as a consequence of the simple rules of inheritance (in which an individual receives one complete set of chromosomes from each parent, with random selection of which copy of a given chromosome in a parent is passed on), no such rules for sharing of environmental factors among family members exist. Some exposures are shared among members of a household (household effect), cultural factors characteristic of a society may be shared among the members of the entire population, while other environmental factors are not shared within families but are unique to an individual.

Another important distinction between genetic and environmental factors is that, while the rules of inheritance impose a correlation structure on the genetic factors, such that virtually all of them can be statistically assessed for having a potential influence on the trait of interest (see more on this point below), no such correlation structure exists for the environmental factors, such that it is not possible to investigate all components of the environment. The investigator thus has to decide beforehand which aspects of the environment are to be measured, and for many complex traits knowledge of which environmental factors may be relevant is inadequate. From a gene-mapping perspective, all genetic and environmental factors except for the one gene to be located in the genome (denoted as major gene 1 in Figure 11-1) represent noise. The greater this noise is, the more difficult it is to map and identify any particular gene.

STUDY DESIGN CONSIDERATIONS FOR
COMPLEX TRAIT GENE IDENTIFICATION

The description of the contrast between Mendelian and complex, multifactorial traits above highlighted the different degree of determinism in the relationship of genotype and phenotype. The probability of displaying a specific phenotype given an underlying genotype, or P(phenotype | genotype), is referred to as the penetrance. For Mendelian diseases, P(phenotype | genotype) $\rightarrow$ 1, that is, a defect in one or several genes will cause disease, and the disease will be absent otherwise. For complex diseases, the relationship is not as deterministic. In the extreme case, P(phenotype | genotype) $\rightarrow$ P(phenotype), that is, alleles of a gene have virtually no influence on the phenotype.

Penetrances describe the unidirectional biological relationship between genes and traits: Genotype determines phenotype, but the reverse generally does not hold. As biological quantities, however, penetrances are not under the control of the investigator, although they do depend on the genetic and environmental background. The reverse relationship, often referred to as detectance, measures how well the observed phenotype predicts the underlying trait locus genotype, that is, P(genotype | phenotype). This quantity varies greatly with the design of a study. The power of a gene-mapping study is a function of the detectance (Weiss and Terwilliger, 2000), and a key goal of designing a powerful study is to select a design in which the observed phenotype predicts the underlying trait locus genotype as accurately as possible, that is, P(genotype | phenotype) $\rightarrow$ 1.

As a demonstration of the importance of study design on the power for gene localization, we previously computed the sample sizes to detect genes for retinitis pigmentosa (RP) under several different popular study designs based on different family sampling units (Terwilliger and Goring, 2000). RP is a serious eye disease (Bhatti, 2006) that is often considered an example of a Mendelian disorder. However, the disease is by no means monogenic, and there is considerable variation in the mode of inheritance (Rivolta et al., 2002). In some families, the disease is found in multiple generations, consistent with dominant inheritance. In other families, RP is observed in only a single sibship or in offspring of consanguineous "marriages" (i.e., the parents are related to each other), which is characteristic of recessive inheritance. Besides autosomal segregation, both dominant and recessive forms of sex-linked segregation are also found.

What the different types of RP have in common is a high penetrance, that is, defects in various different genes cause the disease with high certainty. However, the detectance is fairly low because of substantial locus heterogeneity (i.e., defects in many different genes can cause the disease).

In a single affected individual, in particular if the case is viewed in isolation, it is not clear which gene is responsible, and thus P(genotype at gene X | phenotype) is small.

On the basis of the information provided in Heckenlively and Daiger (1996) and on several simplifying assumptions, we computed the sample size to detect significant evidence for a RP locus using linkage analysis on affected sib pairs (Penrose, 1935) (assuming that one does not attempt to stratify them by the mode of inheritance, which is generally not possible for complex traits). The required sample sizes run on the order of many hundreds to several thousands for recessive loci and tens of thousands for dominant loci. We also computed the sample sizes to detect the gene that is the most common cause of RP (rhodopsin) with trios consisting of parents and an affected offspring. This sampling scheme is popular, and such data are often analyzed using the transmission disequilibrium test (TDT) (Spielman et al., 1993). In this case, the sample sizes range in the tens to the hundreds of thousands.

How is it possible that the sample size requirements for a relatively simple trait such as RP are this high for these two commonly used sampling schemes? The reason is found in the enormous locus and allelic heterogeneity, such that the relationship of disease to causal genes and variants is one to many. As a result, the disease status is a poor predictor of the underlying genotype at any of the many possible RP loci. By collecting extended families segregating the disease, the difficulties posed by the substantial heterogeneity have easily been overcome. By subgrouping these families based on the observed mode of inheritance, or by focusing on a single family of sufficient size, the genetic heterogeneity is greatly reduced, which permitted many different loci to be mapped (Rivolta et al., 2002). Within individual families, it can often be assumed (given the rarity of the disease) that all affected individuals harbor the same underlying genetic defect, that is, there is a one-to-one relationship of disease and causal gene. Hence, the detectance within a family is very high, and linkage analysis should succeed in gene localization.

Hereditary deafness is another good example of substantial locus heterogeneity (Goldfarb and Avraham, 2002). These examples are given here to highlight the substantial impact that study design choices can have. RP and nonsyndromic hearing loss may be extreme examples of locus heterogeneity, perhaps due to the complexity of these sensory organs, but the brain is certainly much more complicated, suggesting that substantial obstacles may lie ahead for gene discovery for predominantly brain-related phenotypes.

Many different strategies can be pursued to increase the power of a gene-mapping study. These strategies are generally not mutually exclusive but can be combined with one another, potentially to great advantage. The following briefly describes some of the available study design choices, yet the list is by no means exhaustive. There is considerable discussion about the relative merits of the various approaches in the scientific community. No approach is accepted as being the best universally. In fact, the choice of study design should be based on the nature of the trait being investigated. The debate about the pros and cons of various approaches is often phrased in terms of statistical methods and software implementations. However, in such a discussion it is important to recognize the critical assumptions that underlie these statistical approaches and computer programs. The chosen study design largely determines the analytical tools to be used in the analysis. And a study should certainly be designed based on fundamental considerations, rather than the ease of analysis with a particular software package.

Population Isolates

One commonly used approach is to focus on an isolated population (Wright et al., 1999). The idea is that, within such a population, the etiological complexity is likely to be substantially reduced, with regard to both genetic variants and environment factors. A population that has received much attention for gene mapping is Finland (Peltonen et al., 1999), and many studies have also been conducted in Iceland, on French Canadians from the province of Quebec, Mormons from Utah, and the Amish and the Mennonite communities in the United States, among others. The genetic etiology may be simpler in such isolates because of a small founding population, such that the genetic material of the entire population goes back to a fairly small number of founder chromosomes, thereby limiting the amount of allelic variation, at least if admixture with other peoples has been absent or minimal since the population was initially established.

For the search for genes influencing rare diseases, these populations have proven to be extremely valuable (Peltonen et al., 1999). For the analysis of complex traits, however, it is not clear whether the founding populations were of sufficiently small size to simplify the allelic architecture to such a degree to make it tractable for genetic dissection (Hovatta et al., 1998, 1999). However, it is clear that these populations should be better suited for gene mapping than mixed populations with many different ethnicities and cultural practices (as are common in many U.S. cities). In addition, some population isolates have very large families, unusual family structures (such as consanguineous parental relationships leading to inbreeding), low rates of nonpaternity, excellent genealogical records, good health care infrastructure, willing study participation, and other features that can be a boon to genetic studies.

Extended Pedigrees

Another strategy is to collect extended pedigrees. This should reduce the genetic complexity because the number of independent founder chromosomes among related individuals is much smaller than the number of independent chromosomes in a set of randomly picked, unrelated individuals. Each founder individual contributes up to two different alleles at a given polymorphism, but, in the absence of mutation, descendants do not further increase the number of different alleles present in a sample, as they merely inherit copies of the founder alleles. Hence, in a case-control study of unrelated individuals, there would be up to twice as many independent allelic forms of each chromosome as there are study participants. This ratio is reduced to 4:3 for parent-offspring trios and 4:(2 + n) for nuclear families consisting of two parents and n siblings. In multigenerational families, the ratio depends on the width of the pedigree (i.e., the sibship sizes) and the depth of the pedigree (i.e., the number of generations). In addition, family members tend to be exposed to more similar environments than unrelated individuals living separately from one another, limiting the environmental complexity.

The reasoning is in direct analogy to the situation of isolated populations whose members tend to be more similar to each other genetically and in environmental exposure than the members of other societies. Furthermore, if genotype data are available across multiple generations, then the segregation of chromosomal segments can be inferred more accurately. For these reasons, extended families tend to have more power to detect linkage per individual than smaller sampling units (Williams and Blangero, 1999; Blangero et al., 2003). And, if families are collected because they segregate some specific trait of interest, then within a family all members showing the trait may have it for the same genetic reason, as described above for RP.

Ascertainment on Phenotype

Especially in the analysis of a rare condition, it is necessary to enrich the sample for the presence of the condition of interest. A "randomly ascertained" sample (i.e., a sample that is collected independently of the trait of interest) would contain few, if any, individuals with the condition. If there is no phenotypic variation in a sample, then genetic factors that may cause variation certainly cannot be identified in that sample. It may also be advantageous to collect families that are densely loaded with affected individuals. The presence of multiple cases in a family generally increases the probability that genetic factors are a major determinant when compared with singleton individuals with the condition

(Terwilliger et al., 2002). Furthermore, the affected individual within a family may well share the same genetic risk factors, as was argued above in the example of RP.

Phenotypic Subtypes

Another consideration may be to focus on subtypes of the trait of interest. For example, there may be considerable variation in the symptoms of individuals afflicted with a particular disorder, including age of onset, severity, or combination of symptoms. Rather than lumping all cases together, it may be advantageous to focus on a particular subclass, even if this reduces the available sample size. For example, in the analysis of Alzheimer disease, studies focusing on early-onset forms of the disease have been much more successful in localization of susceptibility genes (Tilley et al., 1998). In the case of RP, grouping families by detailed symptoms and also the apparent mode of inheritance certainly makes a lot of sense.

Eliminating Effects of Known Risk Factors

Another consideration is to reduce the importance of known risk factors, which represent noise when attempting to localize new genes. In some cases, this can be done as part of the ascertainment scheme. For example, when attempting to identify genetic factors for lung cancer, one may focus on individuals or families with lung cancer despite the absence of smoking. Such individuals may be contrasted with lung cancer–free individuals despite heavy smoking. Alternatively, one may obtain information on exposure to known risk factors and account for them by statistical means, such as by adjusting a quantitative trait for the effects of, say, sex, age, and other variables of known importance. This approach can also be used to account for previously identified genetic risk factors (Blangero et al., 1999; Martin et al., 2001).

Biomarkers

One may focus gene discovery efforts on intermediate phenotypes rather than the trait of ultimate interest (such as clinical disease end points). These intermediate phenotypes are often referred to as biomarkers or endophenotypes. As shown in Figure 11-2, the conceptual basis of this approach is the concern that the trait of interest may be too far removed from the action of individual genes to make it a useful phenotype for gene mapping. However, the trait may be influenced by intermediate phenotypes that have a tractable genetic basis (Blangero et al., 2003).

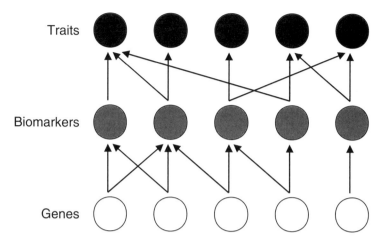

FIGURE 11-2 Utility of biomarkers for gene mapping.
NOTE: The trait of interest may be too far removed from the action of individual genes to allow gene localization and identification. However, the influence of genes on the trait may be mediated through measurable biomarkers (also called endophenotypes) that are much simpler in genetic etiology and thus permit gene discovery. Note that one gene may influence different traits and that different genes may influence the same trait. In this figure, the average biomarker is influenced by 1.8 genes (with a range of 1 to 3), while the average trait is influenced by 3.2 genes (with a range of 2 to 6).

For such intermediate phenotypes to be useful biomarkers, they must be (genetically) correlated with the trait of interest. In addition, they must be "upstream" of the trait of interest rather than "downstream", that is, the endophenotype must influence the trait rather than the other way around.

A problem for genetic research of behaviors, emotional and psychological phenotypes, and psychiatric diseases in particular is that current understanding of the brain is so rudimentary that there are few promising endophenotypes, although this situation is now beginning to change as a result of new techniques, such as improved imaging. Knowledge of other dimensions of human physiology is much more advanced, and many more biomarkers have been discovered and validated. For example, while cardiovascular disease is highly complex, such that the existence of a "stroke gene" or a "heart attack gene" that explains most of these events in the general population is unlikely, many risk factors for cardiovascular disease are known—such as blood pressure, various cholesterol subfractions, inflammatory markers, oxidative stress markers, blood clotting factors, etc.—and most genetic investigations that succeeded in localizing genes have focused on these biomarkers (Comuzzie et al., 1997) rather than the clinical disease end points.

Quantitative Traits

Many endophenotypes are quantitative in nature. Continuous traits are attractive targets for genetic investigations for several reasons (Blangero, 2004). They often provide greater power for gene localization (Blangero et al., 2000). The dichotomization of a naturally continuous phenotype—such as categorizing individuals as obese or nonobese based on body mass index or grouping individuals into hypertensive and non-hypertensive groups based on blood pressure—is generally a poor idea, as dichotomization discards information on the underlying trait locus genotypes (Blangero et al., 2001). For example, a severely hypertensive person with a systolic/diastolic blood pressure of, say, 200/120 mm Hg is more likely to harbor genetic variants of substantial elevating effect on blood pressure than a marginally hypertensive individual with a blood pressure of, say, 140/90 mm Hg. "Random ascertainment," that is, selection of study participants independently of the trait of interest, is a common design of genetic studies of quantitative traits.

The reason this is a suitable strategy is that there is ample variation in the quantitative phenotype in such a sample. This stands in contrast to the investigation of rare diseases in which random samples would contain exclusively or mostly healthy individuals. While nonrandom ascertainment may clearly be used for continuous phenotypes, such as by enriching the sample for individuals with extreme phenotypes from one or both tails of the quantitative distribution, the advantage of such a protocol over random ascertainment is less clear, and attempting to collect larger families may yield greater benefit.

Besides the concepts discussed above, many additional strategies may be pursued in collecting samples for complex trait gene mapping. The common logic behind these approaches is to make the genetic etiology tractable. The goal is to convert, through smart study design, a complicated trait with many etiological factors into a simple trait controlled by a few factors that have substantial importance and are individually identifiable. In a way, in studies of a nonexperimental organism such as humans, we attempt to take advantage of naturally occurring events (such as unusual phenotypes, families, or populations) to get as close as possible to the simplicity of experimental animal models, for which the experimenter has control over both genetic and environmental factors.

Many complex trait gene-mapping studies employ nonrandom ascertainment protocols, such as the preferential collection of families with multiple affected individuals, as described above. As a result, such samples are not reflective of the population as a whole, and the effect size of any locus, gene, and polymorphism may differ between such nonrandom samples and the general population. While this would be a serious drawback for many types of investigations, researchers attempting to identify genetic factors tend to accept this as a condition of having any power for gene localization and identification. In addition, the fact that a sample ascertained on a specific phenotype is different from the wider population is not crucial, because, even in randomly ascertained samples, it is nearly impossible to simultaneously identify genetic factors and estimate their effect size, because of what is sometimes referred to as the winner's curse. Complex trait gene-mapping studies tend to have fairly low power. Many different pointwise tests are conducted when attempting to localize genetic factors in our genome. As a result, considerable luck is required for successful gene localization, and the effect sizes at peaks of the mapping statistic tend to be greatly inflated (Utz et al., 2000; Goring et al., 2001). While various investigators have attempted to correct these biases (Allison et al., 2002; Siegmund, 2002; Sun and Bull, 2005), additional samples will often be required anyway to obtain more accurate estimates of the frequency and impact of genetic variants. Furthermore, given the genetic differences between different populations, multiple samples must be analyzed to establish the consistency of findings across populations and ethnicities.

GENE EXPRESSION LEVELS AS AN EXAMPLE OF EXTREME ENDOPHENOTYPES

Due to recent technological advances, the concentrations of proteins, RNA molecules, and metabolites are increasingly being used as targets in gene-mapping experiments. Transcript abundance and protein levels may be viewed as extreme endophenotypes that are presumably much closer to the action of individual genes than complex human characteristics. The analysis of gene expression levels is now quickly becoming routine, because of the recent commercial availability of microarrays containing probes for vast numbers of transcripts, making it possible to investigate nearly the entire known transcriptome in a single experiment. For recent reviews of genetic analysis of expression profiling, see de Koning and Haley (2005); Gibson and Weir (2005); Pastinen et al. (2006). Expression profiles have recently been generated for lymphocyte samples from 1,240 Mexican American participants in the San Antonio Family Heart Study (Mitchell et al., 1996), the goal of which is to analyze the genetic underpinnings of atherosclerosis in Mexican Americans. We have recently published an initial paper on the genetic regulation of gene expression

(Göring et al., 2007). This example is mentioned here to caution against overly optimistic views on the prospects of complex trait gene-mapping experiments.

A total of 19,648 unique probes detected substantial abundance of the autosomal transcript being targeted and were subjected to statistical genetic analyses. Using a variance components-based approach and the software package SOLAR (Almasy and Blangero, 1998), we observed that, at a 5 percent false discovery rate (Benjamini and Hochberg, 1995), 85 percent of the expression phenotypes are significantly (additively) heritable (i.e., genetic factors in aggregate account for some proportion of the phenotypic variance). This is perhaps not unexpected, given that gene expression phenotypes are about as close to gene action as possible. However, the heritability estimates of many transcripts were modest (46, 68, and 87 percent of transcripts have heritability estimates less than 20, 30, and 40 percent, respectively), hinting at a substantial influence of the environment or physiological state of the individual at the time of blood draw. In an effort to localize major genetic factors influencing the expression levels of individual transcripts, we performed genome-wide variance components-based linkage analysis (Amos, 1994; Almasy and Blangero, 1998).

Figure 11-3 contains a scatterplot of the transcript-specific maximum linkage statistic by the estimated heritability. The figure is included here to make two points: First, no genes were reliably localized in linkage analysis of many transcripts. In fact, using the customary threshold of a lod score of 3 as the criterion (the lod score, short for logarithm of odds score, is the logarithm to base 10 of the likelihood ratio statistic that compares the likelihood of the alternative hypothesis of linkage and the null hypothesis of no linkage; a lod score of 3 thus corresponds to a likelihood ratio of 1,000:1; asymptotically, the genome-wide significance corresponding to this criterion is ~0.05 in humans), only ~10 percent of transcripts had significant linkages. This points out that the etiology of even such seemingly simple traits can be quite complicated, highlighting the challenges faced by investigators attempting to identify the genes underlying complex human conditions and behaviors. Although measurement error in transcript abundance undoubtedly reduces power, the vast majority of expression phenotypes are nonetheless substantially heritable, suggesting that measurement error alone is not the main reason for why no locus was identified for many transcripts.

Second, the heritability estimate is a poor predictor of "mappability." This is illustrated in Figure 11-4. Heritability assesses the influence of genetic variation in the aggregate on phenotypic variation. However, a substantial heritability does not indicate the existence of major genes. While there is a positive correlation between the heritability estimate

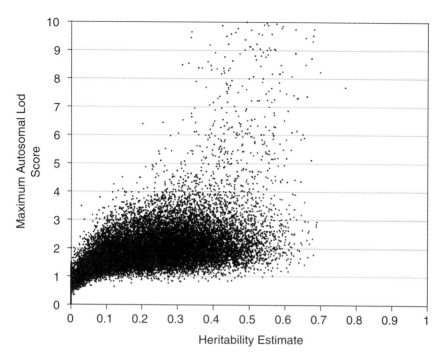

FIGURE 11-3 Relationship of heritability estimate and maximum lod score. NOTE: The figure shows the maximum lod score (obtained in variance components-based linkage analysis across the autosomal chromosomes) as a function of the estimated heritability for the expression levels of 19,648 autosomal transcripts. The expression phenotypes were generated on lymphocyte samples from 1,240 Mexican American family members, participants in the San Antonio Family Heart Study (Mitchell et al., 1996). The expression phenotypes were normalized by an inverse Gaussian transformation and were adjusted for the overall RNA levels in a sample and for the effects of sex and age. The vertical axis is truncated at 10 (the maximum obtained lod score was > 50).

and the probability that a major locus exists, this relationship is very imprecise.

Another example that dramatically illustrates this point is human height. We have investigated normal variation in adult stature in nine extended pedigree samples from several different ethnic groups and countries (Göring et al., 2004a, 2004b). The samples comprised nearly 7,500 phenotyped and genotyped individuals in total. After accounting for the sexual dimorphism in height and for the effects of age, height was found to be highly heritable (with heritability estimates ranging from 63

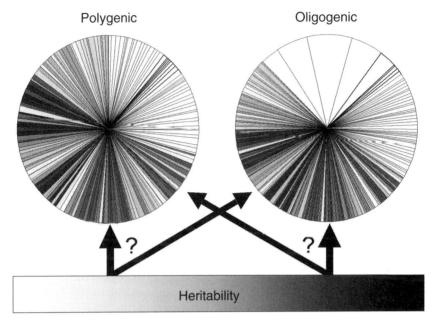

FIGURE 11-4 Heritability and genetic architecture.
NOTE: The heritability estimate of a trait is a poor predictor of whether the trait has a "polygenic" or "oligogenic" genetic architecture. In polygenic inheritance, the genetic factors individually have a small influence on a trait of interest, making their identification impossible by statistical genetic approaches. In oligogenic inheritance, one or several genes have substantial phenotypic effect, potentially allowing them to be localized and identified.

to 92 percent in these samples). Nonetheless, most samples failed to yield a significant lod score in genome-wide linkage analysis, and the few lod score peaks that exceeded the threshold of 3 were inconsistent between the different samples. Other groups have conducted similar studies, but on a smaller scale (Hirschhorn et al., 2001; Perola et al., 2001; Sammalisto et al., 2005). While some interesting linkage results were obtained, in my estimation the findings are consistent with the view that height in the general population, while highly heritable, is a good example of a largely polygenic trait that is quite intractable using standard gene-mapping approaches.

SOCIAL SCIENCE SURVEYS:
A TOOL FOR GENE DISCOVERY OR VALIDATION?

The etiological complexity of multifactorial traits poses an enormous challenge for efforts to localize and identify the underlying genes. The chance of success will be greatly improved if studies focus on biological characteristics that are amenable to genetic investigations and if studies are designed in order to reduce the complexity as much as possible. In the preceding sections, some strategies for simplifying the etiological architecture have been presented.

Social science surveys come in many shapes and sizes, and it is thus difficult to provide general comments on their utility for gene localization, identification, and validation. Whether or not a particular cohort may be useful, these purposes depend on the many specific features of a particular survey. The remarks here are very general. They do not do justice to the uniqueness of individual surveys and do not highlight the many exceptions and caveats. For a more nuanced perspective, the reader is referred to the detailed descriptions of specific surveys that are provided in various chapters of this book.

Social, political, economic, and demographic surveys have very different goals from gene-mapping studies, and thus they are typically designed in a very different and even diametrically opposed manner. Often surveys are conducted in a way that ensures that they are as representative of the surveyed population as possible. This design goal tends to enrich the diversity of environmental factors as well as genetic factors that influence a characteristic of interest. In many cases, surveys conducted in the United States include individuals from many different ethnic groups, due to the ethnic diversity of this country and the desire to understand the importance of racial identity on social conditions. This additional phenotypic variance due to both genetic and environmental factors is generally best avoided in gene-mapping studies, although "admixture mapping" is a gene-mapping approach that seeks to take advantage of any existing allele frequency differences between the founder populations comprising an admixed society (Smith and O'Brien, 2005).

The large size of most surveys necessitates the use of questionnaires or measurements that can be done quickly and cheaply. There often is no time or money for in-depth assessment of characteristics. Ethical concerns may also limit what types of measurements can be conducted on a population-wide scale. Surveys in the social sciences are often primarily interested in behaviors, characteristics, and phenomena that are likely far removed from the influence of individual genes. There is no reason to believe, based on evolutionary considerations, practical gene-mapping experience, and common sense, that a characteristic such as, say, socioeconomic status is amenable to dissection by statistical genetic approaches.

The multitude of factors—such as intelligence, talent, likes, and ambition, as well as family environment, parental socioeconomic status, quality of schooling, type of local economy, availability of jobs, and many others—appears too large to make gene-mapping strategies appear likely to succeed. The brain is so complex that it is extremely difficult to identify the genetic factors underlying even gross disturbances, such as in severe psychiatric disease. The large size of many surveys is wonderful attribute. However, it is unlikely that an increase in sample size alone can readily overcome the challenges of truly complex phenotypes and the drawbacks of study designs that are suboptimal for gene mapping. It appears likely that genotyping technologies will continue to improve so quickly that it will soon be feasible to obtain the entire DNA sequence of all study participants. However, even if complete sequence data were available for all 300 million U.S. residents or even all 6 billion humans on earth, the identification of causal genetic factors influencing human behaviors and similarly complex traits would still be very challenging. In all likelihood, subsetting strategies would be employed that mimic what is now being done through the selection of suitable populations and specific ascertainment criteria. For these reasons, social science surveys are probably a poor tool for gene localization and identification in general.

However, this conclusion does not address the question whether or not such surveys are suitable for "replication" of candidate genes previously identified elsewhere. One reason for the lack of power in gene-mapping experiments is the extremely large hypothesis space. The human genome spans about 3 billion base pairs of DNA and harbors about 30 thousand different genes (although this latter number is quite uncertain and depends on the definition of a gene). The rules of genetic inheritance impose a high degree of correlation among these many factors, making it possible to exhaustively evaluate all hypotheses. No specific prior hypotheses are thus required, and gene-mapping experiments may be better viewed in the framework of estimation (i.e., in which the genes influencing a trait are located) than in the framework of testing (i.e., if a gene influencing the trait of interest is located at a specific position). However, the number of independent tests used to span the genome is still substantial: approximately 500 independent tests are required to span the genome with linkage analysis, and many more are required for genome-wide association analysis. The large number of "tests" substantially reduces the power of genome-wide mapping studies. In contrast, once a specific factor has been identified as being a likely contributor to some trait of interest, a specific hypothesis is at hand that can be tested.

No multiple testing applies in this situation, greatly improving power. Such validation is important for several reasons:

- Given the lack of power of many complex trait gene-mapping studies, many of the published results may represent false positive findings. Replication by others is thus necessary in order for such findings to be accepted.
- As mentioned above, it is next to impossible to map genes and at the same time estimate their effect sizes reliably. Naïve estimators tend to be greatly inflated, for reasons of selection bias (Göring et al., 2001). While procedures for bias reduction are actively being developed, the estimation of effect size in independent samples remains crucial.
- Samples for gene-mapping studies are often not representative of the population as a whole. They are generally restricted to one ethnic group or are greatly enriched for the biological condition of interest. Thus, it is of great interest to assess the generality of findings in the wider population and in different population strata and ethnic groups.

Some characteristics that are typical of many social science surveys suggest that they may be highly suitable for examining individual genetic variants with substantial prior support. The large size of many such surveys should in principle allow for very precise estimation of allelic effect size. And the fact that the sample is typically collected to be representative of the wider population should in principle improve effect size estimation when compared with deriving such estimates from samples that are ascertained based on the trait of interest, as no correction for ascertainment is necessary (which is notoriously difficult). If surveys cover individuals from many different ethnic groups, then the replication of previously identified genetic variants can establish the generality of the findings in many different ethnicities within one survey cohort, assuming that reliable information on ethnic origin is available.

While social science surveys may in principle be useful for validating previously identified genetic variants, survey cohorts are by no means universally useful for that purpose, nor are they the only or necessarily the best type of sample available for replication. In many instances, investigations of the genetic etiology of human traits are conducted by many research groups simultaneously, either in collaboration or in competition with one another. Hence, once one of these groups identifies a particular gene and/or variant and announces the results, replication studies are often done very rapidly, because many other cohorts in which the trait of interest has been measured are at hand. Given the international nature of human gene identification efforts, these cohorts are likely to come from many countries and ethnic groups, such that the importance of the identified variants can be established for many human populations. In this situation, the need for and the utility of social science survey cohorts as further replication samples may then well be limited or nonexistent. And while large surveys can in principle be used for validation of specific genetic variants by association analysis, these cohorts will often not be useful for replicating findings of heritability or linkage, which require information on familial relationships that are missing from many social science surveys.

If social science surveys are to be used for gene validation, two conditions must be met: The phenotype of interest must have been assessed in the survey, and DNA samples must have been collected. To increase the range of phenotypes that can be used for gene validation, it may potentially be useful to collect, as part of such surveys, biological specimens (such as saliva, blood, and urine) that can be obtained easily, cheaply, and at minimal risk to survey participants. A vast number of biomarkers underlying complex diseases have been detected and characterized in these readily available tissues (for details, see the chapters devoted to this topic). If DNA and/or biological specimens are to be collected in social science surveys, they should be designed for this purpose from the outset rather than retrofitted later. Thus, the pros and cons should be carefully weighed beforehand. These considerations should take into account both the many practical issues (such as cost, speed, effect on study participation, transport, storage, etc.) and also the expected value of a given survey for later genetic investigations. For example, if a survey is conducted in a population that is rarely investigated, then the collection of DNA may be more useful than in a well-studied population. And the more expensive or difficult a trait or its associated environmental risk factors are to measure, the greater the potential value of DNA samples collected concurrently. Finally, it is paramount that ethical concerns are given careful consideration.

CONCLUSION

The etiological architecture of complex traits presents a formidable challenge to attempts to identify the underlying genetic components. To overcome this difficulty, gene-mapping studies should be designed in a manner that reduces the etiological complexity as much as possible. Enormous sample sizes, large-scale genotyping and sequencing efforts, sophisticated statistical approaches, and fast computers cannot compensate for a trait that is poorly chosen or for a study that is poorly designed. In general, social science surveys tend to focus on characteristics that are likely to be far removed from the action of individual genes. In addition, these surveys are designed with different goals in mind. For these reasons, it appears that social science surveys will often be poorly suited for complex trait gene discovery.

At the same time, the large size of many social science surveys, and the fact that by design the survey sample is often representative of the target population, make such surveys potentially useful for gene validation. This usage would permit "replication," that is, confirming the role of specific genetic variants in specific traits, allow for a more accurate estimation of the effect size of specific genes and functional variants therein, and allow gene effect size comparisons between different populations and ethnicities. In order to be used in this manner, surveys would need to collect tissue samples that allow DNA extraction for genotyping and potentially also biological specimens permitting phenotyping. Careful considerations must be given to ethical concerns that may arise if genetic data or other information are generated or analyzed by individuals who are not involved in the survey. These issues should be addressed before a survey is conducted, and appropriate informed consent must be obtained.

This chapter has focused mostly on the potential utility of social science surveys for complex trait gene discovery and validation. However, social science and biological or genetic research obviously influence and aid each other in a myriad of ways. For example, if a genetic factor is known to influence some human condition that is of interest in a particular social science survey, then genotyping this genetic factor may be of great help when attempting to identify other explanatory variables underlying the condition. Conversely, known environmental risk factors, including those detected and characterized in social science studies, should certainly be measured and accounted for in genetic studies attempting to identify novel genetic risk factors. After all, the vast majority of human traits are influenced by both genetic and environmental factors, and we should not let ignorance or dogma (including whether a study should still be called a social science survey if it also collects biological specimens) get in the way of using any relevant information to advance the research.

From a more distant perspective, the overarching question regarding the use of genetic indicators in social science surveys is whether it is more useful to have a larger number of smaller and more specialized studies or fewer studies that are larger and more general. As is so often the case, the correct answer probably is "it depends."

REFERENCES

Allison, A.C. (1954). Protection afforded by sickle cell trait against subtertian malarian infection. *British Medical Journal, 1,* 290-294.

Allison, A.C. (1961). Genetic factors in resistance to malaria. *Annals of the New York Academy of Sciences, 91,* 710-729.

Allison, D.B., Fernandez, J.R., et al. (2002). Bias in estimates of quantitative-trait-locus effect in genome scans: Demonstration of the phenomenon and a method-of-moments procedure for reducing bias. *American Journal of Human Genetics, 70*(3), 575-585.

Almasy, L., and Blangero, J. (1998). Multipoint quantitative-trait linkage analysis in general pedigrees. *American Journal of Human Genetics, 62*(5), 1198-1211.

Amos, C.I. (1994). Robust variance-components approach for assessing genetic linkage in pedigrees. *American Journal of Human Genetics, 54*(3), 535-543.

Benjamini, Y., and Hochberg, Y. (1995). Controlling the false discovery rate: A practical and powerful approach to multiple testing. *Journal of the Royal Statistical Society, 57*(1), 289-300.

Bhatti, M.T. (2006). Retinitis pigmentosa, pigmentary retinopathies, and neurologic diseases. *Current Neurology and Neuroscience Reports, 6*(5), 403-413.

Blangero, J. (2004). Localization and identification of human quantitative trait loci: King harvest has surely come. *Current Opinion in Genetics & Development, 14*(3), 233-240.

Blangero, J., Williams, J.T., et al. (1999). Oligogenic model selection using the Bayesian Information Criterion: Linkage analysis of the P300 Cz event-related brain potential. *Genetic Epidemiology, 17*(Suppl 1), S67-S72.

Blangero, J., Williams, J.T., et al. (2000). Quantitative trait locus mapping using human pedigrees. *Human Biology, 72*(1), 35-62.

Blangero, J., Williams, J.T., et al. (2001). Variance component methods for detecting complex trait loci. *Advances in Genetics, 42,* 151-181.

Blangero, J., Williams, J.T., et al. (2003). Novel family-based approaches to genetic risk in thrombosis. *Journal of Thrombosis and Haemostasis, 1*(7), 1391-1397.

Caspi, A., Sugden, K., et al. (2003). Influence of life stress on depression: Moderation by a polymorphism in the 5-HTT gene. *Science, 301*(5631), 386-389.

Comuzzie, A.G., Hixson, J.E., et al. (1997). A major quantitative trait locus determining serum leptin levels and fat mass is located on human chromosome 2. *Nature Genetics, 15*(3), 273-276.

Cox, D.R. (1984). Interaction. *International Statistical Review, 51*(1), 1-31.

de Koning, D.J., and Haley, C.S. (2005). Genetical genomics in humans and model organisms. *Trends in Genetics, 21*(7), 377-381.

Gibson, G., and Weir, B. (2005). The quantitative genetics of transcription. *Trends in Genetics, 21*(11), 616-623.

Goldfarb, A., and Avraham, K.B. (2002). Genetics of deafness: Recent advances and clinical implications. *Journal of Basic and Clinical Physiology and Pharmacology, 13*(2), 75-88.

Göring, H.H.H., Curran, J.E., Johnson, M.P., Dyer, T.D., Charlesworth, J., Cole, S.A., Jowett, J.B.M., Abraham, L.J., Rainwater, D.L., Comuzzie, A.G., Mahaney, M.C., Almasy, K.L., MacCluer, J.W., Kissebah, A.H., Collier, G.R., Moses, E.K., and Blangero, J. (2007). Discovery of expression QTLs using large-scale transcriptional profiling in human lynphocytes. *Nature Genetics, 39*(10), 1208-1216.

Göring, H.H.H., Duggirala, R., et al. (2004a). *Genome-wide linkage analyses of human stature in pedigree samples from different ethnicities.* Paper presented at the 73rd Annual Meeting of the American Association of Physical Anthropologists.

Göring, H.H.H., Duggirala, R., et al. (2004b). *Localization of genetic factors influencing height by genome-wide linkage analysis in large pedigree samples.* Paper presented at the Xth International Congress of Auxology, Firenze, Italy.

Göring, H.H.H., Terwilliger, J.D., et al. (2001). Large upward bias in estimation of locus-specific effects from genomewide scans. *American Journal of Human Genetics, 69*(6), 1357-1369.

Heckenlively, J.R., and Daiger, S.P. (1996). Hereditary retinal and choroidal degenerations. In D.L. Rimoin, J.M. Connor, and R.E. Pyeritz (Eds.), *Emory and Rimoin's principles and practice of medical genetics* (pp. 2555-2576). Edinburgh, Scotland: Churchill-Livingstone.

Hirschhorn, J.N., Lindgren, C.M., et al. (2001). Genomewide linkage analysis of stature in multiple populations reveals several regions with evidence of linkage to adult height. *American Journal of Human Genetics, 69*(1), 106-116.

Hovatta, I., Lichtermann, D., et al. (1998). Linkage analysis of putative schizophrenia gene candidate regions on chromosomes 3p, 5q, 6p, 8p, 20p and 22q in a population-based sampled Finnish family set. *Molecular Psychiatry, 3*(5), 452-457.

Hovatta, I., Varilo, T., et al. (1999). A genomewide screen for schizophrenia genes in an isolated Finnish subpopulation, suggesting multiple susceptibility loci. *American Journal of Human Genetics, 65*(4), 1114-1124.

Kulkarni, P.S., Butera, S.T., et al. (2003). Resistance to HIV-1 infection: Lessons learned from studies of highly exposed persistently seronegative (HEPS) individuals. *AIDS Revealed, 5*(2), 87-103.

Martin, L.J., Comuzzie, A.G., et al. (2001). The utility of Bayesian model averaging for detecting known oligogenic effects. *Genetic Epidemiology, 21*(Suppl 1), S789-S793.

Mitchell, B.D., Kammerer, C.M., et al. (1996). Genetic and environmental contributions to cardiovascular risk factors in Mexican Americans. The San Antonio Family Heart Study. *Circulation, 94*(9), 2159-2170.

Pastinen, T., Ge, B., et al. (2006). Influence of human genome polymorphism on gene expression. *Human Molecular Genetics, 15*(Spec. No. 1), R9-R16.

Peltonen, L., Jalanko, A., et al. (1999). Molecular genetics of the Finnish disease heritage. *Human Molecular Genetics, 8*(10), 1913-1923.

Penrose, L.S. (1935). The detection of autosomal linkage in data which consist of pairs of brothers and sisters of unspecified parentage. *Annals of Eugenics 6*, 133-138.

Perola, M., Ohman, M., et al. (2001). Quantitative-trait-locus analysis of body-mass index and of stature, by combined analysis of genome scans of five Finnish study groups. *American Journal of Human Genetics, 69*(1), 117-123.

Rivolta, C., Sharon, D., et al. (2002). Retinitis pigmentosa and allied diseases: Numerous diseases, genes, and inheritance patterns. *Human Molecular Genetics, 11*(10), 1219-1227.

Sammalisto, S., Hiekkalinna, T., et al. (2005). A male-specific quantitative trait locus on 1p21 controlling human stature. *Journal of Medical Genetics, 42*(12), 932-939.

Siegmund, D. (2002). Upward bias in estimation of genetic effects. *American Journal of Human Genetics, 71*(5), 1183-1188.

Smith, M.W., and O'Brien, S.J. (2005). Mapping by admixture linkage disequilibrium: Advances, limitations, and guidelines. *Nature Reviews Genetics, 6*(8), 623-632.

Spielman, R.S., McGinnis, R.E., et al. (1993). Transmission test for linkage disequilibrium: The insulin gene region and insulin-dependent diabetes mellitus (IDDM). *American Journal of Human Genetics, 52*(3), 506-516.

Sun, L., and Bull, S.B. (2005). Reduction of selection bias in genomewide studies by resampling. *Genetic Epidemiology, 28*(4), 352-367.

Terwilliger, J.D., and Göring, H.H. (2000). Gene mapping in the 20th and 21st centuries: Statistical methods, data analysis, and experimental design. *Human Biology, 72*(1), 63-132.

Terwilliger, J.D., Göring, H.H.H., et al. (2002). Study design for genetic epidemiology and gene mapping: The Korean Diaspora Project. *Shengming Kexue Yanjiu (Life Science Research), 6*, 95-115.

Tilley, L., Morgan, K., et al. (1998). Genetic risk factors in Alzheimer's disease. *Molecular Pathology, 51*(6), 293-304.

Utz, H.F., Melchinger, A.E., et al. (2000). Bias and sampling error of the estimated proportion of genotypic variance explained by quantitative trait loci determined from experimental data in maize using cross validation and validation with independent samples. *Genetics, 154*(3), 1839-1849.

Weiss, K.M., and Terwilliger, J.D. (2000). How many diseases does it take to map a gene with SNPs? *Nature Genetics, 26*(2), 151-157.

Wexler, N.S., Lorimer, J., et al. (2004). Venezuelan kindreds reveal that genetic and environmental factors modulate Huntington's disease age of onset. *Proceedings of the National Academy of Sciences, USA, 101*(10), 3498-3503.

Williams, J.T., and Blangero, J. (1999). Power of variance component linkage analysis to detect quantitative trait loci. *Annals of Human Genetics, 63*(Pt 6), 545-563.

Wright, A.F., Carothers, A.D., et al. (1999). Population choice in mapping genes for complex diseases. *Nature Genetics, 23*(4), 397-404.

12

Overview Thoughts on Genetics: Walking the Line Between Denial and Dreamland, or Genes Are Involved in Everything, But Not Everything Is "Genetic"

Kenneth M. Weiss

". . . You know what M. de Talleyrand said, 'If you don't know how to play whist you are laying up for yourself a desolate old age.'"
Honoré de Balzac, *Lost Illusions*

The word "genetic" is used in different ways by different people or by the same people at different times. There are three basic meanings that should be kept clear. The application of "genetics" to social survey problems depends on which meanings are being used. Indeed, the concept of a gene itself is much more complex than protein-coding, a fact that adds underappreciated nuance and complications to proper genetic studies.

"Genetic" can refer to *mechanism*, that is, what genes do and how they work, a kind of stereotypic biology. When we refer to the fact that p53 is a gene involved in cell cycle regulation in the development of normal tissue, we assume some generality in "normal" people or a standard assay system. "Genetic" is also a population concept that refers to the correlation between phenotype variation and inherited variation in populations. Inherited variation in the p53 gene is associated with abnormal tissue architecture and cancer. Finally, "genetic" refers to *somatic* change that occurs postinheritance—due to mutation in individual body cells—when the changes are inherited by their descendant cells during development and later life (including aging).

But somatic changes that don't occur in the germ line are not inherited from parent to offspring. Genetics in the form of somatic mutation generates variation in mechanism within individuals, and it is fundamental in the etiology of cancer and a few other diseases, probably including cognitive aging, and may have much greater importance than has yet been documented (Weiss, 2005a, 2005b). Somatic mutation in p53 affects individual cells that can transform to found a clone of cells that constitute a life-threatening cancer. Some somatic changes can in various ways affect gene expression rather than gene sequence itself, and these are known as "epigenetic" effects. A number of epigenetic mechanisms are known and new ones are rapidly being discovered.

The "genetic architecture" of a biological trait refers to the number of genes that contribute to it, the way those genes interact with each other and with the environment, their relative contribution to the trait, and the role of their variation on the trait. It involves all three aspects of genetics. The genetic architecture of traits in multicellular organisms, including human traits of any complexity, is far from completely known. Indeed, "complex" is in the eye of the beholder, and even "simple" traits turn out not to be so simple on close inspection (Scriver and Waters, 1999). Most human genetic epidemiology is largely black-box genetics that searches for statistical correlation between inheritance and trait, initially without knowing (or at the discovery stage even caring) about its mechanism. The situation is made more complex by the expanding definition of "gene" to recognize many functions beyond protein-coding, and these are still being discovered.

The genetic architecture of any trait is the product of its evolution. Evolution is the population-historic process that generates the genomes that construct or affect phenotypes. Evolution involves population size, mating patterns, chance, migration, geographic and social distribution, differential reproduction, and the like. The variation that results depends on mutation rates, the size of the mutation target (number and length of relevant genetic units in the genome), and the effects of reproductive fitness (natural selection). These factors are mediated through demographic history. Evolution helps account for and predict the general trait patterns that we see today, and because humans are a globally distributed species, it is especially useful in explaining the amount and geographic distribution of genetic variation affecting human traits.

The 20th century began with the recognition of Mendel's work on the particulate (discrete-factor) nature of inheritance. Discretely varying (e.g., presence/absence) traits initially seemed inconsistent with the more prevalent quantitative variation (e.g., height, weight) seen in nature and that seemed to be the working material for evolution. But early in the 20th century it was realized that the combined action of many independently segregating particulate genes could, through combinatorial contributions, lead to a quantitative phenotype (to which environmental variation contributes as well). The key fact of that realization (as also noted elsewhere in this volume) is *many to many* causation: many different genotypes generate essentially the same phenotype. The genes, known generally as *polygenes*, have been modeled as a homogeneous mix of infinitely many genes, each with two alleles (variant sequence states) of equal frequency, that additively contribute dose-like effects to the trait. The genetic variation in such a model easily generates the typical Gaussian-like trait distributions in a population and the trait-value correlations among relatives, leading to a consistent, complete theory of both variation and inheritance.

This was a fundamental finding, and it can be a lesson difficult to digest for anyone hungering for simple genetic solutions to their favorite traits. It implied that the genetic contributions could not be individually identified. Nonetheless, the basic ideas have been systematically confirmed with ever-increasing depth of understanding during the 20th century. Gene mapping (observational and epidemiological studies to identify genes affecting a trait of interest) and experimental studies have documented these basic points. However, recent genetic and evolutionary research has added important and well-replicated characteristics to the general picture, which dehomogenize polygenic causation to some extent. A considerable amount of data and evolutionary theory show that the distributions of (1) allele frequency, (2) effects of individual alleles on traits they affect, and (3) the effects of alleles on evolutionary "fitness" (natural selection) are skewed. Most alleles in a given gene and population of inference have low frequency, small effect on traits, and small effect on fitness. Since the frequencies of the alleles at a locus must sum to 100 percent, there cannot be too many with high frequency. Most known high-frequency susceptibility alleles have small individual, probabilistic effects on risk and/or risks substantially mediated by the environment, broadly defined and usually only partly known. There is usually a substantial tail of generally rare, strong phenotypic, and/or major fitness effects. The clearest segment of this is the aggregate of lethal mutations, in which the effects are clear and essentially deterministic. The practical upshot is that many complex traits can be called "oligogenic" or "multilocus," in that embedded in a polygenic background are at least some specific loci with alleles frequent enough and/or effects strong enough that standard methods can detect them with adequate sample design—and new instances are continually being reported. But what fraction of genetic architecture comprises this low-hanging fruit in relation to common disease risk is a contentious question. The definition of "common" is a moving target in the eye of the beholder, and a potential source of bias in interpretation, reporting, or accepting or rejecting study results. Such tractable alleles attract scientific, commercial, and media attention, can satisfy vested interests, and can drive sampling and analytic strategies to find them.

There are, of course, clear examples where alleles with substantial effect on disease are at relatively high frequency.

Complicating matters is that genes typically harbor tens to hundreds or more alleles. Most variants are individually rare, so that a given copy of a gene may differ by only one or a few variants relative to the human genome reference sequence. Thus, polygenic contributions involve many genes as well as many alleles within each gene, each with its own population-specific frequency. And because we are diploid, the effects of this variation are genotype-dependent as well.

Because each allele arises by new mutation in some specific individual, most rare alleles are recent and geographically localized, and the more frequent the allele, the older and more geographically dispersed it tends to be. The frequency and/or presence of a given allele varies from place to place. The spatial gradient of frequency is affected by many stochastic demographic factors, but it generally forms a correlation between geographic distance and genetic difference, which applies to individuals and hence to populations. These differences are important for medicine because patients in different populations have different distributions of variation related to a given disease. But the pattern of difference or similarity is probabilistic and subtle, and distant individuals can be more closely related (more similar) at a given gene than individuals from the same population (Witherspoon et al., 2007).

The frequency of an allele and its effect on a given trait are commonly statistically confounded. This is because traits—like disease or behavior—are often defined in terms of deviations from the mean (from normal), and so they can be rare almost by definition: if the frequency of an allele contributing strongly to such a state were common, the mean would be near that allele's effect. An allele could be common in the population and still have a major effect if it results from diversifying selection, such as seen in the immune system, frequency-dependent selection that reduces fitness of any allele that becomes too common, or other demographic or selective situations of that type. Generally with such selection the trait distribution in the population is not unimodal. Sex and sex-related traits are examples.

However, we don't always measure a trait strictly from the mean but sometimes use some desirability standard, such as healthy versus diseased. Hypertension refers to blood pressure elevated enough to pose a presumed universal risk factor for cardiovascular disease per se, not simply relative to the average in the population. In other words, everyone in the population is presumed to be at higher risk with higher blood pressure. Such traits can be common, as for example obesity is in the United States, and certainly genetic risk for obesity could be common. For example, it might be that archetypal human metabolism deposits fat and calcium and so on, such that excessive consumption would elevate blood pressure in anybody regardless of genotype. Such a situation may involve genetic mechanism but not be mappable in observational research: if we all share the genetic mechanism, there may be insufficient or genetic variation contrasts that are too heterogeneous for statistical inference. This usually points to exogenous causal factors, like diet, and does not mean the trait is genetic in the sense that variation in genetic susceptibility is a major cause of variation in risk. Obesity and hypertension are extensively studied traits that are still not well accounted for genetically, but they fit the general many-to-many nature of the polygenic model.

Whether or not a variant affects fitness in the health sense, if it has had little effect on *selective* fitness (that is, on differential reproduction in the natural selection sense) its frequency in a population depends on the chance aspects of reproducing during its sojourn on the population, and hence it can have any frequency. It can be common and yet manifest major harmful effects if there has been a rapid change in environment, so that what was once benign or even helpful is deleterious in today's environments. This may include many chronic diseases of modernization, like heart disease and diabetes. The new harmful effects can, but need not, also confer a new selective disadvantage to the allele.

These facts are all relevant to the objectives of this volume, and other chapters in this book consider them. They are entirely consistent with what we know about evolution and the mechanistic as well as demographic aspects of genetics. They are consistent with the tens of thousands of papers that have been published on the genetics of human disease, both of single-gene and of complex—multigenic—traits. They are consistent with what we know about behavior as well. These facts are easily documented on the web (e.g., OMIM or PubMed at http://www.ncbi.nlm.nih.gov, and many other places).

LEARNING FROM THE FACTS

Strong, rare effects segregate in families according to the principles Mendel discovered in his pea plant experiments, because like Mendel's carefully selected traits, the effects are closely tied to the action of individual genes. They are replicable among relatives within and between families that carry and transmit them. They are the flags by which single-gene diseases were mapped (that is, the gene's chromosomal location identified so the gene itself could then be found in various ways). And they are the flags by which individual genes contributing substantially to quantitative traits (known as quantitative trait loci, or QTL) have been found. There are hundreds if not thousands of success stories, and mapping results based on a variety of study sizes and designs pour forth new results daily, documenting the elements of causal genetic architecture that I have described (e.g., Wellcome Trust Case Control Consortium, 2007). Usually, however, once the gene is found, many more mutations are discovered whose effects are too small to have flagged the gene in a mapping study, but the aggregate of these effects can be an important, subtle source of trait variation in the population (Weiss and Buchanan, 2003).

However, mapping studies typically account for only a fraction of the familial correlation or aggregation of risk. As one moves down the effect-size scale from strong toward weak, or up the frequency scale from rare toward common, the detectability of the genetic signal often reduces, due to a multiplicity of competing factors of similar strength or frequency, measurement error, and environmental factors. Replicability also declines for most of the lesser-effect signals. The farther along this spectrum one goes, the more tightly designed a study may have to be to characterize or even detect their individual effects, because they may be too small or too embedded in heterogeneous complicating factors like environments. A typical finding of the dehomogenization of polygenic risk concepts, referred to above, is that the more restricted or precisely focused the phenotype definition, the more "mappable" it is to identifiable loci or alleles, and vice versa. But then the results apply to only a small subset of the overall trait. This is the case for such behavioral traits as autism, schizophrenia, even Down syndrome and the like.

A potential complicating factor is a third aspect of genetic causation mentioned earlier, that of somatic mutation. We don't yet know how important this may be, but somatic mutation is probably responsible for cases of behavior disorders related to epilepsy and may be a sleeping giant in behavioral biology that will be silent until attacked explicitly, which is difficult to do with today's technologies (Weiss 2005a, 2005b). The scope of epigenetic effects, including those on behavioral traits, is probably much greater than has been thought (Wong, Gottesman, and Petronis, 2005; Petronis, 2006).

The effects of a given allele (which may be defined as a single nucleotide variant or a pattern of them on a single chromosome, known as a haplotype) usually have large variances. We know from many kinds of data, including from experimental animals, that this variance includes the effects of variation in the genomic background (the rest of the genome) of individuals with the same test variant, as well as the environmental context.

Most of our functional genome involves genes whose action is to regulate the expression of other genes. Consequently, much or most genetic variation involves context, timing, and level of expression. These are *quantitative* aspects of gene function. Perhaps especially for chronic disease or age-related traits, genotype-specific effects are manifest as dose-response effects on the age-specific hazard function, rather than proteins made dysfunctional by mutations that change their amino acid sequence.

Overall, to varying but not precisely known extents, about 30-60 percent of the variance in most complex traits, including chronic disease, generally called the heritability, is due to aggregate effects of genetic variation that can be estimated statistically by familial aggregation of risk, monozygotic versus dizygotic twin contrasts, and other common measures. Family-based estimates are probably somewhat inflated because of undetected confounding between environmental and genetic correlations (various study designs, such as adoptee studies, are often suggested to try to disentangle these), so that heritability is a relative measure applicable to a particular sample or population. For example, the common practice of doing genetic studies on multiply affected families may help mapping significance, but at the expense of overestimates of risk (e.g., Begg, 2002; Terwilliger and Weiss, 2003). Nonetheless, despite the many important technical issues associated with heritability as an indicator, it seems indisputable that the overall genetic contribution to most traits is not trivial and often substantial.

However, the substantial heritability of most traits can lead to a kind of mirage of genetic simplicity. Variation in mapped, specifically identified genes usually accounts for only a fraction, usually a small fraction, of the total amount of familial aggregation (and a consequently smaller amount of overall variation). This is because most of the heritable variation appears to be in the form of polygenic effects, whose aggregate is comprised of effects of numerous genes that are individually too small to be reliably or replicably mapped, or they will be mappable in one but not other populations or samples, because while genetic in the sense of inherited variation, in each individual they will be due to a different set of particular genetic variants.

This substantial fraction of apparent genetic etiology is an epistemological mirage in the sense that its individual components cannot be identified with practicable samples or study designs. To a substantial extent, cases of the trait may be causally so heterogeneous, for reasons described above, that reductionist approaches may be impossible in principle, since they are based on the replicability of observation. These instances of the trait can be characterized as *noninferrably genetic*, that is, putatively caused by genetic variation but in a way that can't be inferred with practicable or even achievable samples or sample sizes. Causal assertion in such instances may perhaps be true but may border on the mystic.

Note also that heritability refers to the fraction of trait variance that is statistically accounted for by genetic variance. How they do it is an entirely separate question. Thus $h^2 = 0.40$ doesn't imply that 40 percent of cases are caused by a gene (or even by an aggregate of genes) while the remaining 60 percent have nothing to do with genes.

Matters are made more complex—and, importantly, more ephemeral—by the "environmental" contributions, a term I put in quotes because we typically know very little about them except in broad or aggregate terms (e.g., stress, excess dietary intake, etc.). This is a source of major frustration in biomedical genetics today, but it is the expectable product of evolution in which many genes contribute to complex biological phenotypes in contextually dependent ways, and natural selection is tolerant of variation (Weiss and Buchanan, 2003; Buchanan, Weiss, and Fullerton, 2006a, 2006b).

I've tried to illustrate these points heuristically in Figure 12-1. The genetic objective is to predict outcome from inherited genotype. Part A shows a symbolic DNA sequence with several alleles in fonts proportional to their inherent effect on some trait of interest (inherent effect is a rather epistemologically dubious notion itself, but we'll let it pass). Surrounding the sequence are representatives of the factors that interpose between inheritance at the time of fertilization and eventual outcome. Part B shows how these factors overlay each other during life and the consequent difficulty we should expect in reliably inferring the underlying sequence

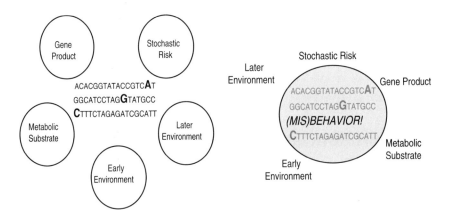

FIGURE 12-1 How darkly through a glass?

(Left). At the center, an inherited genotype with nucleotide font size proportional to an assumed inherent effect. Around this are several contributing factors that intervene between the genotype and the effect.

(Right). An important objective of genetic epidemiology is to be able to infer the genetic cause through these layers.

effects from the final phenotype. Each individual differs in these various components (e.g., their degree of transparency or grayness in this heuristic), at least to some extent, even if we somehow know what they are and measure them accurately. The immediate genetic epidemiological problem, the inverse of the causal process itself, is to infer the underlying genotype from a sample of outcomes, and from that to be able to make risk predictions. The figure shows metaphorically why it is problematic to think that genes code "for" complex traits.

A common strategy to pull the causal rabbit out of the uncertainty hat is meta-analysis, the aggregation of different studies. This approach is taken when existing studies vary widely in their conclusions or in their estimates of a putative risk factor, or it is thought that individual studies may have insufficient statistical power to detect a putative effect. Frequently, as the aggregate study size increases, the net overall estimated effect (e.g., relative risk) becomes smaller. While some substantial effects have been confirmed by meta-analysis, it must be recognized that there has been a continuing flow of meta-analyses finding that what initially seemed to be clear-cut risk factors turn out to have less or no effect. Along with causal and sampling heterogeneity, a likely reason for the latter outcomes is the bias in the design of samples that are responsible for the initial findings, referred to above. It remains difficult to find the rabbit, or to determine if it is the same rabbit in different hats. This makes the assessment of risk very problematic, even for frank disease, for which, unlike many aspects of social behavior, there is no controversy about the desire for risk assessment or intervention.

SNATCHING VICTORY FROM FAILURE

We do not yet know clearly where on the spectrum from monogenic to polygenic control the genetic components of complex traits lie, and this clearly varies among traits and to some extent among different populations because of their different population history.

There are two possible explanations for the frustration we have faced in attempting to dissect complex traits into their genetic and environmental causes or to identify the factors when they interact. Some think (or fear) that the traits of interest really do lie mainly toward the polygenic range of the etiological spectrum. Alternatively, others think (or hope) that such traits are more likely to be oligogenic (involve only a few major genes) and that the appearance of complexity is due to poor trait definition or identification of environmental factors that affect the phenotypic variance around a modest number of genotypes.

An important point worth repeating is that there are strong reasons to view the genetic contributions as environmentally contextual. The environment can be the physical, psychological, genomic, or ecological setting in which each genotype uniquely finds itself. It may be difficult or epistemologically impossible to separate out interacting factors, but we do have enough evidence from many known secular epidemiological trends that major changes in risk occur rapidly in populations whose gene pool is essentially unchanged, in traits with substantial heritability, and even in specific genotypes that are among the strongest and clearest of genetic risk factors, such as the major mutations associated with breast cancer (e.g., Begg, 2002; King, Marks, and Mandell, 2003). When environments are not known, not measured, or change unpredictably, risk estimation on which prospective intervention is based may be more art than science.

It is a sign of scientific victory, not failure, that we basically understand the nature of the uncertainties I have described, even aspects of risk we know that we don't know, or perhaps *can't* know with current conceptual approaches. Could complex traits really be genetically simple but we're missing something? That seems unlikely to be the general case, after the diversity of approaches that have been taken to the genetics of such traits.

Science should be guided by this victory of understanding rather than decrying it as a source of failure or distress because it doesn't lead to a tidy reduction of causation to a manageable number of clear causal elements. Current knowledge raises the epistemological and methodological question of what to do in the face of a substantial fraction of polygenic control, rather than hoping, against the bulk of the existing evidence, that we can identify a large fraction of individual contributing genes with expensive technological fixes, such as more elaborate statistics, more extensive genotyping technology, and faster computers.

DENIAL VERSUS DREAMING

The substantial debate about the nature and extent of genetic control of complex traits becomes more heated when the subject is social or nonpathogenic behavioral traits. Peoples' views are inextricably tied to their social politics, whether that fact is stated or tacit. Many wish to deny that nature controls traits they would rather see as facultative, as manifestations of free will, or as avoidable by lifestyle and societal change. Genetic determinism is a threat to a person's inherent value, something history shows that society can and often does abuse.

Others take the opposite world view. Basically, they see the living world in selfish gene terms, as the genetically determined, materialistic, reducible results of inevitable molecular interactions. They view natural selection as prescriptive, so that our traits today have limited genetic variation and high genetic control, although usually offering politically correct denials that they are not being genetic determinists or essentialists. But since natural selection affects only genes that determine a trait, Darwinian arguments are inescapably deterministic.

Where between these views—one denying genetic involvement in social behavioral traits, and the other dreaming of a simple causal world—does the truth lie? From the mechanistic sense of "genetic," behavioral traits, like any other traits, clearly involve genes. For behavioral traits, the mechanisms are likely to be very indirect, since genes don't talk, rape, shudder, cheat, or decide what pensions to invest in. But the same is true of diabetes, since genes don't eat, avoid exercise, etc.

In addition to mechanism, we also have every reason to assume that genetic variation affects variation in essentially every trait, social behavior included. This doesn't imply simple genetic architecture, however, and again, as with complex disease traits, behavioral phenotype correlations among relatives do not imply that environments or their nonlinear interaction with genes are unimportant, or the causal genes few. Each case will differ, and the central issues involve the *degree* to which the latter applies, and to which behavior, no matter what inherent "tendencies" we may wish to ascribe to it, is malleable by culture, just as we ask whether obesity is malleable by diet or exercise. The situation is clearly complex and heavily nuanced (Ryff and Singer, 2005). However, genetics is a particularly strong attractant in social behavior research these days, and it is important to try to understand why this might be.

Some nonscientific factors are at work, including subjective, sociopolitical, status, and desperational views ("nothing else is working") that are widespread in behavioral and social science. Funding and other entirely human considerations lead genetics to have a strong gravitational attraction for research, not unlike a black hole in space.

From a strictly scientific viewpoint, reductionist causation is convenient, manipulable. That suits our technophilic, interventionist yearning for simple and hence reassuring explanations. A driving force is the familial aggregation of many serious pathological behavioral (psychiatric) traits, even when specific genes have not yet been found. Also, some specific genetic variation with statistical effects on the nature or risk of behavioral traits have been found, as mentioned by various chapters in this book. As with other disease traits, such findings fire enthusiasm for continued searching. But there is a tendency to extend these findings to nonpathological aspects of behavior—to search for genetic determinants of normal social interactions.

I think it is very instructive to recall that Durkheim and the other founders of social science knew that social structure is the emergent result of superorganic interactions, not specific genotypes (basically everyone can speak Chinese, eat with a fork, be a Hindu, etc.) and that social facts exist, change, and are to be explained mainly in terms of their natural level of organization, that is, in terms of other social facts. Those social thinkers were only beginning and certainly didn't get everything right, but they had the central idea.

One way to see the point is perhaps to recognize that just because tallness (and its genetic basis—after all, stature is 90 percent heritable) is correlated with basketball, it does little to explain basketball. But a technophilic excitement over discovering the "basketball genes" that might account for the extremes of height, and so truly could account for a fraction of what we know about contemporary basketball, would distract from the more challenging effort to understand the phenomenon itself. And assuming that some positions of today's style of basketball are optimally played by people with genes for tallness (in a rich nutritional environment) would ignore the facts that basketball originated and flourished when most people were short and tall people were gawky, and the game can be played by people in wheelchairs.

The founders of social science wrote to resist the growing late 19th-century reductionist *zeitgeist* to explain culture in terms of psychology, and they were right. It remains relevant to ask what the scientific rationale is that leads social scientists today to hunger yet again for rescue by aliens, in this case molecular rescue by genes, another reductionist approach that is too far below the emergent level of the phenomena to fully explain them, and a diversion from trying to fix the problems in social science itself.

Why would social scientists essentially abandon the notion that social facts are to be understood in their own terms and instead seek to reduce these facts to molecular terms? I think the attraction, when it is scientific rather than one of funding politics and the like, rests understandably on two basic, often implicit assumptions.

First, a human being develops from a single cell, and all the information in the cell that affects development is ultimately genetic. Children look like their parents, twins highly resemble each other, and rabbits never give birth to mice. Ergo, genes must determine traits. That's genes as mechanism. Second is the idea that evolution occurs through heritable variation, and since natural selection molds variation adaptively, what we are must be what nature intended us to be, which is controlled by genes.

These premises are substantial but not complete truths. Many factors affect development, most traits of importance to this volume develop gradually with age and are context-dependent, stochasticities of countless sorts demonstrably affect biological processes and organisms. Natural selection is far more tolerant of diversity than one would gather from the melodramatic accounts of selfish genes and ruthless nature that one sees in the media (and often in the literature itself). To a great extent, what behavior is, and evolved "for," is assessing unpredicted and unclear situations and responding to them. Facultative responses cannot be too highly hard-wired. This should be especially true for humans, who are the social, *facultative* species par excellence. We should be trying to understand how that works, not to make it into its robotic opposite.

BIOETHICS

Relevant ethical issues go beyond the routine mechanics of informed consent, and these are especially complex and sensitive when applied to behavioral or social traits rather than outright disease. This is especially true when the causal epistemology is shaky. Genetic findings relate to the subjects involved but also give some information about their relatives. Topics discussed in this volume concern genetic contributions to behavioral traits that may be ultimately related to aspects of health but only indirectly so. Behavioral genetic findings also apply to those that the subject interacts with. Knowledge that you are an inherent gambler or intolerant to stress can be relevant to the lives of your affinal relatives, work cohorts, neighbors, and the like, not to mention those who might wish to exploit your characteristics commercially in one way or another. Let's look before we leap into prenuptial genetic testing to see who is genetically suited for the state of marital bliss, or to tolerate child abuse best (or provoke others to abuse them), or whether you'll be a liberal or conservative.

When social behavior affects health, it can be a legitimate concern of the National Institutes of Health, but genetic variation cuts many ways. It is usually thought that genetic variation responds to the social context, but to a considerable extent it may be that genetic variation contributes to the construction of that context (Odling-Smee, Laland, and Feldman, 2003; Weiss and Buchanan, 2004). As a society, we even hungrily modify our behavior when we're told what our genes make us do, changing the causal landscape on the fly. This is poorly understood territory that deserves more attention before leaping into assumptions of genetic causation.

An important ethical issue is informed consent. Subjects should be informed of the true nature of what can be psychologically or socially intrusive as well as physically invasive research, and also of its risks and benefits, many of which one might only be able to guess at. Informed consent should be frank and clear, and testing should be done to ensure that subjects understand that the investigators are going to be trying to find genetically inherited reasons why they are without adequate pensions, poor, poorly paid, divorced, bad investors, or whatever. Geneticists are aware of the damage that some of their work on disease per se can do if, for example, confidentiality is breached. Employment, insurance, or other forms of discrimination can result. While known abuses may be rare, in the information age the increasing potential for unauthorized data transfer keeps these issues on the table, and society's guardians should always press legislators to ensure that the laws keep up. Probably the greatest risk is in the damage that could be done to people by screeners who don't understand, but make decisions based on, the problematic probabilism of most genotypic risk assessment. What does informed consent actually inform study subjects of?

What do we intend to do with the information we'll collect on genetic variation and behavior? Even at the individual level, discussion quickly slides from abstract investigation to the topic of intervention and prevention. We should acknowledge our historical and clearly resurgent tendency to extend from the individual to the group, which usually includes racial categorization. There is often a disclaimer that a given kind of social genetics is just exploration of the nature of nature, to build models of basic mechanism that can't be done through mouse experiments, and that the work is not intended to identify individuals for policy reasons, nor is understanding mechanism necessarily a step toward genetic determinism in the variation sense. But read carefully: do the same authors—even the same paper—eventually relate their findings to points of intervention? Intervention by whom, and at whose discretion?

Behavioral genetic studies often categorize groups, such as with/ without a test genotype, male/female, or by "race" and report statistically significant differences in some measured outcome. Even when the p-values are convincing and the result replicable, the actual differences are often very modest. Yet focus is naturally drawn to the "significant" mean difference and away from the variance and the much greater overlap between categories. A focus on the mean has potential for unintended social interpretation. Are women different from men in, say, spatial or mathematical ability? Does one "race" have a higher IQ than another? And how sensitive are the distributions to environment? These questions can easily be given less attention than they deserve, because they are less newsworthy and more problematic.

From the late 1800s through the end of World War II, developed societies experienced many aspects of eugenics. The nominal purpose was beneficent, to prevent the human species from deteriorating by accumulating too much defective genetic variation. This was an obvious extension of Darwinian value judgments, since society has always to some extent or other cared for those in need. But the new idea was the notion that modern society protected the unfit in a deeper evolutionary sense when it allowed them to reproduce. Genetic knowledge can identify harmful variants, and they can thus be eliminated.

Eugenics deservedly became a dirty word and is easy to confuse with attempts to help those with life-limiting impairment. Individual eugenics is also routine and voluntary. It is eugenic for parents to decide whether to have children who inherit variation from them that might lead to disease, or to select the sex of their next child, or simply to pick mates with traits they like. Screening for deleterious genetic variation by amniocentesis, and other means, is possible for many different traits, and when the trait is clearly devastating *and* clearly genetic, as in Tay Sachs disease or thalassemia, this may not be controversial. But the definition of "harmful," or the threshold of acceptable harm can open a slippery slope of subjective judgment, especially when the causal epistemology is weak.

What about selective abortion based on genetic results for behavioral traits? How undesirable a behavioral trait might be involves subjective judgments or values, which can delve into senses of personal worth, and there are countless ways in which societal discrimination could be imposed on those deemed inherently less worthy. The potential to interpret culturally defined worth as being what was mandated by our history of natural selection (and hence "good") is a deterministic Darwinian excess to which there is great temptation.

I invite readers who think this isn't dangerous and that we could never plausibly revisit the venomous abuses into which the eugenic movement morphed—that fed the Nazi era in Europe—to read some of the eugenic and human genetic literature of the time (e.g., the main source on human genetics for 20 years is Baur, Fischer, and Lenz, 1931, or see Proctor, 1989; Carlson, 2001). Filter out dated terms and knowledge and you will see that the logic is remarkably similar to what is being widely offered today. Eugenic societal policy, racial hygiene, scientific racism, and social Darwinism were propounded by respected scientists, often the leading scientists, and often in the name of medicine. The traits of interest then have uncanny resemblance to the targets of social and behavioral genetics now.

What starts as "objective" and disinterested study of the facts of life as they are in the raw can lead even well-meaning scientists to accommodate for self-interested reasons to policies that are directly harmful, or even horrific, for many people. What may be benign in normal times turns ugly in societies under stress. To think that those days are over is to be oblivious not just to history but to trends in our own society today. Feelings are heated and opinions sharply divided when it comes to how, when, or whether to study what is "in our genes" regarding heavily culture-dependent aspects of human life and behavior. The risk of societal abuse may be small, but it can't be said to be zero. It's inappropriate to dismiss these concerns as "just political." This kind of science is by its very nature inherently political: social behavior *is* politics.

CONCLUSION

A character in Balzac's *Lost Illusions* says that "ideas are bipolar." He means that clever people can find ways to argue opposite points of view from the same data. So far as we understand it today, the genetic architecture of diverse traits has the characteristics I've outlined above. I have tried to show why it has been so difficult to identify in specific genetic terms more than a modest fraction of the control and variation of complex traits or even of their familial aggregation.

Work presented in this volume is consistent with that characterization. There are positive findings that may help in the understanding of genetic mechanisms affecting social behavior, but even classic examples of success, like ApoE, have subtleties that may be important to take into account. Extensive data from research on complex disease traits, which one might think would be more tractable than behavioral traits, suggests that a lot of effort, time, distraction, and resources will be committed to chasing down false positive results if the studies are not properly designed to detect genetic factors or if due attention is not paid to tempering results. Given the subtleties of environments in the social-behavioral context, really good design may be hard to come by.

Balzac's epigram at the beginning of this commentary illustrates the bipolar nature of ideas. It's relevant at this stage to keep them in mind. Some people may be born with a propensity to gamble at cards, but that propensity may only be enabled under peer pressure, and peer pressure is a societal structure that can coerce resistant genotypes to gamble as well. Is the cause of the economic disaster of a desolate old age, with its negative health consequences, in the gamblers' genotypes? The gambling system? In the causes of the peer pressure? Or lack of societal guarantees? To build genetic essentialism carelessly into our thinking is to gamble with everyone's future.

REFERENCES

Baur, E., Fischer, E., and Lenz, F. (1931). *Human heredity.* New York: Macmillan.

Begg, C.B. (2002). On the use of familial aggregation in population-based case probands for calculating penetrance. *Journal of the National Cancer Institute, 94*(16), 1221-1226.

Buchanan, A.V., Weiss, K.M., and Fullerton, S.M. (2006a). Dissecting complex disease: The quest for the philosopher's stone? *International Journal of Epidemiology, 35*(3), 562-571.

Buchanan, A.V., Weiss, K.M., and Fullerton, S.M. (2006b). On stones, wands, and promises. *International Journal of Epidemiology, 35*(3), 593-596.

Carlson, E. (2001). *The unfit: History of a bad idea.* Cold Spring Harbor, NY: Cold Spring Harbor Press.

King, M.C., Marks, J.H., and Mandell, J.B. (2003). Breast and ovarian cancer risks due to inherited mutations in BRCA1 and BRCA2. *Science, 302*(5645), 643-646.

Odling-Smee, F.J., Laland, K.N., and Feldman, M.W. (2003). *Niche construction: The neglected process in evolution*. Princeton; NJ: Princeton University Press.

Petronis, A. (2006). Epigenetics and twins: Three variations on the theme. *Trends in Genetics, 22*(7), 347-350.

Proctor, R. (1989). *Racial hygiene: Medicine under the Nazis*. Cambridge, MA: Harvard University Press.

Ryff, C.D., and B.H. Singer. (2005). Social environments and the genetics of aging: Advancing knowledge of protective health mechanisms. *Journals of Gerontology Series B: Psychological Sciences and Social Sciences, 60*(Spec No. 1), 12-23.

Scriver, C.R., and Waters, P.J. (1999). Monogenic traits are not simple: Lessons from phenylketonuria. *Trends in Genetics, 15*(7), 267-272.

Terwilliger, J.D., and Weiss, K.M. (2003). Confounding, ascertainment bias, and the blind quest for a genetic "fountain of youth." *Annals of Medicine, 35*(7), 532-544.

Weiss, K.M. (2005a). Cryptic causation of human disease: Reading between the (germ) lines. *Trends in Genetics, 21*(2), 82-88.

Weiss, K.M. (2005b). Genetics: One word, several meanings. In C. Scriver (Ed.), *Online metabolic and molecular basis of inherited disease* (Ch. 2.1). New York: McGraw-Hill.

Weiss, K.M., and Buchanan, A.V. (2003). Evolution by phenotype: A biomedical perspective. *Perspectives in Biology and Medicine, 46*(2), 159-182.

Weiss, K.M., and Buchanan, A.V. (2004). *Genetics and the logic of evolution*. Hoboken, NJ: Wiley.

Wellcome Trust Case Control Consortium. (2007). Genome-wide association study of 14,000 cases of seven common diseases and 3,000 shared controls. *Nature, 447*, 661-678.

Witherspoon, D.J., Wooding, S., Rogers, A.R., Marchani, E.E., Watkins, W.S., Batzer, M.A., and Jorde, L.B. (2007). Genetic similarities within and between human populations. *Genetics, 176*(1), 351-359.

Wong, A.H.C., Gottesman, I.I., and Petronis, A. (2005). Phenotypic differences in genetically identical organisms: The epigenetic perspective. *Human Molecular Genetics, 14*(Spec No. 1), R11-R18.

Part III:
New Ways of Collecting, Applying, and Thinking About Data

13

Minimally Invasive and Innovative Methods for Biomeasure Collection in Population-Based Research

Stacy Tessler Lindau and *Thomas W. McDade*

A new laboratory for health research has emerged at the intersection of the social and biomedical sciences. Sociological, demographic, and economic inquiries that have traditionally relied on self-reported health data are increasingly complemented by objective physical measures at the individual level. Biomedical investigations constrained by biases derived from clinic-based samples can now pursue questions about health disparities between social groups and about mechanisms linking social conditions to health and disease. Data from large probability samples can also provide valuable reference data for clinical diagnosis. Broad advances in information and biomedical technology, combined with emphasis on interdisciplinary research from the National Institutes of Health (2005), the National Academy of Sciences (2004), and others, have given momentum to new approaches to primary collection of individual-level health data that span subjective and objective social, health, and biophysiological domains. Hybridization of gold standard social science methodologies with minimally invasive techniques for biophysiological data collection in the home or in vivo defines a new frontier for health research.

In large part, these advances derive from rich traditions of field-based research that have valued the integration of multilevel data toward an understanding of human health and development. In the last several years, anthropometric measures (e.g., height, weight, waist and hip circumference) have replaced self-report or subjective estimates in many population-based studies and have proven highly clinically relevant.

Similarly, clinical measures, such as blood pressure sphygmomanometry, spirometry, and bone densitometry, have been successfully incorporated into epidemiological and other population-based health research. Investigators seeking to incorporate biophysiological measures must regularly choose between tried and true measures with established validity, reliability, and predictive power and experimental or cutting-edge measures, such as those that may have unknown clinical utility or are newly adapted to home-based collection procedures.

For this reason, in our review of innovative biophysiological methods suitable for population-based research, we include (1) methods with a track record of successful implementation in population-based research, (2) established clinical methods amenable to use in population settings, and (3) emerging and experimental methods with promise for future population research. This review is oriented toward experienced population researchers with interest in health but with limited experience in the collection of objective measures of biological function. We emphasize procedures and rationale for collection of biological measures that can be reasonably implemented in field settings and that can be meaningfully integrated with survey research.

ISSUES IN THE APPLICATION OF
MINIMALLY INVASIVE METHODS

Biological measures collected in the population setting can include direct measures of physical or physiological characteristics (e.g., hip circumference, blood pressure), functional testing (e.g., cognitive function, balance, grip strength), or collection of specimens that require laboratory processing in order to generate analyzable data. Such data may also be generated via experimental design (e.g., neuropsychiatric or olfactory testing). In traditional survey research, such constructs are approximated using self-report or subjective assessment by the study subject or the data collector (or both).

Translation of clinical or other laboratory methods for data and specimen collection to the population setting can occur by replication or adaptation. For example, investigators may choose to replicate the clinical encounter by sending a clinician, nurse, or phlebotomist to the home to conduct a physical examination or venipuncture for blood, or to bring participants into a mobile clinic close to where they live or work. Alternatively, adaptation of clinical or experimental laboratory methods using minimally invasive strategies and nonmedically trained interview personnel may enhance the feasibility of data collection and prove more cost-effective (Rockett, Buck, Lynch, and Perreault, 2004). Furthermore, the thrust for minimal invasiveness in population-based health research encourages technological innovation (e.g., development of easily portable medical equipment, adaptation of venous blood assays for use with dried blood spots) that may contribute to improvements in clinical medical services, particularly in remote and resource-poor areas. Data from the largest ongoing population-based health study in the United States, the National Health and Nutrition Examination Survey (NHANES), uses a replicative approach (e.g., state-of-the-art mobile clinics) and provides a wealth of gold standard measures against which adapted methods can be benchmarked.

As with any survey measure, several key considerations should guide decisions regarding implementation and application of biomeasures in population research. Test performance characteristics, such as reliability, validity, sensitivity and specificity, in combination with information about the expected distribution of the measure of interest in the population, require close attention. Issues particularly relevant to biological measures include the relationship of the instrument and measure to the clinical or laboratory gold standard, focused training of data collectors who may have limited experience with these methods, and quality control at the levels of data collection, transportation, and sample analysis. Changes in the availability of instrumentation and laboratory reagents may pose challenges for comparability across time and studies. Knowledge about the physiological and biological mechanisms pertinent to the system of interest, as well as variations in these across populations, is critical; this informs the sample design and size, timing, interpretation, consideration of relevant confounders, and the full range of environmental and contextual factors that may influence the measure and the process of measurement.

In the context of population-based research, several criteria must be met in order to achieve the goals of minimal invasiveness (Box 13-1). Minimally invasive methods aim to minimize burden to and maximize the safety of research subjects and data collectors and to contain research costs (York, Mahay, and Lindau, 2004; Mack, 2001).

BIOMEASURES AND TECHNOLOGIES WITH ESTABLISHED USE IN THE POPULATION SETTING

Anthropometrics

External measures of physical dimensions can be used to assess body size and composition as indicators of energy balance and nutritional history (Gibson, 2005; World Health Organization, 1995). Anthropometric measures can be performed quickly with portable equipment at minimal cost and therefore represent a set of objective health measures with low

---

BOX 13-1
Principles of Minimal Invasiveness in
Population-Based Biomeasure Collection

1. Compelling rationale: high value to individual health, population health, or scientific discovery.
2. In-home collection is feasible.
3. Cognitively simple.
4. Can be self-administered or implemented by single data collector during a single visit.
5. Affordable.
6. Low risk to participant and data collector.
7. Low physical and psychological burden.
8. Minimal interference with participant's daily routine.
9. Logistically simple process for transport from home to laboratory.
10. Validity with acceptable reliability, precision, and accuracy.

---

barriers to implementation. The availability of standardized methods and reference data improve the precision, reliability, and comparability of these measures (Lohman, Roche, and Martorell, 1988).

Commonly obtained anthropometric measures include height and weight, as well as circumferences and skinfold thicknesses taken from various locations on the body (Lohman et al., 1988). Raw measures are often converted to indexes or compared with age- and sex-specific reference values. For children, particularly in low-income settings, standardized scores for height-for-age, weight-for-age, and weight-for-height have long been used to identify short- and long-term growth faltering that increases risk for subsequent morbidity and mortality (World Health Organization, 1995). With growing awareness of an impending epidemic of overweight/obesity—both in the United States and internationally—body mass index (BMI, also called Quetelet's index; weight in kg/height in m^2) is frequently used as a tool for assessing weight relative to height in both children and adults.

However, BMI cannot differentiate lean tissue weight from weight due to body fat, nor does it reveal the pattern of fat distribution (e.g., central versus peripheral) that may be more predictive of disease risk. Waist-to-hip ratio, waist circumference, and strategically placed skinfold measurements provide more direct indicators of the quantity and distribution of fat and have therefore been implemented in a large number of epidemiological studies (Gibson, 2005). Better estimates of body fat also allow more accurate determinations of lean mass, which may have implications for bone loss and glucose metabolism. Recently leg length in adulthood has received attention as a potential indicator of nutritional quality in early childhood, when leg growth is most rapid and may be compromised by adverse environments (Wadsworth, Hardy, Paul, Marshall, and Cole, 2002).

More precise estimates of fat-free mass, fat mass, and relative body fat can be obtained through biological impedance analysis (BIA), using portable instruments that measure the impedance of a small electrical current passed through the body (National Institutes of Health, 1996). Computerized tomography, magnetic resonance imaging, and dual energy X-ray absorptiometry (DEXA) provide highly precise measures of body composition in clinical and laboratory settings, but they are too costly and cumbersome for field settings. However, relatively portable DEXA and ultrasound densitometry instruments have been successfully used in community settings to measure peripheral bone mass as a predictor of fracture risk (Andersen, Boeskov, Holm, and Laurberg, 2004; Wear and Garra, 1998).

Grip Strength

Several population-based studies, particularly studies on aging, have incorporated measures of hand grip strength as a biomarker of general muscle strength. Grip strength offers a relatively simple, minimally invasive measure of motor performance that has been shown to correlate with health status and a variety of health outcomes, including general physical function, bone density, mobility, and long-term mortality (Schaubert and Bohannon, 2005; Bohannon, 2001). Early work using grip strength in a French population study demonstrated progressive loss in strength with age (Clement, 1974).

Handheld dynamometry has been used as a research tool since 1916. In a thorough review of the technology, four classes of instruments commonly used for measuring grip strength in the clinical setting are described: (1) hydraulic dynamometers, (2) sphygmomanometers, (3) the vigorimeter (manometer with tubing and rubber ball), and (4) computerized dynamometry (Shechtman, Gestewitz, and Kimble, 2005). Hydraulic dynamometers, adapted from the gold standard Jamar™ dynamometer developed in 1954, have been the most widely used instrument in the population setting. This instrument uses a hydraulic mechanism to register static grip strength in pounds (kilograms) of force and is manufactured in portable, handheld models. Although protocols vary, testing typically involves two or three repeated trials of hand contraction separated by periods of rest with the subject in a comfortable seated position (for clinical protocol detail, see Shechtman et al., 2005).

Related measures that have been successfully used in population studies include lower extremity dynamometry, body weight, body mass index, chair stand, and timed-up-and-go (Schaubert and Bohannon, 2005). Computer-based, portable dynamometry for hand grip, pinch strength, and lower extremity strength may significantly enhance measurement and ease of integration with survey data collected in population studies.

Accelerometry

Physical activity is a central part of energy balance, but it is difficult to measure through self-report methods. Accelerometry provides objective measures of the frequency, intensity, and duration of physical activity in everyday settings (Crouter, Clowers, and Bassett, 2006; Chen and Bassett, 2005; Welk, Schaben, and Morrow, 2004). Several watch-sized monitors are commercially available and are worn on the wrist, ankle, or waist. These monitors record acceleration of the body in one, two, or three dimensions and collect data for hours, days, or even weeks at a time. Some units have event markers, as well as light and temperature sensors. Data on acceleration events per unit time are downloaded and analyzed to provide estimates of physical activity and energy expenditure. At this point, data reduction algorithms are not standardized, and different approaches have significant implications for outcome variables (Masse et al., 2005).

Accelerometry has been widely used in the exercise sciences and is increasingly applied to research on obesity and sleep (Treuth, Hou, Young, and Maynard, 2005; Ancoli-Israel et al., 2003; Yngve, Sjöström, and Ekelund, 2002). An accelerometry module was added to NHANES in 2003. Heart rate monitoring has also been successfully used in field settings to measure physical activity and energy expenditure (Wareham, Hennings, Prentice, and Day, 1997; Leonard, Katzmarzyk, Stephen, and Ross, 1995), and improved estimates may be possible with the integration of accelerometry and heart rate data (Strath, Brage, and Ekelund, 2005).

Dried Blood Spots

Population-based health research often seeks to define the reciprocal effects of health and sociodemographic factors. Until recently, the measurement of biological measures in blood specimens was the exclusive domain of clinical or laboratory research. Many key biomarkers of health and physiological function are accessible only through serum or plasma, but venipuncture is a relatively invasive procedure that has served as an impediment to the application of biomeasures to population-level research. Dried blood spots—drops of capillary whole blood collected on filter paper following a simple prick of the finger—represent a viable

TABLE 13-1 Recent Applications of Dried Blood Spots (DBS) in Large Population-Based Studies

Study	N	Biomarkers in DBS
Great Smoky Mountains Study	1,071	Androstenedione, DHEAS, EBV antibodies, estradiol, FSH, LH, testosterone
Health and Retirement Study	7,000[a]	CRP, HbA1c, total cholesterol, HDL
Los Angeles Family and Neighborhood Survey	5,000[a]	CRP, EBV antibodies, HbA1c, total cholesterol, HDL
National Longitudinal Study of Adolescent Health	17,000[a]	CRP, EBV antibodies, HbA1c
National Social Life, Health, and Aging Project	1,945	CRP, EBV antibodies, HbA1c, hemoglobin
Work and Iron Status Evaluation	18,000	CRP, transferrin receptor
Tsimane' Amazonian Panel Study	600[b]	CRP, EBV antibodies, transferrin receptor, leptin

[a]Projected; final plans for analyses are to be determined.
[b]Panel study; 3 waves of dried blood spots.
NOTES: DHEAS = dehydroepiandrosterone sulfate; EBV = Epstein-Barr virus; FHS = follicle stimulating hormone; LH = luteinizing hormone; CRP = C-reactive protein; HDL = high-density lipoproteins; HbA1c = hemoglobin A1C.

alternative to venipuncture and have recently been incorporated into a number of population-based studies. Table 13-1 summarizes the range of analytes currently being measured in blood spots among these studies. For additional information, see McDade, Williams, and Snodgrass (2007). This follows on the success of previous applications in remote settings internationally, including lowland Bolivia, Samoa, Kenya, Papua New Guinea, and Nepal (McDade et al., 2005; Shell-Duncan and McDade, 2004; Panter-Brick, Lunn, Baker, and Todd, 2001; McDade, Stallings, and Worthman, 2000; Worthman and Stallings, 1997).

Sample collection is relatively straightforward, and can be implemented by nonmedical interviewers: (1) the participant's finger is pricked using a sterile, single-use disposable lancet; (2) up to five drops of blood are spotted onto filter paper; (3) samples are allowed to dry (four hours to overnight); and (4) they are shipped via express or standard mail to the laboratory for freezer storage (most analytes are stable in dried blood spots at normal room temperatures for at least two weeks). A small disc of dried whole blood is punched from the filter paper blood spot and is placed in solution to create a sample of reconstituted blood, analyzable in a similar manner as a serum or plasma sample. Since these filter papers were originally developed to facilitate the collection of blood samples from neonates as part of an ongoing national screening program, they are certified to performance standards for sample absorption and lot-to-lot consistency (Mei, Alexander, Adam, and Hannon, 2001).

Protocols for over 100 analytes have been validated, including important indicators of endocrine, immune, reproductive, and metabolic function, as well as measures of nutritional status and infectious disease (McDade et al., 2007). Many of these biomarkers have been applied clinically and may be used in survey research to determine risk for the development of disease or to gain insight into the impact of psychosocial or behavioral contexts across multiple physiological systems. Recent innovations in immunoassay technology now make it possible to simultaneously quantify multiple analytes in one sample, rather than measuring one analyte at a time (Bellisario, Colinas, and Pass, 2000).

Advantages of dried blood spots include the minimally invasive sample collection procedure, simplified field logistics associated with sample processing, transport, and storage, and long-term stability under freezer storage that allows for future analyses as new biomarkers of interest, such as genetic markers, emerge (McDade et al., 2007). Disadvantages derive primarily from the fact that biomarker measurement in serum or plasma represents the gold standard in clinical assessment. Protocols must therefore be validated specifically for use with dried blood spots, and results may not be directly comparable with serum or plasma methods. In addition, the relatively small quantity of sample collected with blood spots may be an insurmountable limitation for some analytes that require large volumes of blood.

Urine

Among biospecimens amenable to home collection, urine provides a relatively simple platform for a very wide array of analyses. These include measures of renal, neuroendocrine, and sex hormone function, urine chemistry and cytology, nutritional measures, human chorionic gonadotropin (pregnancy hormone), microbes including sexually transmitted agents (e.g., chlamydia, N. gonorrhea, trichomonas, HIV) uropathogens, toxic substances, drugs, and drug metabolites (Rockett et al., 2004). Assays for the measurement of reproductive hormones or neuroendocrine metabolites commonly require collection of multiple urine voids over time (e.g., overnight, 12- or 24-hour periods, or daily) and sometimes involve food restriction, but most other assays can be performed on a single urine specimen. For some assays, and with younger participants, a first morning specimen can approximate an 8-12 hour multiple void collection. Measurement of urinary creatinine, a metabolic correlate of muscle mass that is normally found in relatively constant concentrations in urine, can be used to adjust for variability in urinary volume and concentration (Garde, Hansen, Kristiansen, and Knudsen, 2004; Naranayan and Appleton, 1980).

In addition to data obtained via laboratory processing, physical characteristics of urine, such as color, odor and temperature, can be useful. Urine temperature measured via an adhesive thermometer strip applied to the specimen cup can provide a close approximation of core body temperature. Decisions about specimen collection media depend on the volume of urine to be collected, the desired assays, and transportation issues.

In most cases, urine is collected in plastic specimen cups or jugs. However, innovative collection methods include filter paper, diapers, and commode collection pans. Although filter paper collection is limited by the paucity of validated assays, further development could substantially improve ease of transportation and storage. For most assays, urine must either be refrigerated or frozen within two hours (Simerville, Maxted, and Pahira, 2005) of collection and therefore requires cold storage or packaging. This method has been widely used and found acceptable in population studies with younger and older samples, men and women, and in a variety of cultural settings (Auerswald, Sugano, Ellen, and Klausner, 2006; Zheng et al., 2005; Serlin et al., 2002; Wawer et al., 1998). Cell-free genetic information (DNA molecules) may also be obtained and amplified from urine specimens (Botezatu et al., 2000).

Saliva

For many biomarkers, saliva is an attractive alternative to blood sampling since collection is noninvasive and can be successfully performed by participants in their homes or as they go about their normal daily routines. Furthermore, many assays that can be collected via blood can also be obtained from saliva. Repeat sampling is possible, and saliva can be collected from infants and children with minimal difficulty. Most analytes are stable at room temperature for up to a week—much longer for some analytes and collection devices—and saliva samples can therefore be stored and shipped without refrigeration for limited periods of time (Hofman, 2001). For these reasons, biobehavioral research has a track record of success with saliva sampling, both in experimental and naturalistic settings (Nepomnaschy et al., 2006; Gunnar and Donzella, 2002; Beall et al., 1992).

Assays for physiological indicators of stress, immune function and infectious disease, reproductive function, and drug use have been validated for use with saliva (Hofman, 2001; Granger, Schwartz, Booth, and Arentz, 1999; Nishanian, Aziz, Chung, Detels, and Fahey, 1998; Kirschbaum and Hellhammer, 1994; Ellison, 1988). In most cases, salivary measures are of value only if they reflect circulating concentrations in serum. This may be an insurmountable obstacle for some analytes. For others (e.g., steroid hormones), concentrations in saliva are free of binding proteins and may provide a better estimate of the active fraction in circulation (Vining, McGinley, and Symons, 1983).

Despite the noninvasive nature of saliva sampling, there are a number of important issues in the collection and processing of saliva that may significantly affect assay results. For example, concentrations of some analytes are affected by salivary flow rate, as well as contamination of saliva with blood or food (Kivlighan et al., 2004; Kugler, Hess, and Haake, 1992). The use of oral stimulants to promote saliva production and the absorption of saliva into cotton rolls to facilitate sample collection have been shown to modify assay results for some, but not all, analytes (Shirtcliff, Granger, Schwartz, and Curran, 2001; Schwartz, Granger, Susman, Gunnar, and Laird, 1998). The composition of the containers (e.g., glass, polystyrene) in which saliva is collected and stored can also affect some results and should be evaluated for each analyte (Ellison, 1988). These issues, as well as the application of different laboratory protocols, can lead to difficulties in interpretation and comparison across studies (Hofman, 2001).

Currently, salivary cortisol is frequently measured in naturalistic settings as a biomarker of stress, reflecting activation of the hypothalamic-pituitary-adrenal axis (Adam, 2006; Cohen et al., 2006; Kirschbaum and Hellhammer, 1994). Normal diurnal rhythms in cortisol production provide the opportunity to investigate concentrations at different times of day, patterns of change across the day, and overall levels of cortisol exposure. However, this variation poses significant challenges to measurement, with little consensus on sampling protocols beyond recognition of the importance of collecting multiple samples per day, preferably across multiple days. The timing of sampling is critical, particularly in the morning, and some studies have used timers or tracking devices to ensure compliance with collection protocols (Broderick, Arnold, Kudielka, and Kirschbaum, 2004).

TRANSLATION OF ESTABLISHED CLINICAL TECHNOLOGIES TO THE POPULATION SETTING

Ambulatory Electrocardiogram

Holter monitors provide the possibility of measuring heart rate variability continuously as participants go about their normal daily activities. A compact monitor is worn on the waist or over the shoulder, and electrodes attached to the chest record heart rate activity for 24 to 72 hours. Analysis of electrocardiogram data can be used to assess cardiac arrhythmias and vagal tone, both of which have been analyzed in relation to psychosocial factors and cardiovascular risk (Cacioppo, Tassinary, and Berntson, 2000; Kawachi, Sparrow, Vokonas, and Weiss, 1995). Holter monitoring is widely used clinically and has considerable potential for population-based research.

Spirometry

Pulmonary function testing is a mainstay of pulmonary medicine and plays an important role in the monitoring of such common diseases as asthma and obstructive pulmonary diseases. Spirometry has been used in several population studies, providing measures of lung volume (e.g., forced expiratory volume in one second or FEV_1) and expiratory flow (e.g., peak expiratory flow or PEF), as well as estimates of lung capacity (e.g., forced vital capacity or FVC).

More recently, spirometry has been incorporated into home-based studies as an indicator of vitality or disability and has been found to be a useful predictor of functional status and decline in the elderly. Data from large, population studies demonstrate that reduced lung function as measured by FEV_1 is associated with systemic inflammation (e.g., elevated C-reactive protein) and cardiovascular mortality independent of smoking (Sin, Wu, and Man, 2005), and it may be an independent predictor of long-term mortality (Schunemann, Dorn, Grant, Winkelstein, and Trevisan, 2000). A variety of portable spirometry devices are available, including handheld peak flow meters and computer-based devices (most clinical centers in developed nations use computerized spirometry). Interpretation of spirometry data for diagnostic purposes requires comparison with an appropriate reference group. An excellent review provides spirometry testing and interpretation standards as well as an overview of available equipment (Ruppel, 1997). Data from NHANES III provide clinic-based spirometry reference values for whites, African Americans, and Mexican Americans ages 8-80 (Hankinson, Odencrantz, and Fedan, 1999).

Evaluation of portable spirometers against laboratory spirometry shows acceptable agreement (Ezzahir, Leske, Peiffer, and Trang, 2005; Korhonen, Remes, Kannisto, and Korppi, 2005; Mortimer, Fallot, Balmes, and Tager, 2003; Koyama, Nishimura, Ikeda, Tsukino, and Izumi, 1998), but poor intermodel agreement (Koyama et al., 1998). This implies that longitudinal studies using peak flow meters should avoid changing meter models from one wave to another and highlights the importance of working with an established and reliable proprietor. Use of spirometry or peak flow meters in population research, particularly with nonmedically trained interviewers, requires careful training and a very detailed protocol. Similar to Holter monitoring, development of portable spirometers that have the capacity for longitudinal data storage and phone or computer line transmission (i.e., telemetry) may facilitate longitudinal home-based data collection (Giner and Casan, 2004; Wagner et al., 1999).

Hearing

Technological advances have significantly facilitated the transfer of audiometrics from the clinical to the population setting; in the near future, handheld and laptop-based platforms for hearing testing are likely to replace gold standard audiometry equipment even in clinical settings. For population studies, integration of audiometry software with laptop-based questionnaires may significantly enhance ease of implementation. Hearing testing has been widely used in population studies, particularly with older adults (e.g., NHANES, Sin et al., 2005) and the Leiden, Netherlands, 85+ Study (Gussekloo, de Craen, Oduber, van Boxtel, and Westendorp, 2005). Self-report is a poor indicator of hearing function (Newman, 1990).

Several minimally invasive instruments are currently available for hearing testing. The most widely used is the portable audiometer, a hand-held device that performs physiological audiometry. A hybrid device, the audioscope, combines audiometry with an otoscope (allowing for direct inspection of the ear canal and tympanic membrane). Portable audiometry in the frequency ranges of human speech (between 500 and 4,000 Hz at 25-40 dB) demonstrates excellent sensitivity against the gold standard (pure tone audiometry with sound booth isolation) and good specificity, but because it provides a physiological rather than a functional test, it may detect individuals with asymptomatic hearing loss. In contrast, it may fail to detect individuals with functional hearing impairment. These are appropriate for home use and can be handled by lay personnel.

Currently, the high cost of portable audiometry has limited its application in population studies. The Hearing Handicap Inventory for the Elderly-Screening Version (HHIE-S) offers a low-cost, widely used alternative to objective physiological measurement of hearing. This 10-item,

2-5 minute test screens for functional hearing, with scores on a 0-40 point scale indicating a probability of hearing impairment. It has been widely used in population studies with acceptable performance compared with audiometry in detecting hearing loss. Older measures used to screen for hearing loss aim to detect threshold sensitivity to whispered voice, the sound of the examiner's fingers rubbing together, or vibratory sensation with a tuning fork. Concerns about the subjectivity and particularly reliability of these measures limit their use (Pirozzo, Papinczak, and Glasziou, 2003).

Vision

Similar to hearing loss, ocular disorders are strongly associated with psychosocial impairment (Burmedi, Becker, Heyl, Wahl, and Himmelsbach, 2002), and have been related to increase in mortality (Knudtson, Klein, and Klein, 2006). Protocols appropriate for home-based vision testing, particularly acuity assessment, can be informed to some degree by clinical procedures and large-scale clinic-based studies that use population samples, such as the Beaver Dam Eye Study (Klein, Klein, Linton, and De Mets, 1991), the Baltimore Eye Survey (Tielsch, Sommer, Witt, Katz, and Royall, 1990), the Blue Mountains Eye Study (Attebo, Mitchell, and Smith, 1996), and the Salisbury Eye Evaluation Study (Muñoz et al., 2000). NHANES has also performed vision testing with population samples in a mobile clinic setting (Vitale, Cotch, and Sperduto, 2006). Together, these studies provide useful population prevalence estimates of vision impairment and blindness (see Table 2 in Xu et al., 2006).

However, these studies typically bring individuals from a population sample into a clinical examination setting. Home-based acuity testing is limited in part by the size and rigidity of eye charts, the need for standardized lighting, and, in small residential spaces, the chart distance requirements. An excellent review of visual acuity assessment summarizes the most widely used charts and scales for visual acuity testing of adults and children, such as the Snellen eye chart, the Early Treatment Diabetic Retinopathy Study Charts (ETDRS), and symbol charts that do not require English language or literacy skills (Kniestedt and Stamper, 2003). Studies in developing countries have perhaps made greatest use of field-based visual acuity testing (Gouthaman et al., 2005; Amansakhatov, Volokhovskaya, Afanasyeva, and Limburg, 2002), using adaptations of the Snellen tumbling "E" chart or the ETDRS chart in ambient indoor lighting, sunlight, or portable lighting. Similar to developments with auditory testing, researchers are investigating the performance of computerized acuity testing (Rosser, Murdoch, Fitzke, and Laidlaw, 2003), although this has not yet been widely implemented. Also on the horizon, fundus photography and other imaging instruments that can be used without retinal dilation (Wang, Tielsch, Ford, Quigley, and Whelton, 1998) may provide innovative opportunities for population-based research. Cost and ease of application by nonmedical personnel will determine the usefulness of these technologies in the field setting.

Point-of-Care Devices

Recent technological advances have led to the development of small, portable "point-of-care" devices that analyze clinically important bio-markers in real time (Glazer, 2006). As with blood spot sampling, a sterile lancet is used to stimulate the flow of capillary whole blood, and a drop of blood is placed into a cartridge or cuvette, which is then inserted into the instrument. Results are available onsite seconds or minutes later. The HemoCue™ instrument for analysis of hemoglobin has been widely used in international and domestic settings, and other portable devices are currently available, such as those that measure glucose, HbA1c, CRP, lipids, blood chemistry, or liver function in a single drop of blood. Devices vary in weight, ease of setup, whether they can operate on batteries, and the conditions under which cartridges must be stored prior to use. A single finger prick can provide capillary whole blood for onsite analysis, as well as for collection on filter paper as dried blood spots. By combining these procedures, results can be shared with participants, while blood spots can be assayed in the lab for a broader range of measures. In some cases, this may provide valuable health screening services and act as an incentive for research participants.

TRANSLATION OF EMERGING AND EXPERIMENTAL METHODS TO THE POPULATION SETTING

Given rapidly growing interest in the integration of biomeasures into population-based data collection efforts, acceleration in the translation of new technology from clinical and experimental laboratory research is expected. As an example, the National Social Life, Health, and Aging Projects (NSHAP) study recently completed collection of a broad panel of sensory function data using lay interviewers in the home setting with excellent cooperation from field staff and respondents. This required adaptation of experimental methods used to quantify olfactory and gustatory function in the laboratory setting as well as use of a simple and affordable clinical measure of tactile sensation.

Tactile Sensation

Measures of tactile sensation have not yet been widely implemented in broad population studies. Aside from use by hand surgeons, therapists, and neurologists to quantify neurological deficit and recovery, for example before and after hand surgery, population studies are limited to military personnel or workers using vibratory equipment (Wild et al., 2001) or method validation studies (Kaneko, Asai, and Kanda, 2005). NSHAP, a study of the interactions between social and biological factors at older ages, incorporates tactile sensation as part of a global evaluation of sensory function. Variations in touch sensibility may indicate microvascular disease, as seen in diabetes, or they may correlate with social characteristics, such as occupation or social contact with others. Cognitive function may also influence touch sensibility measures (Lundborg and Rosen, 2004). In a study of volunteers, thresholds for digital pressure perception appeared to increase with age (Kaneko et al., 2005).

Sensation to touch, or hand sensibility, has long been estimated in the clinical setting using the two-point discrimination test (2PD), described originally by Weber in 1835 (Lundborg and Rosen, 2004). This test measures the distance (in millimeters) between two points necessary for the individual to feel two distinct contacts, indicating peripheral innervation density. 2PD instruments range from a simple paper clip or compass (aesthesiometer, about $65 U.S.) to a multisided handheld instrument with graded intraprong distances (e.g., Dellon's Disk-Criminator™, about $100 U.S.) to pressure-specifying devices with force transducers in order to standardize application pressure. Despite concerns regarding standardization of 2PD equipment and protocols, particularly with regard to application of pressure with the prongs, 2PD has been demonstrated to be a valid measure of chronic nerve compromise (Dellon and Keller, 1997). Semmes-Weinstein nylon filaments provide greater standardization in sensation and pressure perception testing (Mayfield and Sugarman, 2000) but are relatively expensive for population use. Computer-assisted sensorimotor technologies are also available and may offer better performance, but they have not yet been adapted for nonclinical use (Dellon and Keller, 1997).

Taste

Gustatory function plays an important role in nutrition and quality of life and serves as a defense against ingestion of noxious or toxic substances. Because social interactions and cultural traditions frequently involve eating, impairment of gustatory sensibility can be associated with social alienation or withdrawal. The vast majority of gustatory function

testing occurs in the clinical or laboratory settings, with important applications to diagnosis and prognosis of common otolaryngological and
neurological disorders. Commonly used medications can interfere with
gustatory function. Aging can result in diminished sensitivity to some
tastes, salty and sweet for example, and therefore increased intake of
foods that can have negative health consequences (Schiffman, 1997). Oral
mucosal changes in postmenopausal women may contribute to changes in
gustatory function with age (Delilbasi, Cehiz, Akal, and Yilmaz, 2003).

Work led largely by German researchers has resulted in advancement of minimally invasive, easily portable, and well-performing methods for gustatory testing appropriate for population use. The latest of
these innovations involves use of tastant-impregnated paper strips (sour,
sweet, salty, bitter) in varying concentrations for evaluation of taste identification ability and threshold sensibility. The strips are designed so that
both sides of the tongue can be tested separately. Long shelf life, ease of
administration, and a relatively brief protocol offer major advantages over
liquid-based drop tests commonly used for clinical purposes (Mueller et
al., 2003). NSHAP has just completed olfactory testing using this method
with a national U.S. probability sample of more than 3,000 men and
women ages 57-85 with very high cooperation rates. Other technologies
for gustatory testing that could be used in the home include taste disks
and edible taste wafers. As opposed to regional testing accomplished with
the taste strips, these measures provide whole mouth tests of gustatory
function, perhaps a truer measure of everyday taste experiences (Ahne,
Erras, Hummel, and Kobal, 2000). Of course, everyday taste experience
typically involves complex combinations of tastes and aromas. Interdependence between taste and olfactory function limits the value of assessing one without the other (Mueller et al., 2003).

Olfaction

Over the last several years, a growing body of evidence linking olfactory function to cognitive decline at older ages has prompted both public
and research interest in this area and has motivated development of
portable, minimally invasive, efficient olfactory function tests. Decline in
olfactory function appears to occur with age (50 percent between ages 65
and 80 suffer significant loss of function—Frank, Dulay, and Gesteland,
2003) and some common medical illnesses and medication use, and it
is closely related to taste sensibility. Compromised olfaction presents
a risk for injury or other hazardous events, such as inability to detect
smoke or other noxious fumes and consumption of spoiled food (Santos,
Reiter, DiNardo, and Costanzo, 2004). Olfactory function testing shares
common principles with psychophysical testing of gustatory and other

chemosensory abilities; odorants are presented and the subject is assessed on the ability to detect, identify, and discriminate these (Doty, 2006). Although most authors agree that no gold standard olfactory assessment test exists, many are compared with the widely used 40-item University of Pennsylvania Smell Identification Test (UPSIT) (Doty, Shaman, and Dann, 1984), a comprehensive, self-administered, odor identification test using a scratch-and-sniff technique. Although probably too time-consuming for most population studies, this measure has been used in over 50,000 individuals and is validated in several languages (Doty, 2006).

A recent review article (Doty, 2006) describes the broad range of olfactory function measures available and provides a concise primer on the relative benefits and drawbacks of each. Shortened versions of the UPSIT have been successfully implemented in some clinical settings. In addition to the scratch-and-sniff technology, several other modalities offer options for population studies. These include "Sniffin' Sticks," which use a penlike odor device (Mueller and Renner, 2006; Hummel, Konnerth, Rosenheim, and Kobal, 2001; Kobal et al., 2000) and the odor stick identification test from Japan (Hashimoto et al., 2004). A laptop-based device is used for the sniff magnitude test, which assesses olfactory ability by comparing the nature of a person's sniff in response to air versus sniff with odors. Although used primarily in the clinical research setting, the laptop-based operation and data capture offer unique advantages over other olfactory test methods. The NSHAP study protocol for olfactory function testing in a large probability sample of older adults is an adaptation of the short Sniffin' Sticks olfactory function screening method (Mueller and Renner, 2006) and takes about 3-5 minutes to administer with very high cooperation rates (> 90 percent).

Vaginal Self-Sampling

Vaginal self-sampling provides population and clinical researchers a minimially invasive method for obtaining data historically restricted to the clinical setting in which a gynecologic pelvic examination could be performed. In addition to microbial pathogen data, quantification of the normal vaginal flora and mucosal inflammation provide data points of relevance to health, sexual behavior and function, and systemic hormonal and inflammatory processes. In comparison to the methodologies summarized here, vaginal self-sampling imposes relatively high respondent burden, albeit much less so than a gynecological examination. The implementation of this method, which requires a respondent to self-collect vaginal material in a private setting by inserting and rotating one or more small, sterile cotton- or Dacron™-tipped swabs, a small cytobrush, or a tampon in the vagina, derives primarily from experience in remote field settings where clinical pelvic examination is infeasible. Recent introduction of flexible menstrual collection devices provides another method for self-sampling to obtain cervicovaginal secretions (Boskey, Moench, Hees, and Cone, 2003) and biomeasures of menstruation (Koks, Dunselman, de Goeij, Arends, and Evers, 1997).

Once the samples are obtained, they are placed into transport media appropriate to the assays of interest and typically must be frozen and delivered to a laboratory for analysis. Established clinical protocols for microbial testing are appropriate for specimens collected in the home; the NSHAP study is also developing protocols for home-based vaginal sampling appropriate for cytological analysis. The vaginal self-swabbing method has demonstrated acceptability to study participants in a variety of settings (e.g., Bradshaw, Pierce, Tabrizi, Fairley, and Garland, 2005; Chernesky et al., 2005; Nelson, Bellamy, Gray, and Nachamkin, 2003; Serlin et al., 2002) and has performed favorably in comparison to urinary and clinical pelvic examination protocols for testing of sexually transmitted infections (e.g., Harper, Noll, Belloni, and Cole, 2002; Knox et al., 2002).

Vaginal self-sampling requires careful selection and training of field staff to maximize both respondent and data collector comfort with and understanding of the rationale and steps for the sampling procedures. Cervical self-sampling can also be accomplished using similar techniques and can approximate Papanicalou smear findings. However, interpretation of assays, designed to detect dysplastic or malignant conditions, requires expert or expertly supervised personnel. Furthermore, anticipatory guidance for research subjects must include clear information about the physical effects of self-sampling (e.g., transient irritation or a small amount of discharge or blood are not uncommon following sampling in older women) and whether the specimen will or will not be used for cancer screening. In almost all cases, implementation of vaginal sampling in population studies requires a professional results reporting and counseling mechanism, such as that offered by the American Social Health Association.

Magnetic Resonance Imaging

Magnetic resonance imaging (MRI) is a critical diagnostic tool that provides a window into the inner workings of the human body, but one that is not easily adapted to population research due to technical and logistical constraints. Brain imaging is emerging as a fundamental tool for social neuroscience inquiry that aims to locate and map the neurological pathways through which social stimuli may influence health and health outcomes. However, MRI systems contained in mobile trailers must be transported to centralized locations for community-based research. In addition, a portable MRI device has recently been developed that can provide images of the extremities, although, at a weight of approximately 100 pounds, it is still too large for household surveys. Additional innovations in MRI and functional MRI are likely to lead to more portable instrumentation in the near future.

ADJUNCT METHODS

Widely used by psychiatric and psychological researchers, experience sampling methods do not directly measure biological or physiological parameters, but instead serve as an adjunct to such measurements by providing documentation of real-time health-related information, feelings, symptoms, reflections, and thoughts or events pertinent to the individual psychosocial context, such as stress, mood, emotion, and well-being. These include paper-and-pencil diary, thought sampling (Hurlburt, 1997), and ecological momentary assessment (Stone and Shiffman, 1994) and are uniquely home- or field-based. However, most examples in the literature involve convenience, rather than probability, samples. These methods aim to reduce reporting bias caused by recall and the constraints of close-coded instruments used in traditional designs. Although the use of paper-and-pencil diaries for research dates to the 1940s (Bolger, Davis, and Rafaeli, 2003), thought sampling and experience sampling methods appear to have emerged nearly simultaneously in the mid-1970s (Hurlburt, 1997). Recently, Kahneman and colleagues reported on the Day Reconstruction Method, a hybrid of experience sampling methods and time-budget measurement (Kahneman, Krueger, Schkade, Schwarz, and Stone, 2004).

Vigorous debate in the recent literature about optimal diary methods suggests that no single method is superior for all study designs (Green, Rafaeli, Bolger, Shrout, and Reis, 2006; Broderick, Stone, Calvanese, Schwartz, and Turk, 2006; Bolger, Shrout, Green, Rafaeli, and Reis, 2006) and that these methods may, in some cases, be limited in their superiority to retrospective self-report (Takarangi, Garry, and Loftus, 2006). However, there is broad enthusiasm for advancement beyond paper-and-pencil methods toward augmentation with signaling devices (such as beepers, watch alarms, and phone calls) and for replacement of paper and pencil by electronic data collection using handheld devices, tablet personal computers, and electronic mail or web-based entries. The electronic devices offer the advantage of time- and date-stamping to corroborate subject compliance with the research protocol and to provide response-time data. Such devices may also enhance privacy, minimize the risk of data loss, allow for closer monitoring by researchers, and, combined with signaling, allow for dynamic flexibility in the intervals between entries. In addition to cost, some downsides of high-technology diary methods may include higher respondent burden due to disruptiveness, logistics of interacting with and transporting the equipment, or unfamiliarity with computer technology. A clinic-based study of patient compliance showed significantly higher compliance with an electronic versus paper-and-pencil diary (Stone, Shiffman, Schwartz, Broderick, and Hufford, 2003).

Rapid innovations in diary research methods include real-time electronic interaction with participants, voice recording and recognition technology for verbal entries, behavioral medicine technologies that integrate self-report of subjective mood, and cardiovascular and physical activity indices with ambulatory monitoring. One can imagine the combination of these innovative technologies with global satellite positioning; the value of the data versus potential infringements on subjects' privacy will have to be carefully weighed.

Medication use records provide another important adjunct to biomeasures, particularly in studies of aging. Self-report either by interview or self-administered questionnaire substantially limits the usefulness of such data. Direct observation of medication containers by in-home data collectors with immediate data entry is likely to improve data quality (Landry et al., 1988), but lack of unique identifiers for pharmaceuticals presents a major challenge with regard to coding and analysis.

CONCLUSION

Major advances in minimally invasive and portable biomedical technology and growing collaborations between social, biomedical, and life scientists, combined with state-of-the art survey technology, offer a tremendous opportunity for new kinds of health-related discovery. Many of the methods described here are innovative by virtue of bringing them into the field or home setting. Others implement novel inventions motivated by a desire to reach generalizable or remote samples. In the case of biological specimens, such as blood spots, saliva, vaginal samples, and urine, the power of the method is limited not by cooperation of the research participant, but by the availability and translation of suitable assays. The importance of high-quality, well-trained data collectors who embrace the rationale for biomeasure collection and can convey this to research participants cannot be underestimated. Innovations for population research that accomplish minimal invasiveness may facilitate cooperation in the clinical setting and diagnosis and treatment of disease in populations who otherwise would or could not access medical care.

REFERENCES

Adam, E.K. (2006). Transactions among adolescent trait and state emotion and diurnal and momentary cortisol activity in naturalistic settings. *Psychoneuroendocrinology, 31*(5), 664-679.

Ahne, G., Erras, A., Hummel, T., and Kobal, G. (2000). Assessment of gustatory function by means of tasting tablets. *Laryngoscope, 110*(8), 1396-1401.

Amansakhatov, S., Volokhovskaya, Z.P., Afanasyeva, A.N., and Limburg, H. (2002). Cataract blindness in Turkmenistan: Results of a national survey. *British Journal of Ophthalmology, 86*(11), 1207-1210.

Ancoli-Israel, S., Cole, R., Alessi, C., Chambers, M., Moorcroft, W., and Pollak, C.P. (2003). The role of actigraphy in the study of sleep and circadian rhythms. *Sleep, 26*(3), 342-392.

Andersen, S., Boeskov, E., Holm, J., and Laurberg, P. (2004). Feasibility of dual-energy X-ray absorptiometry in arctic field studies. *International Journal of Circumpolar Health, 63*(Suppl. 2), 280-283.

Attebo, K., Mitchell, P., and Smith, W. (1996). Visual acuity and the causes of visual loss in Australia. The Blue Mountains Eye Study. *Ophthalmology, 103*(3), 357-364.

Auerswald, C.L., Sugano, E., Ellen, J.M., and Klausner, J.D. (2006). Street-based STD testing and treatment of homeless youth are feasible, acceptable and effective. *Journal of Adolescent Health, 38*(3), 208-212.

Beall, C.M., Worthman, C.M., Stallings, J., Strohl, K.P., Brittenham, G.M., and Barragan, M. (1992). Salivary testosterone concentration of Aymara men native to 3600 m. *Annals of Human Biology, 19*(1), 67-78.

Bellisario, R., Colinas, R.J., and Pass, K.A. (2000). Simultaneous measurement of thyroxine and thyrotropin from newborn dried blood-spot specimens using a multiplexed fluorescent microsphere immunoassay. *Clinical Chemistry, 46*(9), 1422-1424.

Bohannon, R.W. (2001). Dynamometer measurements of hand-grip strength predict multiple outcomes. *Perceptual Motor Skills, 93*(2), 323-328.

Bolger, N., Davis, A., and Rafaeli, E. (2003). Diary Methods: Capturing life as it is lived. *Annual Review of Psychology, 54*(1), 579-616.

Bolger, N., Shrout, P.E., Green, A.S., Rafaeli, E., and Reis, H.T. (2006). Dear diary, is plastic better than paper? I can't remember: Comment on Green, Rafaeli, Bolger, Shrout, and Reis (2006). *Psychological Methods, March, 11*(1), 123-125.

Boskey, E.R., Moench, T.R., Hees, P.S., and Cone, R.A. (2003). A self-sampling method to obtain large volumes of undiluted cervicovaginal secretions. *Sexually Transmitted Diseases, 30*(2), 107-109.

Botezatu, I., Serdyuk, O., Potapova, G., Shelepov, V., Alechina, R., Molyaka, Y., Anan'ev, V., Bazin, I., Garin, A., Narimanov, M., Knysh, V., Melkonyan, H., Umansky, S., and Lichtenstein, A. (2000). Genetic analysis of DNA excreted in urine: A new approach for detecting specific genomic DNA sequences from cells dying in an organism. *Clinical Chemistry, 46*(8), 1078-1084.

Bradshaw, C.S., Pierce, L.I., Tabrizi, S.N., Fairley, C.K., and Garland, S.M. (2005). Screening injecting drug users for sexually transmitted infections and blood borne viruses using street outreach and self collected sampling. *Sexually Transmitted Infections, 81*(1), 53-58.

Broderick, J.E., Arnold, D., Kudielka, B.M., and Kirschbaum, C. (2004). Salivary cortisol sampling compliance: Comparison of patients and healthy volunteers. *Psychoneuroendocrinology, 29*, 636-650.

Broderick, J.E., Stone, A.A., Calvanese, P., Schwartz, J.E., and Turk, D.C. (2006). Recalled pain ratings: A complex and poorly defined task. *The Journal of Pain, 7*(2), 142.

Burmedi, D., Becker, S., Heyl, V., Wahl, H.-W., and Himmelsbach, I. (2002). Emotional and social consequences of age-related low vision. *Visual Impairment Research, 4*(1), 47.

Cacioppo, J.T., Tassinary, L.G., and Berntson, G.G. (2000). *Handbook of psychophysiology.* New York: Cambridge University Press.

Chen, K.Y., and Bassett, D.R., Jr. (2005). The technology of accelerometry-based activity monitors: current and future. *Medicine and Science in Sports and Exercise, 37*(11), S490-S500.

Chernesky, M.A., Hook, E.W., III, Martin, D.H., Lane, J., Johnson, R., Jordan, J.A., Fuller, D., Willis, D.E., Fine, P.M., Janda, W.M., and Schachter, J. (2005). Women find it easy and prefer to collect their own vaginal swabs to diagnose Chlamydia trachomatis or Neisseria gonorrhoeae infections. *Sexually Transmitted Diseases, 32*(12), 729-733.

Clement, F.J. (1974). Longitudinal and cross-sectional assessments of age changes in physical strength as related to sex, social class, and mental ability. *Journal of Gerontology, 29*(4), 423-429.

Cohen, S., Schwartz, J.E., Epel, E., Kirschbaum, C., Sidney, S., and Seeman, T. (2006). Socioeconomic status, race, and diurnal cortisol decline in the Coronary Artery Risk Development in Young Adults (CARDIA) Study. *Psychosomatic Medicine, 68*(1), 41-50.

Crouter, S.E., Clowers, K.G., and Bassett, D.R., Jr. (2006). A novel method for using accelerometer data to predict energy expenditure. *Journal of Applied Physiology, 100*(4), 1324-1331.

Delilbasi, C., Cehiz, T., Akal, U.K., and Yilmaz, T. (2003). Evaluation of gustatory function in postmenopausal women. *British Dental Journal, 194*(8), 447-449; discussion 441.

Dellon, A.L., and Keller, K.M. (1997). Computer-assisted quantitative sensorimotor testing in patients with carpal and cubital tunnel syndromes. *Annals of Plastic Surgery, 38*(5), 493-502.

Doty, R.L. (2006). Olfactory dysfunction and its measurement in the clinic and workplace. *International Archives of Occupational and Environmental Health, 79*(4), 268-282.

Doty, R.L., Shaman, P., and Dann, M. (1984). Development of the University of Pennsylvania smell identification test: A standardized microencapsulated test of olfactory function. *Physiology and Behavior, 32*(3), 489-502.

Ellison, P.T. (1988). Human salivary steroids: Methodological considerations and applications in physical-anthropology. *Yearbook of Physical Anthropology, 31*, 115-142.

Ezzahir, N., Leske, V., Peiffer, C., and Trang, H. (2005). Relevance of a portable spirometer for detection of small airways obstruction. *Pediatric Pulmonology, 39*(2), 178-184.

Frank, R.A., Dulay, M.F., and Gesteland, R.C. (2003). Assessment of the sniff magnitude test as a clinical test of olfactory function. *Physiology and Behavior, 78*(2), 195-204.

Garde, A.H., Hansen, A.M., Kristiansen, J., and Knudsen, L.E. (2004). Comparison of uncertainties related to standardization of urine samples with volume and creatinine concentration. *The Annals of Occupational Hygiene, 48*(2), 171-179.

Gibson, R.N. (2005). *Principles of nutritional assessment.* Oxford, England: Oxford University Press.

Giner, J., and Casan, P. (2004). Spirometry at home: Technology within the patient's reach. *Archivos de Bronconeumologia, 40*(1), 39-40.

Glazer, W.M. (2006). Point-of-care tests in behavioral health. *Behavioral Healthcare, 26*, 37-39.

Gouthaman, M., Raman, R.P., Kadambi, A., Padmajakumari, R., Paul, P.G., and Sharma, T. (2005). A customised portable LogMAR chart with adjustable chart illumination for use as a mass screening device in the rural population. *Journal of Postgraduate Medicine, 51*(2), 112-124, discussion 115.

Granger, D.A., Schwartz, E.B., Booth, A., and Arentz, M. (1999). Salivary testosterone determination in studies of child health and development. *Hormones and Behavior, 35*(1), 18-27.

Green, A.S., Rafaeli, E., Bolger, N., Shrout, P.E., and Reis, H.T. (2006). Paper or plastic? Data equivalence in paper and electronic diaries. *Psychological Methods, 11*(1), 87-105.

Gunnar, M.R., and Donzella, B. (2002). Social regulation of the cortisol levels in early human development. *Psychoneuroendocrinology, 27*(1-2), 199-220.

Gussekloo, J., de Craen, A.J., Oduber, C., van Boxtel, M.P., and Westendorp, R.G. (2005). Sensory impairment and cognitive functioning in oldest-old subjects: The Leiden 85+ Study. *American Journal of Geriatric Psychiatry, 13*(9), 781-786.

Hankinson, J.L., Odencrantz, J.R., and Fedan, K.B. (1999). Spirometric reference values from a sample of the general U.S. population. *American Journal of Respiratory and Critical Care Medicine, 159*(1), 179-187.

Harper, D.M., Noll, W.W., Belloni, D.R., and Cole, B.F. (2002). Randomized clinical trial of PCR-determined human papillomavirus detection methods: Self-sampling versus clinician-directed–biologic concordance and women's preferences. *American Journal of Obstetrics and Gynecology, 186*(3), 365-373.

Hashimoto, Y., Fukazawa, K., Fujii, M., Takayasu, S., Muto, T., Saito, S., Takashima, Y., and Sakagami, M. (2004). Usefulness of the odor stick identification test for Japanese patients with olfactory dysfunction. *Chemical Senses, 29*(7), 565-571.

Hofman, L.F. (2001). Human saliva as a diagnostic specimen. *The Journal of Nutrition, 131*(5), 1621S-1625S.

Hummel, T., Konnerth, C.G., Rosenheim, K., and Kobal, G. (2001). Screening of olfactory function with a four-minute odor identification test: Reliability, normative data, and investigations in patients with olfactory loss. *Annals of Otolaryngology, Rhinology, and Laryngology, 110*(10), 976-981.

Hurlburt, R.T. (1997). Randomly sampling thinking in the natural environment. *Journal of Consulting and Clinical Psychology, 65*(6), 941-949.

Kahneman, D., Krueger, A.B., Schkade, D.A., Schwarz, N., and Stone, A.A. (2004). A survey method for characterizing daily life experience: The day reconstruction method. *Science, 306*(5702), 1776-1780.

Kaneko, A., Asai, N., and Kanda, T. (2005). The influence of age on pressure perception of static and moving two-point discrimination in normal subjects. *Journal of Hand Therapy, 18*(4), 421-424, quiz 425.

Kawachi, I., Sparrow, D., Vokonas, P.S., and Weiss, S.T. (1995). Decreased heart rate variability in men with phobic anxiety. *American Journal of Cardiology, 75*, 882-885.

Kirschbaum, C., and Hellhammer, D.H. (1994). Salivary cortisol in psychoneuroendocrine research: Recent developments and applications. *Psychoneuroendocrinology, 19*(4), 313-333.

Kivlighan, K.T., Granger, D.A., Schwartz, E.B., Nelson, V., Curran, M., and Shirtcliff, E.A. (2004). Quantifying blood leakage into the oral mucosa and its effects on the measurement of cortisol, dehydroepiandrosterone, and testosterone in saliva. *Hormones and Behavior, 46*(1), 39-46.

Klein, R., Klein, B.E., Linton, K.L., and De Mets, D.L. (1991). The Beaver Dam Eye Study: Visual acuity. *Ophthalmology, 98*(8), 1310-1315.

Kniestedt, C., and Stamper, R.L. (2003).Visual acuity and its measurement. *Ophthalmology Clinics of North America, 16*(2), 155-170.

Knox, J., Tabrizi, S.N., Miller, P., Petoumenos, K., Law, M., Chen, S., and Garland, S.M. (2002). Evaluation of self-collected samples in contrast to practitioner-collected samples for detection of Chlamydia trachomatis, Neisseria gonorrhoeae, and Trichomonas vaginalis by polymerase chain reaction among women living in remote areas. *Sexually Transmitted Diseases, 29*(11), 647-654.

Knudtson, M.D., Klein, B.E., and Klein, R. (2006). Age-related eye disease, visual impairment, and survival: The Beaver Dam Eye Study. *Archives of Ophthalmology, 124*(2), 243-249.

Kobal, G., Klimek, L., Wolfensberger, M., Gudziol, H., Temmel, A., Owen, C.M., Seeber, H., Pauli, E., and Hummel, T. (2000). Multicenter investigation of 1,036 subjects using a standardized method for the assessment of olfactory function combining tests of odor identification, odor discrimination, and olfactory thresholds. *European Archives of Otorhinolaryngology, 257*(4), 205-211.

Koks, C.A., Dunselman, G.A., de Goeij, A.F., Arends, J.W., and Evers, J.L. (1997). Evaluation of a menstrual cup to collect shed endometrium for in vitro studies. *Fertility & Sterility, 68*(3), 560-564.

Korhonen, H., Remes, S.T., Kannisto, S., and Korppi, M. (2005). Hand-held turbine spirometer: Agreement with the conventional spirometer at baseline and after exercise. *Pediatric Allergy & Immunology, 16*(3), 254-257.

Koyama, H., Nishimura, K., Ikeda, A., Tsukino, M., and Izumi, T. (1998). Comparison of four types of portable peak flow meters (Mini-Wright, Assess, Pulmo graph and Wright Pocket meters). *Respiratory Medicine, 92*(3), 505-511.

Kugler, J., Hess, M., and Haake, D. (1992). Secretion of salivary immunoglobulin A in relation to age, saliva flow, mood states, secretion of albumin, cortisol, and catecholamines in saliva. *Journal of Clinical Immunology, 12*(1), 45-49.

Landry, J.A., Smyer, M.A., Tubman, J.G., Lago, D.J., Roberts, J., and Simonson, W. (1988). Validation of two methods of data collection of self-reported medicine use among the elderly. *The Gerontologist, 28*(5), 672-676.

Leonard, W.R., Katzmarzyk, P.T., Stephen, M.A., and Ross, A.G. (1995). Comparison of the heart rate-monitoring and factorial methods: Assessment of energy expenditure in highland and coastal Ecuadoreans. *The American Journal of Clinical Nutrition, 61*(5), 1146-1152.

Lohman, T.G., Roche, A.F., and Martorell, R. (1988). *Anthropometric standardization reference manual.* Champaign, IL: Human Kinetics Books.

Lundborg, G., and Rosen, B. (2004). The two-point discrimination test—Time for a reappraisal? *Journal of Hand Surgery [Br], 29*(5), 418-422.

Mack, M.J. (2001). Minimally invasive and robotic surgery. *Journal of the American Medical Association, 285*(5), 568-572.

Masse, L.C., Fuemmeler, B.F., Anderson, C.B., Matthews, C.E., Trost, S.G., Catellier, D.J., and Treuth, M. (2005). Accelerometer data reduction: A comparison of four reduction algorithms on select outcome variables. *Medicine and Science in Sports and Exercise, 37*(11), S544-S554.

Mayfield, J.A., and Sugarman, J.R. (2000). The use of the Semmes-Weinstein monofilament and other threshold tests for preventing foot ulceration and amputation in persons with diabetes. *Journal of Family Practice, 49*(11 Suppl.), S17-S29.

McDade, T.W., Leonard, W.R., Burhop, J., Reyes-Garcâia, V., Vadez, V., Huanca, T., and Godoy, R.A. (2005). Predictors of C-reactive protein in Tsimane' 2 to 15 year-olds in lowland Bolivia. *American Journal of Physical Anthropology, 128*(4), 906-913.

McDade, T.W., Stallings, J.F., and Worthman, C.M. (2000). Culture change and stress in Western Samoan youth: Methodological issues in the cross-cultural study of stress and immune function. *American Journal of Human Biology, 12*(6), 792-802.

McDade, T.W., Williams, S., and Snodgrass, J.J. (2007). What a drop can do: Dried blood spots as a minimally-invasive method for integrating biomarkers into population-based research. *Demography.*

Mei, J.V., Alexander, J.R., Adam, B.W., and Hannon, W.H. (2001). Use of filter paper for the collection and analysis of human whole blood specimens. *The Journal of Nutrition, 131*(5), 1631S-1636S.

Mortimer, K.M., Fallot, A., Balmes, J.R., and Tager, I.B. (2003). Evaluating the use of a portable spirometer in a study of pediatric asthma. *Chest, 123*(6), 1899-1907.

Mueller, C., Kallert, S., Renner, B., Stiassny, K., Temmel, A.F., Hummel, T., and Kobal, G. (2003). Quantitative assessment of gustatory function in a clinical context using impregnated taste strips. *Rhinology, 41*(1), 2-6.

Mueller, C., and Renner, B. (2006). A new procedure for the short screening of olfactory function using five items from the Sniffin Sticks identification test kit. *American Journal of Rhinology, 20*(1), 113-116.

Muñoz, B., West, S.K., Rubin, G.S., Schein, O.D., Quigley, H.A., Bressler, S.B., and Bandeen-Roche, K. (2000). Causes of blindness and visual impairment in a population of older Americans: The Salisbury Eye Evaluation Study. *Archives of Ophthalmology, 118*(6), 819-825.

Naranayan, S., and Appleton, H.D. (1980). Creatinine: A review. *Clinical Chemistry, 26*(8), 1119-1126.

National Academy of Sciences, National Academy of Engineering, and Institute of Medicine. (2004). *Facilitating interdisciplinary research*. Committee on Science, Engineering, and Public Policy. Washington, DC: The National Academies Press.

National Institutes of Health. (1996). National Institutes of Health: Technology Assessment Conference Statement. Bioelectrical impedance analysis in body composition measurement. *American Journal of Clinical Nutrition, 64*, 524S-532S.

National Institutes of Health. (2005). *NIH Roadmap for medical research: Accelerating medical discovery to improve health*. Available: http://research.musc.edu/nih/NIH2005/NIH%20Roadmap.ppt [accessed Sept. 2007].

Nelson, D. B., Bellamy, S., Gray, T.S., and Nachamkin, I. (2003). Self-collected versus provider-collected vaginal swabs for the diagnosis of bacterial vaginosis: An assessment of validity and reliability. *Journal of Clinical Epidemiology, 56*(9), 862-866.

Nepomnaschy, P.A., Welch, K.B., McConnell, D.S., Low, B.S., Strassmann, B.I., and England, B.G. (2006). Cortisol levels and very early pregnancy loss in humans. *Proceedings of the National Academy of Sciences, USA, 103*(10), 3938-3942.

Newman, D. (1990). Assessment of hearing loss in elderly people: The feasibility of a nurse administered screening test. *Journal of Advanced Nursing, 15*(4), 400-409.

Nishanian, P., Aziz, N., Chung, J., Detels, R., and Fahey, J.L. (1998). Oral fluids as an alternative to serum for measurement of markers of immune activation. *Clinical and Diagnostic Laboratory Immunology, 5*(4), 507-512.

Panter-Brick, C., Lunn, P.G., Baker, R., and Todd, A. (2001). Elevated acute-phase protein in stunted Nepali children reporting low morbidity: Different rural and urban profiles. *The British Journal of Nutrition, 85*(1), 125-131.

Pirozzo, S., Papinczak, T., and Glasziou, P. (2003). Whispered voice test for screening for hearing impairment in adults and children: Systematic review. *British Medical Journal, 327*(7421), 967.

Rockett, J.C., Buck, G.M., Lynch, C.D., and Perreault, S.D. (2004). The value of home-based collection of biospecimens in reproductive epidemiology. *Environmental Health Perspectives, 112*(1), 94-104.

Rosser, D.A., Murdoch, I.E., Fitzke, F.W., and Laidlaw, D.A. (2003). Improving on ETDRS acuities: Design and results for a computerized thresholding device. *Eye, 17*(6), 701-706.

Ruppel, G.L. (1997). Spirometry. *Respiratory Care Clinics of North America, 3*(2), 155-181.

Santos, D.V., Reiter, E.R., DiNardo, L.J., and Costanzo, R.M. (2004). Hazardous events associated with impaired olfactory function. *Archives of Otolaryngology: Head & Neck Surgery,* *130*(3), 317-319.

Schaubert, K.L., and Bohannon, R.W. (2005). Reliability and validity of three strength measures obtained from community-dwelling elderly persons. *Journal of Strength & Conditioning Research, 19*(3), 717-720.

Schiffman, S.S. (1997). Taste and smell losses in normal aging and disease. *Journal of the American Medical Association, 278*(16), 1357-1362.

Schunemann, H.J., Dorn, J., Grant, B.J., Winkelstein, W., Jr., and Trevisan, M. (2000). Pulmonary function is a long-term predictor of mortality in the general population: 29-year follow-up of the Buffalo Health Study. *Chest, 118*(3), 656-664.

Schwartz, E.B., Granger, D.A., Susman, E.J., Gunnar, M.R., and Laird, B. (1998). Assessing salivary cortisol in studies of child development. *Child Development, 69*(6), 1503-1513.

Serlin, M., Shafer, M.A., Tebb, K., Gyamfi, A.A., Moncada, J., Schachter, J., and Wibbelsman, C. (2002). What sexually transmitted disease screening method does the adolescent prefer? Adolescents' attitudes toward first-void urine, self-collected vaginal swab, and pelvic examination. *Archives of Pediatric Adolescent Medicine, 156*(6), 588-591.

Shechtman, O., Gestewitz, L., and Kimble, C. (2005). Reliability and validity of the DynEx dynamometer. *Journal of Hand Therapy, 18*(3), 339-347.

Shell-Duncan, B., and McDade, T. (2004). Use of combined measures from capillary blood to assess iron deficiency in rural Kenyan children. *The Journal of Nutrition, 134*(2), 384-387.

Shirtcliff, E.A., Granger, D.A., Schwartz, E., and Curran, M.J. (2001). Use of salivary biomarkers in biobehavioral research: Cotton-based sample collection methods can interfere with salivary immunoassay results. *Psychoneuroendocrinology, 26,* 165-173.

Simerville, J.A., Maxted, W.C., and Pahira, J.J. (2005). Urinalysis: A comprehensive review. *American Family Physician, 71*(6), 1153-1162.

Sin, D.D., Wu, L., and Man, S.F. (2005). The relationship between reduced lung function and cardiovascular mortality: A population-based study and a systematic review of the literature. *Chest, 127*(6), 1952-1959.

Stone, A.A., and Shiffman, S. (1994). Ecological momentary assessment (EMA) in behavioral medicine. *Annals of Behavioral Medicine, 16*(3), 199-202.

Stone, A.A., Shiffman, S., Schwartz, J.E., Broderick, J.E., and Hufford, M.R. (2003). Patient compliance with paper and electronic diaries. *Controlled Clinical Trials, 24*(2), 182-199.

Strath, S.J., Brage, S., and Ekelund, U. (2005). Integration of physiological and accelerometer data to improve physical activity assessment. *Medicine and Science in Sports and Exercise, 37*(11), S563-S571.

Takarangi, M.K., Garry, M., and Loftus, E.F. (2006). Dear diary, is plastic better than paper? I can't remember: Comment on Green, Rafaeli, Bolger, Shrout, and Reis (2006). *Psychological Methods, 11*(1), 119-122; discussion 123-125.

Tielsch, J.M., Sommer, A., Witt, K., Katz, J., and Royall, R.M. (1990). Blindness and visual impairment in an American urban population. The Baltimore Eye Survey. *Archives of Ophthalmology, 108*(2), 286-290.

Treuth, M.S., Hou, N., Young, D.R., and Maynard, L.M. (2005). Accelerometry-measured activity or sedentary time and overweight in rural boys and girls. *Obesity Research, 13*(9), 1606-1614.

Vining, R.F., McGinley, R.A., and Symons, R.G. (1983). Hormones in saliva: Mode of entry and consequent implications for clinical interpretation. *Clinical Chemistry, 29*(10), 1752-1756.

Vitale, S., Cotch, M.F., and Sperduto, R.D. (2006). Prevalence of visual impairment in the United States. *Journal of the American Medical Association, 295*(18), 2158-2163.

Wadsworth, M.E., Hardy, R.J., Paul, A.A., Marshall, S.F., and Cole, T.J. (2002). Leg and trunk length at 43 years in relation to childhood health, diet, and family circumstances: Evidence from the 1946 National birth cohort. *International Journal of Epidemiology, 31*(2), 383-390.

Wagner, F.M., Weber, A., Park, J.W., Schiemanck, S., Tugtekin, S.M., Gulielmos, V., and Schuler, S. (1999). New telemetric system for daily pulmonary function surveillance of lung transplant recipients. *Annals of Thoracic Surgery, 68*(6), 2033-2038.

Wang, F., Tielsch, J.M., Ford, D.E., Quigley, H.A., and Whelton, P.K. (1998). Evaluation of screening schemes for eye disease in a primary care setting. *Ophthalmic Epidemiology, 5*(2), 69-82.

Wareham, N.J., Hennings, S.J., Prentice, A.M., and Day, N.E. (1997). Feasibility of heart-rate monitoring to estimate total level and pattern of energy expenditure in a population-based epidemiological study: The Ely Young Cohort Feasibility Study 1994-5. *The British Journal of Nutrition, 78*(6), 889-900.

Wawer, M.J., Gray, R.H., Sewankambo, N.K., Serwadda, D., Paxton, L., Berkley, S., McNairn, D., Wabwire-Mangen, F., Li, C., Nalugoda, F., Kiwanuka, N., Lutalo, T., Brookmeyer, R., Kelly, R., and Quinn, T.C. (1998). A randomized, community trial of intensive sexually transmitted disease control for AIDS prevention, Rakai, Uganda. *AIDS, 12*(10), 1211-1225.

Wear, K.A., and Garra, B.S. (1998). Assessment of bone density using ultrasonic backscatter. *Ultrasound in Medicine and Biology, 24*(5), 689-695.

Welk, G.J., Schaben, J.A., and Morrow, J.R., Jr. (2004). Reliability of accelerometry-based activity monitors: A generalizability study. *Medicine and Science in Sports and Exercise, 36*(9), 1637-1645.

Wild, P., Massin, N., Lasfargues, G., Baudin, V., Unlu, D., and Donati, P. (2001). Vibrotactile perception thresholds in four non-exposed populations of working age. *Ergonomics, 44*(6), 649-657.

World Health Organization. (1995). World Health Organization. Physical status: The use and interpretation of anthropometry report of a WHO Expert Committee. *World Health Organization Technical Report Series, 14.*

Worthman, C.M., and Stallings, J.F. (1997). Hormone measures in finger-prick blood spot samples: New field methods for reproductive endocrinology. *American Journal of Physical Anthropology, 104*(1), 1-21.

Xu, L., Cui, T., Yang, H., Hu, A., Ma, K., Zheng, Y., Sun, B., Li, J., Fan, G., and Jonas, J.B. (2006). Prevalence of visual impairment among adults in China: The Beijing Eye Study. *American Journal of Ophthalmology, 141*(3), 591-593.

Yngve, A., Sjöström, M., and Ekelund, U. (2002). Amount and pattern of health-related physical activity among adults in relation to current recommendations. *Medicine & Science in Sports & Exercise, 34*(5: Supplement 1), S265.

York, E., Mahay, J., and Lindau, S.T. (2004). 7 criteria for the advancement of biomarker collection technology in population-based research. In S.T. Lindau and J. Mahay, Eds., *Proceedings of the 2004 Chicago Workshop on Biomarkers in Population-Based Health and Aging Research* (pp. 11-14). Chicago, Center on Demography and Economics of Aging, National Opinion Research Center at the University of Chicago.

Zheng, W., Chow, W.H., Yang, G., Jin, F., Rothman, N., Blair, A., Li, H.L., Wen, W., Ji, B.T., Li, Q., Shu, X.O., and Gao, Y.T. (2005). The Shanghai Women's Health Study: Rationale, study design, and baseline characteristics. *American Journal of Epidemiology, 162*(11), 1123-1131.

14

Nutrigenomics

John Milner, Elaine B. Trujillo,
Christine M. Kaefer, and *Sharon Ross*

Belief in the preventable nature of many chronic diseases coupled with rising health care costs has propelled consumers to seek more information about the quality of their diet and how dietary change might influence their life. This increased interest in the medicinal uses of foods or their components is not a new concept but has been handed down for generations. In fact, Hippocrates is often quoted as suggesting almost 2,500 years ago "to let food be thy medicine and medicine be thy food." In the United States, approximately 80 percent of adults ages 65 and older have at least one chronic health condition, and at least half of this population has two or more chronic health conditions that contribute to disability, decreased quality of life, and increases in health care costs (Goulding, 2003). Not surprisingly, the estimated health care costs for those ages 65 and older in developed countries typically range from three to five times greater than the health care costs associated with younger people. In addition, health care expenditures in the United States are almost double the amounts spent on health care in many other countries (Goulding, 2003; Organisation for Economic Co-operation and Development, 2005). Many consumers believe lifestyle, which includes dietary habits, must be a contributor to the higher health care expenditures in the United States. Since more than 75 percent of U.S. health care costs are linked with one or more chronic diseases, and diet is linked to 5 of the 10 major causes of death for Americans, there is ample justification for such widespread interest.

The significance of dietary components to health is not limited to the

United States, since nearly 60 percent of deaths worldwide are diet related, and there is increased recognition that inappropriate nutrition contributes to increasing health care costs in most settings (World Health Organization, 2004). Some of the most common causes of death both globally and domestically include cardiovascular disease, cancer, and diabetes, which are intimately linked to eating behaviors (Centers for Disease Control and Prevention, 2004; World Cancer Research Fund and American Institute for Cancer Research, 1997; World Health Organization, 2004, 2005). The 2003 report from a Joint World Health Organization/Food and Agriculture Organization Expert Consultation on Diet, Nutrition, and the Prevention of Chronic Diseases reviewed a large body of scientific evidence and found that up to 80 percent of coronary heart disease, 90 percent of type 2 diabetes, and one-third of cancers may be prevented by healthy eating practices, maintenance of a normal weight, and regular physical activity (Nishida, Uauy, Kumanyika, and Shetty, 2004).

Public health messages have centered on optimizing health for populations. The approach has considerable merit, although it is clear that not all individuals respond identically to treatment, whether by dietary change, lifestyle change, or drug therapy. There is much that needs to be learned regarding what accounts for individual differences, including the interactions occurring between genetics and the environment. As noted in a 2003 World Health Organization report on chronic diseases, genes help "define opportunities for health and susceptibility to disease, while environmental factors, including diet, determine which susceptible individuals are most likely to develop illness" (World Health Organization, 2003). Each stage of life is influenced by manmade and environmental factors that affect chronic disease risk. The interaction of genes with environmental influences throughout the life span creates multiple opportunities for tailored behavioral health interventions.

"Genome" is a term that refers to the entire DNA sequence of an organism, and "genomics" is the scientific discipline of mapping, sequencing, and analyzing the genome (Mathers, 2004). The human genome is an undeniably amazing blueprint of information. While all DNA consists of four simple bases, their sequence has a pronounced effect on health. Advances in molecular biology have already begun to reveal a wealth of information about human growth and development. Fundamental to genomics is the transcription of DNA to RNA (transcriptomics), which is subsequently translated to proteins (proteomics). Proteins bring about changes in cellular structure and small molecular weight compounds (metabolomics), which can influence one or more biological processes that ultimately determine a person's phenotype (Figure 14-1).

It is important to note that the relationship between transcriptomics, proteomics, and metabolomics is not linear. That is to say, there is not

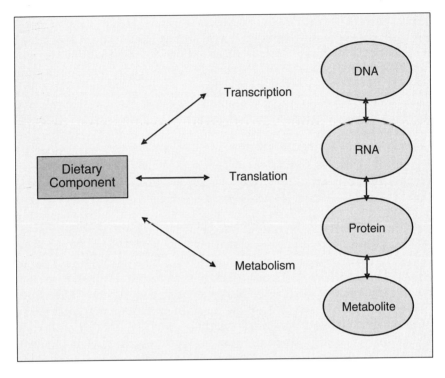

FIGURE 14-1 Dietary components can modify DNA transcription, translation, and metabolism.

necessarily a direct link between the amount of expression of a particular mRNA and the level of protein expressed. If the resulting protein is an enzyme, the resulting metabolite concentrations are not always proportional. Thus, complex and poorly understood regulatory processes exist that can influence overall phenotype. Regardless, it is safe to conclude that the more knowledge gained about the human genome, the more there will be to learn about the prevention and treatment of human disease. The discovery of about 3 billion chemical coding units in human DNA has already opened unexpected opportunities for understanding in greater detail what commonalities are shared among all humans and what constitutes individuality.

Differences in gene pool may account for population differences in the risk of developing such diseases as diabetes, heart disease, and cancers. Single-nucleotide polymorphisms (SNPs) differ in a single base from the generally accepted sequence, which occurs in at least 1 percent of the general population. SNPs are the commonest form of genetic variability. Humans have about 30,000 genes and roughly 5 to 8 million SNPs (Harland, 2005). Several gene polymorphisms have been identified as screening tools for predicting disease risk, including the HFE gene for hereditary hemochromatosis and the E-4 allele of the apolipoprotein (ApoE) gene for hypercholesterolemia and Alzheimer disease (Motulsky, 1999). Recently, the Genetic Association Information Network (GAIN) was created to support a series of genome-wide association studies designed to identify specific points of DNA variation associated with the occurrence of common diseases (http://www.genome.gov/19518664). In addition, in 2006, the National Institutes of Health (NIH) established the Genes and Environment Initiative (GEI) to support research that will lead to the understanding of gene-environment interactions in common diseases (http://www.genome.gov/19518663).

NUTRIGENOMICS

It has long been recognized that humans have an individual responsiveness to the foods they consume. Phenotypic variation to foods can be as subtle as sensitivity to bitterness, as reflected by the response to compounds like phenylthiocarbamide, or as gross as obesity, as reflected by differences in energy utilization (Clement, 2005; Wooding et al., 2004). Collectively, the scientific study of the way foods or their components interact with genes to influence phenotype is referred to as "nutrigenomics" or "nutritional genomics" (Davis and Milner, 2004; Trujillo, Davis, and Milner, 2006). The science of nutrigenomics is starting to provide greater clarity to the genetic pathways and associated molecular targets that account for the ability of food components to result in a physiologically relevant response. Researchers are beginning to unravel the genetic factors that influence an individual's eating behaviors. Although this area of research is still in its infancy, new analytical approaches, including genome-wide linkage scans and association studies with candidate gene markers and eating behavior are beginning to surface (Rankinen and Bouchard, 2006). This chapter focuses on the role of nutrigenomics in health and disease prevention and related insights on how nutrition-related biomarkers may assist in dissecting individual differences in populations.

The use of genomic technologies in nutrition research is relatively new but can have profound scientific implications for the development of public health messages and policy. Public screening programs that capture genetic information about an individual are already having an impact on nutritional intervention strategies. A perfect example is phenylketonuria (PKU), a rare metabolic disease involving the absence or deficiency of an enzyme needed to process an essential amino acid, which leads to mental retardation without proper treatment. Newborn blood screening for PKU has been part of public health programs in many parts of the world for decades, and the use of a phenylalanine-restricted diet is the accepted standard of care for this disease (National Institutes of Health Consensus Panel, 2001).

Nutrients and Single-Nucleotide Polymorphisms

Vitamin D is a fat-soluble vitamin that is found in food and can be formed in a person's skin in response to sunlight. In the presence of adequate ultraviolet light (UVB) in the wavelength range of 290-315 nm, a dietary intake of vitamin D may not be needed. Since adequate exposure to UVB is not always possible for a variety of reasons, a dietary source of vitamin D is needed to avoid skeletal diseases that weaken bones, such as rickets and osteomalacia. There is also evidence that vitamin D adequacy may play a role in immune function and the regulation of cell growth and differentiation, and therefore vitamin D may be a factor in the development of cancer (Holick, 2006). The vitamin D receptor (VDR), a nuclear hormone receptor, is known to mediate the biological actions of 1,25-dihydroxyvitamin D3 (1,25(OH)2 D3), which is the physiologically active form of vitamin D, by regulating a variety of target genes involved in cell proliferation and differentiation.

There are several known VDR polymorphisms that may affect the response to various dietary components and disease risk. One particular VDR polymorphism is FokI, which results in a VDR protein that is three amino acids longer than the protein produced from individuals carrying the nonvariant F codon. Individuals with the Ff or ff genotype were reported to have a 51 and 84 percent greater risk, respectively, of developing colorectal cancer (Wong et al., 2003). Those consuming a low-calcium or low-fat diet were found to have more than double the risk of colorectal cancer when they carried the ff compared with the FF genotype. Data (Wong et al., 2003) suggest that once the inadequacy of the diet is eliminated, the effect of genotype disappears. Thus, this polymorphism may serve as a predictive marker for those who will benefit most from adequate nutrient intakes.

Two additional vitamin D receptor polymorphisms (Bsm1 and poly A) also have been linked to calcium and vitamin D intake. Interestingly, these polymorphisms were related to energy consumption and the risk for colorectal cancer (Slattery et al., 2004). The occurrence of multiple VDR gene polymorphisms raises questions about the importance of single SNPs in accounting for variation and which diet-allele interactions are the most important determinants of phenotype. At least some evidence suggests that racial differences may exist in regard to the type of variance that occurs and its relationship to disease (Kidd et al., 2005).

Thus, multiple variables, including diet and race, can influence the relationship between VDR and disease risk (Slattery et al., 2004; Wong et al., 2003). Studies are needed to expand the understanding of the molecular and cellular significance of various polymorphisms, copy number variations, and their utility in population studies to detect susceptibility under a variety of environmental conditions (Redon et al., 2006). In fact, large longitudinal cohorts to evaluate both genetic and environmental factors that contribute to disease have been proposed (Manolio, Bailey-Wilson, and Collins, 2006).

A polymorphism in the angiotensinogen gene may determine how an individual's blood pressure responds to dietary fiber (Hegele, Jugenberg, Connelly, and Jenkins, 1997). Angiotensinogen is a liver protein involved with increasing vascular tone and promoting sodium retention, and plasma levels correlate with blood pressure. Individuals with a particular angiotensinogen genotype (TT) had a decrease in blood pressure when provided a diet with increased amounts of insoluble fiber compared with increased amounts of soluble fiber. In contrast, blood pressure in individuals with a different genotype (TM or MM) was not significantly influenced by the type of fiber consumed. Thus, some of the reported discrepancies in the response of blood pressure to dietary fiber may be related to inter-individual genetic differences in response to different types of fiber.

The response to other dietary components, such as caffeine, may also depend on specific SNPs. A study investigating the role of caffeine as a risk factor for bone loss in elderly women found that those with a variant of the vitamin D receptor (tt genotype) and who had caffeine intakes of greater than 300 mg/day had significantly higher rates of bone loss than did women with a different genotype (TT) (Rapuri, Gallagher, Kinyamu, and Ryschon, 2001). Since some individuals will not be receptive to caffeine avoidance, it may be wise to develop alternative strategies for minimizing risk, including providing additional calcium, vitamin D, or both.

Some additional examples of the interrelationship between SNPs and food components illustrate how reported discrepancies in the response to disease may arise from failure to account for interindividual genetic differences. The ApoE polymorphism is probably the most common example of a genetic polymorphism directly affected by a nutrient. ApoE plays an important role in lipid metabolism and its relationship to health and disease. There are three common allele variants, ApoE-2, ApoE-3, and ApoE-4. The ApoE-4 variant is associated with increased levels of total cholesterol and low-density lipoprotein cholesterol, which may increase risk for cardiovascular disease, but the opposite association occurs for carriers of ApoE-2. Some studies report greater plasma lipid responses to dietary manipulation in subjects carrying the ApoE-4 allele, while others fail to do so. A meta-analysis showed that ApoE-4 carriers were hyper-responsive to a low-fat diet. However, it is also true that ApoE-4 carriers may be hyporesponders to other types of interventions, such as dietary fiber (Ordovas et al., 1995; Uusitupa et al., 1992). While there is evidence that some diseases are associated with a SNP, the majority of the chronic diseases are thought to have multigenic roots. Thus, examining a single SNP may not provide sufficient detail to predict risk or appropriate inter-vention strategies.

Diet-gene interactions, including the impact of the ratios of nutri-ents consumed on a particular genotype, may be particularly important. Nuclear receptors, specifically peroxisome proliferator-activated receptors (PPARs), regulate the expression of genes involved in the storage and metabolism of fats. One particular PPAR, PPAR gamma, is recognized for its involvement in regulating insulin resistance and blood pressure. In individuals with a specific polymorphism in PPAR gamma, a low polyunsaturated-to-saturated fat ratio is associated with an increase in body mass index and fasting insulin concentrations. When the dietary ratio of polyunsaturated to saturated fats is high, the opposite is true (Luan et al., 2001). The interaction between ratio and type of dietary fat and PPAR gamma genotype is another example of the complexity found in examining diet-nutrient interactions in conjunction with a single polymorphism.

While there is mounting evidence that the frequency of functional polymorphisms may influence the response to a variety of dietary com-ponents, we need to validate and verify these findings (Davis et al., 2004; Stover, 2006; Trujillo, Davis, and Milner, 2006). Most findings are associ-ated with single observations and therefore need to be substantiated for their relevance and physiological significance in other settings. In addi-tion, attention needs to be given to the interaction of multiple genes in order to understand what is occurring within cells and ultimately being expressed in terms of health outcomes. Because of cost restraints, molecu-lar epidemiological studies have considered only a limited number of polymorphisms that may confer disease susceptibility. The use of hap-lotypes, which are a set of closely linked genetic markers present on one chromosome that tend to be inherited together, may offer a cost-effective solution for screening large populations. Alternatively, the need for low-cost whole genome DNA sequencing is being encouraged by a recent initiative from the National Human Genome Research Institute of NIH in order to reduce costs for sequencing individual genomes (http://grants. nih.gov/grants/guide/rfa-files/RFA-HG-04-003.html).

Nutritional Epigenomics

Epigenetics is the study of heritable changes in gene expression that occur without a change in DNA sequence and provide an extra layer of control in gene expression regulation. These regulatory processes are critical components for normal development and growth of cells. Evidence continues to support the hypothesis that epigenetic abnormalities are causative factors in cancer, genetic disorders, and pediatric syndromes as well as contributing factors in autoimmune diseases and aging.

Three distinct mechanisms are intricately related to epigenetics: DNA methylation, histone modification, and RNA-associated silencing. Abnormal methylation patterns are a nearly universal finding in cancer, as changes in DNA methylation have been observed in many cancer tissues, specifically colon, stomach, uterine cervix, prostate, thyroid, and breast tissues (Ross, 2003). Site-specific alterations in DNA methylation have also been observed in cancer and are thought to play a significant role in gene regulation and tumor behavior. The relationship between aberrant hypermethylation and hypomethylation on the expression of genes and their relationship to disease risk remains an area of active investigation.

Because epigenetic events can be changed, they offer another explanation for how environmental factors, including diet, can influence biological processes and phenotypes. Dietary components have been reported to influence DNA methylation patterns. Food components influence these events in at least four different ways (Ross, 2003). First, dietary factors are important in providing and regulating the supply of methyl groups available for the formation of S-adenosylmethionine (SAM), the universal methyl donor. Second, dietary factors may modify the utilization of methyl groups by processes that include shifts in DNA methyltransferase activity. A third plausible mechanism relates to DNA demethylation activity. Finally, the DNA methylation patterns may influence the response by regulating genes that influence absorption, metabolism, or the site of action for the bioactive food component.

The effect of maternal diet on phenotypic outcome in offspring has been examined in a mouse model, which is important because of its similarity in many ways to humans. The agouti mouse indicates that supplementation of choline, betaine, folic acid, vitamin B12, methionine, and zinc to the maternal diet increases the level of DNA methylation in the promoter region of the agouti gene and causes a change in the color pattern of the hair coat in offspring (Cooney, Dave, and Wolff, 2002). This phenotypic change coincides with a lower susceptibility to obesity, diabetes, and cancer. More recently, dietary genistein supplementation during pregnancy has been found to change the coat color of the offspring and reduce body mass, which were again reported to be related to changes in DNA methylation (Dolinoy, Weidman, Waterland, and Jirtle, 2006). While humans do not have the long-term repeating unit found in these mice, these studies serve as a proof-of-principle that diet can influence epigenetic events and lead to phenotypic change. These types of studies also suggest that *in utero* exposure to dietary components may not only influence embryonic development but also have profound and long-term health consequences.

DNA methylation patterns are being utilized more frequently as biomarkers in population and case-control studies to determine if differences exist between certain exposed groups or between cases and controls. In one such investigation in patients with alcohol dependence, the DNA methylation pattern in the HERP (homocysteine-induced endoplasmic reticulum protein) promoter region was found to be hypermethylated in patients with alcohol dependence when compared with healthy controls (Bleich et al., 2006). Interestingly, this aberrant hypermethylation was also significantly associated with diminished expression of HERP mRNA as well as elevated homocysteine levels. These findings assist in understanding the pathogenesis of a disorder as well as support the importance of epigenetic control on gene expression and the impact of dietary influences on epigenetic control.

Nutritional Transcriptomics

Genomic and epigenomic shifts do not entirely account for the influence that dietary factors can have on a person's phenotype, since changes in the rate of transcription of genes (transcriptomics) can also be exceedingly important (Feder and Walser, 2005). Several bioactive food components have been reported to be important regulators of gene expression patterns both *in vitro* and *in vivo*. Vitamins, minerals, and various phytochemicals have been reported to significantly influence gene transcription and translation in a dose- and time-dependent manner. These changes are likely to be key to the ability of food components to influence one or more biological processes, including cellular energetics, cell growth, apoptosis, and differentiation, all of which are important in regulating disease risk and consequences.

Transcriptomics allows for a genome-wide monitoring of expression for the simultaneous assessment of tens of thousands of genes and their relative expression. While microarray technologies provide an important tool to discover expression changes that are linked to cell processes, it must be remembered that any response may be cellular dependent and may vary between healthy and diseased conditions. Studies using animal models are beginning to identify specific sites of action of bioactive food components. For example, the nuclear factor E2 p45-related factor

2 (Nrf2) and the Kelch domain-containing partner Keap1 are modified by sulforaphane and allyl sulfur (Chen et al., 2004; Gamet-Payrastre, 2006). Sulforaphane is primarily found in cruciferous vegetables, such as broccoli and cabbage. Gene expression profiles from wild-type and Nrf2-deficient mice fed sulforaphane have shown several novel downstream events and thus provide more clues about the true biological response to this food component. The up-regulation of glutathione s-transferase, nicotinamide adenine dinucleotide phosphate:quinone reductase, gamma-glutamylcysteine synthetase, and epoxide hydrolase, occurring because of release of Nrf2 from its cytosolic complex, may explain the ability of sulforaphane to influence multiple processes, including those involving xenobiotic metabolizing enzymes, antioxidants, and biosynthetic enzymes of the glutathione and glucuronidation conjugation pathways.

Mammals are known to adapt to excess exposure to foods and their components through shifts in absorption, metabolism, or excretion. Thus, the quantity and duration of exposure must be considered when evaluating the response in gene expression patterns. Since microarray technologies provide only a single snapshot, overinterpretation of their physiological significance is certainly possible. While mRNA microarray technology continues to provide a powerful tool for examining potential sites of action of food components, their usefulness for population studies remains uncertain. Transcriptomic technologies have been used to examine the relationship between diet and prostate cancer among native Japanese and second-generation Japanese American men as a function of consumption of animal fat and soy (Marks et al., 2004). This technology was able to discriminate between men with cancer and those who were cancer free. Likewise, detectable changes associated with body mass and metabolism were observed (Marks et al., 2004). Weight loss caused by caloric restriction has been reported to be associated with changes in the expression of several inflammatory-related genes as discovered by transcriptomics (Clement et al., 2004; Viguerie et al., 2005).

To date, relatively few human studies have used transcriptomics to characterize the response to specific dietary components, and thus it is hard to make firm conclusions about the utility of this technology. Nevertheless, a recent study demonstrates that dietary intervention with high protein or carbohydrate breakfast cereals can influence gene expression patterns within a few hours after food consumption (van Erk, Blom, van Ommen, and Hendriks, 2006). However, much of the current evidence suggests that mRNA abundance is not always proportional to protein activity and thus cannot substitute for functional and ecological analyses of candidate genes (Feder and Walser, 2005). While the transcriptional profile can be useful in predicting metabolic stress, simpler indicators may suffice. It is possible that more select arrays may be useful if targeted to some cellular process. At this point it seems wise to evaluate carefully the costs and benefits of transcriptomics before including this research approach into large population studies.

Other Factors Determining Risk: Population Differences

There are well-documented differences in health between various populations and geographic areas; however, there is debate over the extent to which health disparities are due to "innate genetic differences," the "biological impact" of racial discrimination and lack of economic resources, or both (Krieger, 2005). Historically and in current scientific literature, a person's race or ethnic group has been used as a way to categorize and compare individuals as well as populations and their risks for developing various health conditions. It has been suggested that many of these labels are quite arbitrary (Krieger, 2005) and that the self-report of a person's race or ethnicity often fails to accurately identify the full range of their genetic makeup (Gonzalez et al., 2005). Ongoing projects are aimed at cataloguing the patterns of human genetic variation throughout populations in Africa, Asia, and the United States to assist in the identification of genes that impact health (Couzin, 2004).

Genetic comparisons provide important information about ancestral links, and geographic distances between various populations may be more closely related to the intermingling of populations from different geographical areas than to the racial or ethnic group to which an individual self-reports (Thomas, Irwin, Shaugnessy, Zuiker, and Millikan, 2005). Nevertheless, migration studies provide interesting clues about the importance of the environment, including diet, in changing the health risk of individuals. For instance, men moving from Japan and China to the United States adopt increased risks of prostate cancer (Brawley, Knopf, and Thompson, 1998). Likewise, women moving from Japan to the United States were found to exhibit an increased risk of breast cancer in their new location (Cole and Cramer, 1977). Today, the incidence of breast and colon cancers is similar in both Japan and the United States, suggesting that environmental factors such as diet are primary determinants of risk.

BIOMARKERS

The use of a biomarker to indicate a biological response to selected foods and food components is critical, since long-term intervention studies are difficult to conduct for a variety of reasons, including cost. Almost any measure that reflects a change in a biochemical process, structure, or function can serve as a useful biomarker. Several biomarkers may be used successfully to distinguish between healthy and diseased states and, in some instances, to predict future susceptibility to disease. While risk factors, including diabetes mellitus, smoking, hypertension, and hyper-cholesterolemia, have been linked to the risk of developing symptom-atic atherosclerosis, more sensitive biomarkers may offer early signals of shifts in risk. For example, lipid metabolism-related biomarkers such as lipoprotein(a) and apolipoprotein A-1 are associated positively and nega-tively, respectively, with premature atherosclerotic disease; however, there may be even earlier signals that are predictive of disease (Ordovas, 2006; Scheuner, 2001). Inflammatory markers, such as C-reactive protein (CRP) and fibrinogen, as well as thrombotic markers, such as fibrin D-dimer (DD) and tissue plasminogen activator, are receiving increased attention for their predictive value. Even nutrition-related factors, such as elevated plasma homocysteine, have been associated with the presence of athero-sclerotic arterial disease and its progression.

It is highly unlikely that one biomarker will be shown to adequately predict disease risk; therefore, several sensitive, reliable, and inexpensive biomarkers are needed to adequately assess the benefits and risks associ-ated with consumption of specific foods and their bioactive components. It is likely that intake, biological effect, and susceptibility biomarkers will be needed to adequately evaluate the effectiveness of foods and their bio-active components (Figure 14-2).

Assessment of dietary intake, performed by using various techniques such as 24-hour dietary recalls and food frequency questionnaires, is

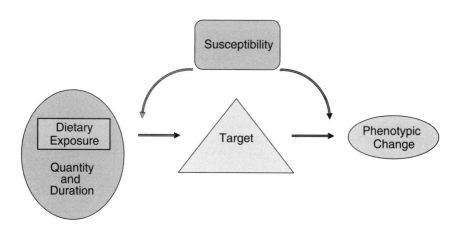

FIGURE 14-2 The quantity and duration of dietary exposures may bring about changes in the target (or biomarker) and result in phenotypic outcomes. Suscep-tibility factors may influence this process.

central for studies of the relationship between diet and health, and thus it is important that these dietary methods give an adequate measure of dietary intake. Given the variation in the content of individual food components, biological intake indicators, as reflected by circulating concentrations or perhaps other obtainable tissue, may be particularly useful for reflecting the amount of bioactive food component or metabolite present in cells, tissues, or body fluids. Assessment of intake indicators is relatively straightforward analytically, but their use is complicated by the need to know the optimal measurement period after consumption and by variability in rates of metabolism and accumulation across tissues and biological fluids (Kohlmeier, 1995). These and other measurement errors can have profound effects on how dietary data are interpreted (Paeratakul et al., 1998).

An interesting example concerning the relationship between dietary questionnaire data and circulating concentrations is highlighted for lycopene, a bioactive food component, in the European Prospective Investigation into Cancer and Nutrition (EPIC) study (Jenab et al., 2005). Serum lycopene concentrations have been reported to be inversely related to prostate cancer risk. EPIC investigators measured the consumption of tomatoes (raw and cooked) and tomato products (sauces, pastes, ketchup) in 521,000 subjects with country-specific dietary questionnaires across 10 countries. Furthermore, these investigators obtained plasma lycopene concentrations in a subgroup of 3,089 subjects from 16 EPIC regions (100 men and 100 women per region). The overall correlation of the total tomatoes and the tomato product intake with plasma lycopene was .33; the within-region correlation coefficient was .23, whereas the among-region correlation was .53.

These modest correlations between the dietary consumption of tomato and the lycopene concentration in blood may be due to the imprecision of dietary measurements as well as variations in the bioavailability and the absorption of lycopene. The cooking and seasoning methods of tomatoes and tomato products and the methods of consumption can affect the bioavailability of lycopene. In addition, these generally low correlations of tomatoes and tomato products with plasma lycopene may be related to the time span between the exposure and the blood sampling, especially in light of the strong seasonal variations in the intake of tomato and tomato products. This example illustrates the complexity of assessing dietary exposure and the need for new approaches. The National Heart, Lung, and Blood Institute, along with the National Cancer Institute and several other institutes and centers at NIH, and the National Science Foundation, recently released a request for applications to develop technologies or biomarkers to measure diet and physical activity in free-living, diverse populations (http://grants.nih.gov/grants/guide/rfa-files/rfa-ca-07-032. html).

Although serum and blood cells have frequently been used to evaluate exposure to bioactive food components, evaluation of bioactive food components may not always be predictive of the target tissue. Surrogate samples, such as exfoliated cells, may offer a noninvasive opportunity to evaluate exposures and physiological responses in target tissues. A source of exfoliated cells for oral or esophageal tissue can presumably be found in saliva. Because of the possible application of tea in the prevention of oral and esophageal cancers, Yang and collaborators (Yang, Lee, and Chen, 1999) measured the salivary levels of tea catechins in six human volunteers after drinking tea. Although exfoliated cells from saliva were not isolated in this study, the results suggest that tea catechins were absorbed through the oral mucosa and that saliva—a source of exfoliated cells—may be another biological source material in which to evaluate dietary exposure of certain bioactive food components and their physiological effects.

Recently serum biomarkers were examined in a randomized controlled trial using matrix-assisted laser desorption/ionization-time of flight mass spectrometry proteomic profiling (MALDI-TOF) and statistical analysis (Mitchell, Yasui, Lampe, Gafken, and Lampe, 2005). In this study, 38 participants ate a diet devoid of fruits and vegetables and one supplemented with cruciferous (broccoli) family vegetables. Interestingly, two significant peaks were detected (m/z values of 2,740 and 1,847) that could classify participants based on diet (control vs. cruciferous) with 76 percent accuracy. The 2,740 m/z peak was identified as the B-chain of alpha 2-HS glycoprotein, a serum protein previously found to vary with diet and be involved in insulin resistance and immune function. Thus, the expanded use of proteomic technologies may provide some important clues not only about consumption patterns (exposure), but also about their biological consequences.

Dietary components are known to have widespread "effects" on various cellular processes associated with health and disease prevention, including carcinogen metabolism, hormonal balance, cell signaling, cell cycle control, apoptosis, and angiogenesis (Davis and Milner, 2004; Trujillo et al., 2006). In addition, the combination of foods or nutrients may incur favorable outcomes. Studies in men have found that soy combined with black or green tea has been reported to synergistically reduce serum prostate-specific antigen concentrations, which is a marker for prostate cancer (Zhou, Yu, Zhong, and Blackburn, 2003). Moreover, certain combinations of foods or nutrients may diminish the magnitude of the response compared to when those foods or nutrients are provided alone. For example, the impact of omega-3 fatty acids on gene expression was not observed when combined with the antioxidant vitamin E in a cell culture system (Aktas and Halperin, 2004). Whether this is true in vivo remains unresolved, but if this blunting effect of vitamin E also occurs in vivo, it would explain some of the inconsistencies about the health benefits of fish that currently exist in the literature.

Effect biomarkers are particularly useful if they can predict a potentially detrimental response long before it occurs. However, few biomarkers are universally accepted as being reliable (Schatzkin and Gail, 2002). The most common biomarkers are body mass index, blood pressure, and cholesterol. However, more sensitive biomarkers are needed to detect subtle change long before disease complications arise. The range of effect biomarkers required is immense because of the need to detect a variety of metabolic events that alter cognitive and physical performance or change the risk of disease. Associating genetic polymorphisms with carcinogen-DNA adduct measurements shows promise for monitoring dietary exposure to cancer-causing agents (Kyrtopoulos, 2006; Warren and Shields, 1997). Many other biomarkers are beginning to emerge that might be used effectively to monitor the impact of dietary habits on growth and development, including platelet-derived growth factor, transforming growth factor, basic fibroblast growth factor, epidermal growth factor, insulin-like growth factor, and hepatocyte growth factor (Giovannucci, 1999; Fletcher et al., 2005). Additional research is needed to unravel the effects of dietary habits on these and other effect biomarkers.

An example of an *effect* biomarker that is influenced by diet comes from studies on the ability of fish oil to suppress tumor necrosis factor (TNF-α) production and mediate the inflammatory response. It is well known that TNF-α mediates inflammation and that high TNF-α production has adverse effects during disease. However, the effect biomarker is also influenced by the genetics of the consumer, since the response is influenced by polymorphisms in the TNF-α genes. Grimble and colleagues (Grimble et al., 2002) found that men with high inherent TNF-α production were more sensitive to the anti-inflammatory effects of fish oil compared with men with lower levels of TNF-α production.

The biological response to a food component is probably not consistent across all tissues. For example, the effect of nutritional zinc-deficiency on the activities of O6-alkylguanine:DNA methyltransferase (AGT) in nine rat tissues, including liver, lung, kidney, spleen, brain, esophagus, forestomach, gastric stomach, and small intestine, was examined by two measures (Fong, Cheung, and Ho, 1988). In both measurements, the activity of AGT was significantly reduced in the esophagus of zinc-deficient rats compared with zinc-sufficient controls, but the other tissues were less affected, suggesting sensitivity of this tissue to zinc deficiency. A human clinical example for this concept comes from a randomized controlled clinical study on the effect of selenium supplementation for prevention of skin carcinoma (Clark et al., 1996). It was found that selenium treatment did not protect against development of basal or squamous cell carcinomas of the skin, but results from secondary end-point analyses supported the hypothesis that supplemental selenium may reduce the incidence of, and mortality from, carcinomas of several sites, including lung, colon, and prostate.

The complexity in understanding the biological consequences of a change in an effect marker is illustrated by studies with the isoflavone genistein, a bioactive compound found in soy and other legumes. Circulating concentrations of estradiol are strongly and positively related to bone health and breast cancer risk in women (Thomas, Gallo, and Thomas, 2004). At low exposures, genistein may reduce the influence of naturally occurring estrogen, but at higher concentrations it may promote estrogenic effects (Gikas and Mokbel, 2005; Reinwald and Weaver, 2006). This dose dependency may explain part of the confusion that exists in the literature (Cassidy, 2003). Likewise, variation in how estradiol is formed and metabolized may influence the response to genistein supplements or dietary soy products.

A recent study investigated the relationship among dietary isoflavone exposure, genotype, and plasma estradiol levels in postmenopausal women using three markers of isoflavone exposure: diet, urine, and serum. There was a strong correlation between isoflavone and plasma estradiol in women with a particular polymorphism in a gene involved in estrogen metabolism. For the women with that polymorphism, the data would translate into a 30 percent reduction in breast cancer risk (Low et al., 2005). These observations raise the interesting possibility of a diet-gene interaction in which the effect of isoflavone exposure may be exceptionally pronounced in women with a particular genotype but is attenuated when women of different genotypes are considered. Thus, to gain a greater understanding of the observations, a biological effect measure must consider genomics.

A host of *susceptibility* biomarkers also will be critical for evaluating the merits of changing dietary habits. These biomarkers should allow the measurement of individual differences associated with genetic background or variation associated with many environmental factors (Milner, 2003; Perera, 1994). The ability of genetic differences to activate or detoxify genotoxic agents is becoming increasingly recognized as an appropriate susceptibility biomarker. Many genes, associated products, or receptors are now under investigation as markers of susceptibility, including those associated with *OB, UCP, erbB-2, ras, myc, p53, BCL-2, Ki-67, and HNF-1-α* (Wiedmann and Caca, 2005). Although several single-gene mutations have been shown to cause problems in experimental animal models, such as those that occur in some animal models of obesity, the situation in humans is likely to be considered more complex. Interactions among several genes, environmental factors, and behavior make the search for appropriate gene markers especially challenging. Nevertheless, new markers are being developed that should offer exciting opportunities for clarification of the effect of genes and the environment.

Recently, the drug herceptin received attention for its novel characteristics in blocking a specific target and thereby susceptibility associated with the increased risk of certain breast cancers (Flaherty and Brose, 2006). This drug has been showcased as an emerging new strategy for dealing with disease, namely a molecular medicine approach. Interestingly, it has been recognized that various dietary components can modify this same process and influence the risk of disease. Evidence, from a variety of sources, points to olive oil, an integral ingredient of the Mediterranean diet, as a deterrent to cancer. Recent evidence suggests the ability of its monounsaturated fatty acid oleic acid (18:1n-9) to specifically regulate overexpression of HER2 (Her2/neu, erbB-2), a well-characterized oncogene that may account for at least part of the observed protection in population studies (Colomer and Menendez, 2006). Interestingly, the efficacy of trastuzumab, a humanized monoclonal antibody binding with high affinity to the ectodomain (ECD) of the Her2-coded p185 (HER2) oncoprotein, was found to be enhanced by oleic acid (Colomer et al., 2006). In addition to olive oil, experimental evidence suggests one of the active components in green tea, epigallocatechin-3-gallate, is effective in inhibiting Her-2/ neu expression and leads to a suppression in the downstream events that it typically would bring about (Masuda, Suzui, Lim, and Weinstein, 2003). Evidence also exists that n-3 fatty acids from fish, as well as flavonoids from various grains, can retard Her2/neu expression (Menendez, Lupu, and Colomer, 2005; Way, Kao, and Kin, 2004).

The field of nutrigenomics and associated biomarkers is helping to illuminate how nutrients modulate processes within human tissues. However, there are many unique challenges that must be faced when dealing with nutritional and genomic research. People generally do not eat one food at a time, and it is often difficult to ascertain how much is consumed, which dietary component(s) brings about a positive or negative health effect, and if food components are working alone or in combination. Information coming from the Women's Health Initiative also emphasizes the need to not only understand quantities of individual food components but also the duration of their intake (Prentice et al., 2006). This was particularly true for the effects of reducing fat intake on breast cancer risk, which did not appear until four to five years after the intervention started. How the genetic background of an individual, as well as that person's age, gender, and lifestyle affect response to nutrients are additional issues that need to be more adequately addressed in the future. Although the task of answering these questions seems daunting, progress is being made and may well continue as newer tools and biomarker research unfolds.

Overall, a variety of biomarkers that can monitor the intake (exposure), molecular target (effect) and variation in response (susceptibility) will be needed to develop a profile for an individual that reflects the effect of diet on overall performance and health. To assess the benefits of foods or their components, additional attention must be paid to examining the variability in response among populations and individuals, the strengths of any association or correlation, the specificity of the relationship, the reversibility of the response, and the biological basis for any proposed benefits. Undeniably, the development and application of validated and sensitive biomarkers have enormous importance not only in improving health but also in assessing the importance of dietary change. While it is difficult to assess the significance of individual types of investigations, it would be wise to develop a model that incorporates various types of data before highlighting the particular benefits or risks of a functional food or bioactive component. While some may argue that a model including epidemiologic evidence (25 percent), intervention studies (35 percent), animal models (25 percent), and mechanism of action (15 percent) might be appropriate, others might shift the proportions markedly. This is in no way a novel approach; it was used in preparing six consensus statements about chronic disease, yet many scientific and lay publications do not use the same criteria when showcasing information about the merits or limitations of nutrition and health studies.

ETHICS

Although many researchers are enthusiastic about the potential to tailor public health products and services for disease prevention and treatment based on an individual's genetic profile, and many consumers are anxious to gain more personal control over their health by learning about their genetic susceptibility for various health concerns, there are risks and benefits that must be taken into consideration. Potential benefits from genetic testing include relief from anxiety, improved ability to plan for the future, and more informed decisions regarding measures that may or may not play a role in disease prevention or treatment. Potential risks include the cost of testing, psychological and emotional reactions to results, disruption to families, and potential for discrimination (Anderlik and Rothstein, 2001).

In the realm of nutrigenomics, it is important for the public to understand that the companies currently in existence that tout "personalized nutrition" tend to base their recommendations on a small number of genes for which there is currently inadequate information to predict an individual's response to specific food components or eating habits (Check, 2003; Kaput et al., 2005). In addition, because of the often complex interactions between genes and environmental factors, it is challenging for researchers, clinicians, and counselors to determine an individual's likelihood of developing a disease. In some cases, behavioral changes will not modify a person's likelihood of developing a given disease. However, in many cases, behavioral changes will modify individual risk, even if it is higher based on their genetics, so it is important that the information is communicated in a manner that does not create deterministic attitudes that might reduce self-efficacy. The fact that some companies exist that provide results directly to individuals without any discussion with a physician or genetics counselor is troubling, since this information may be used incorrectly or inappropriately (Castle, 2003). Regardless of the current concerns related to the accuracy of health claims made by companies marketing personalized nutrition and the manner in which these results are delivered to individuals, it is likely that the future of nutrition includes the ability to create effective individual guidance based on genetic profiles.

A concern regarding the future of nutrigenomics (and genomics in general) is whether the extremely large investments in research infrastructure to support ___omics research will further widen the health disparities gap between the rich and the poor at the individual and population levels (Darnton-Hill, Margetts, and Deckelbaum, 2004; Cannon and Leitzmann, 2005). As described above, the public throughout the world, as well as policy makers, health care providers, and a wide range of researchers need to understand the concept of genomics and how it can have a positive impact on lives in order to make informed decisions about whether or not it will be in their best interest to support the large investments needed for genomics research. When genomics is considered at the population level, the issue of individual health benefit versus the good of the overall community, population, or country comes into question, especially since there are a limited number of health care dollars to treat everyone with medical issues (Anderlik et al., 2001). In order for genomics to improve health in developing countries and wealthier developed countries, technological capacity and training need to be increased regionally to address health concerns and reduce the belief that "90% of research expenditure is dedicated to the health problems of 10% of the world's population" (Singer and Daar, 2001).

The application of nutrigenomics to the field of agriculture has been used already in several developing countries to address some of the regional concerns related to nutritional deficiencies through improvements to crop yield and enhancement of the nutritional profile of staple crops. There has been an ongoing debate for years regarding the safety of genetically modified foods, fueled primarily by issues of trust and public perceptions of risk (Anderlik et al., 2001), and several countries will not allow these products into their food supply. If genomics can be applied as part of the solution to a wide range of health concerns in developing countries, which rank as some of the most populous nations in the world, the return on investment and health benefits need to be carefully considered, along with issues of trust and public understanding of genetics and risk.

In order for nutrigenomics to be applied to disease prevention and treatment, the public must be able to trust those who collect DNA for individual analysis as well as for inclusion into larger research databases, and those who will be providing them with health education based on genetic information. For example, in 1999 and 2000, the National Bioethics Advisory Committee commissioned studies on the ethical issues and policy guidance regarding research involving human biological materials. The reports noted that in the United States, certain population groups that have experienced research abuses in the past, such as African Americans and Jews, had the highest levels of concern related to privacy and trust issues (Anderlik et al., 2001). The 2002 World Health Organization report *Genomics and Health* reviewed how certain cultural practices may also affect confidentiality (World Health Organization, 2002). There are many people who fear there is insufficient security and protection of their personal health data. Some countries, including Iceland and Estonia, which have national projects containing genetic information, have laws to safeguard genetic information, who may use it, and in what manner. However, in the United States there are serious gaps in the areas of privacy protection, the degree of specificity needed in informed consent procedures for research, and in the laws governing privacy of health information and genetic discrimination (Anderlik et al., 2001).

CONCLUSION

Advances in nutrition must be consistent with meeting the needs of populations as well as individuals in subpopulations. Since nutrition is not a pure science, it is ideally suited to build on the social, behavioral, and biological domains (Cannon et al., 2005). Nevertheless, each of these domains introduces a variety of perspectives about what is important and what strategies are critical for bringing about a change in health and disease prevention. In order to determine whether personalized nutrition with tailored risk reduction strategies to improve health through eating behaviors will motivate individuals and populations to change, behavioral and social scientists need to partner with basic researchers in current and future efforts to understand the impact of nutrigenomics and help the field reach its fullest potential. It is clear that not all individuals respond identically to foods or their components. Identifying those who benefit and those placed at risk is the real challenge. While the challenges are not insignificant, the potential societal benefits are truly enormous.

REFERENCES

Aktas, H., and Halperin, J.A. (2004). Translational regulation of gene expression by omega-3 fatty acids. *Journal of Nutrition, 134*(9), 2487S-2491S.

Anderlik, M.R., and Rothstein, M.A. (2001). Privacy and confidentiality of genetic information: What rules for the new science? *Annual Review of Genomics and Human Genetics, 2*, 401-433.

Bleich, S., Lenz, B., Ziegenbein, M., Beutler, S., Frieling, H., Kornhuber, J., and Bonsch, D. (2006). Epigenetic DNA hypermethylation of the HERP gene promoter induces down-regulation of its mRNA expression in patients with alcohol dependence. *Alcoholism: Clinical and Experimental Research, 30*(4), 587-591.

Brawley, O.W., Knopf, K., and Thompson, I. (1998). The epidemiology of prostate cancer part II: The risk factors. *Seminal Urology and Oncology, 16*(4), 193-201.

Cannon, G., and Leitzmann, C. (2005). The new nutrition science project. *Public Health Nutrition, 8*(6A), 673-694.

Cassidy, A. (2003). Potential risks and benefits of phytoestrogen-rich diets. *International Journal for Vitamin and Nutrition Research, 73*(2), 120-126.

Castle, D. (2003). Clinical challenges posed by new biotechnology. *Postgraduate Medical Journal, 79*, 65-66.

Centers for Disease Control and Prevention. (2004). *The burden of chronic diseases and their risk factors: National and state perspectives.* Atlanta, GA: Author.

Check, E. (2003). Consumers warned that time is not yet ripe for nutrition profiling. *Nature, 426*, 107.

Chen, C., Pung, D., Leong, V., Hebbar, V., Shen, G., Nair, S., Li, W., and Kong, A.N. (2004). Induction of detoxifying enzymes by garlic organosulfur compounds through transcription factor Nrf2: Effect of chemical structure and stress signals. *Free Radical Biology and Medicine, 37*(10), 1578-1590.

Clark, L.C., Combs, G.F., Jr., Turnbull, B.W., Slate, E.H., Chalker, D.K., Chow, J., Davis, L.S., Glover, R.A., Graham, G.F., Gross, E.G., Krongrad, A., Lesher, J.L., Jr., Park, H.K., Sanders, B.B., Jr., Smith, C.L., and Taylor, J.R.. (1996). Effects of selenium supplementation for cancer prevention in patients with carcinoma of the skin: A randomized controlled trial. Nutritional Prevention of Cancer Study Group. *Journal of the American Medical Association, 276*(24), 1957-1963.

Clement, K. (2005). Genetics of human obesity. *Proceedings of the Nutrition Society, 64*(2), 133-142.

Clement, K., Viguerie, N., Poitou, C., Carette, C., Pelloux, V., Curat, C.A., Sicard, A., Rome, S., Benis, A., Zucker, J.D., Vidal, H., Laville, M., Barsh, G.S., Basdevant, A., Stich, V., Cancello, R., and Langin, D. (2004). Weight loss regulates inflammation-related genes in white adipose tissue of obese subjects. *Journal of the Federation of American Societies for Experimental Biology, 18*(14), 1657-1669.

Cole, P., and Cramer, D. (1977). Diet and cancer of endocrine target organs. *Cancer, 40*(1 Suppl), 434-437.

Colomer, R., and Menendez, J.A. (2006). Mediterranean diet, olive oil, and cancer. *Clinical and Translational Oncology, 8*(1), 15-21.

Cooney, C.A., Dave, A.A., and Wolff, G.L. (2002). Maternal methyl supplements in mice affect epigenetic variation and DNA methylation of offspring. *Journal of Nutrition, 132*(8 Suppl.), 2393S-2400S.

Couzin, J. (2004). Genomics: Consensus emerges on HapMap strategy. *Science, 304,* 671-673.

Darnton-Hill, I., Margetts, B., and Deckelbaum, R. (2004). Public health nutrition and genetics: Implications for nutrition policy and promotion. *Proceedings of the Nutrition Society, 63,* 173-185.

Davis, C.D., and Milner, J. (2004). Frontiers in nutrigenomics, proteomics, metabolomics, and cancer prevention. *Mutation Research, 551*(1-2), 51-64.

Dolinoy, D.C., Weidman, J.R., Waterland, R.A., and Jirtle, R.L. (2006). Maternal genistein alters coat color and protects Avy mouse offspring from obesity by modifying the fetal epigenome. *Environmental Health Perspectives, 114*(4), 567-572.

Feder, M.E., and Walser, J.C. (2005). The biological limitations of transcriptomics in elucidating stress and stress responses. *Journal of Evolutionary Biology, 18*(4), 901-910.

Flaherty, K.T., and Brose, M.S. (2006). Her-2 targeted therapy: Beyond breast cancer and trastuzumab. *Current Oncology Reports, 8*(2), 90-95.

Fletcher O., Gibson, L., Johnson, N., Altmann, D.R., Holly, J.M., Ashworth, A., Peto, J., and Silva Idos, S. (2005). Polymorphisms and circulating levels in the insulin-like growth factor system and risk of breast cancer: A systematic review. *Cancer Epidemiology Biomarkers & Prevention, 14*(1), 2-19.

Fong, L.Y., Cheung, T., and Ho, Y.S. (1988). Effect of nutritional zinc-deficiency on O6-alkylguanine-DNA-methyl-transferase activities in rat tissues. *Cancer Letters, 42*(3), 217-223.

Gamet-Payrastre, L. (2006). Signaling pathways and intracellular targets of sulforaphane mediating cell cycle arrest and apoptosis. *Current Cancer Drug Targets, 6*(2), 135-145.

Gikas, P.D., and Mokbel, K. (2005). Phytoestrogens and the risk of breast cancer: A review of the literature. *International Journal of Fertility in Women's Medicine, 50*(6), 250-258.

Giovannucci, E. (1999). Insulin-like growth factor-I and binding protein-3 and risk of cancer. *Hormone Research, 51*(Suppl. 3), 34-41.

Gonzalez Burchard, E., Borrell, L.N., Choudhry, S., Naqvi, M., Tsai, H.J., Rodriguez-Santana, J.R., Chapela, R., Rogers, S.D., Mei, R., Rodriguez-Cintron, W., Arena, J.F., Kittles, R., Perez-Stable, E.J., Ziv, E., and Risch, N. (2005). Latino populations: A unique opportunity for the study of race, genetics, and social environment in epidemiological research. *American Journal of Public Health, 95*(12), 2161-2168.

Goulding, M.R. (2003). Public health and aging: Trends in aging, United States and worldwide. *Morbidity and Mortality Weekly Report, 52*(6), 101-106.

Grimble, R.F., Howell, W.M., O'Reilly G., Turner, S.J., Markovic, O., Hirrell, S., East, J.M., and Calder, P.C. (2002) The ability of fish oil to suppress tumor necrosis factor α production by peripheral blood mononuclear cells in healthy men is associated with polymorphisms in genes that influence tumor necrosis factor α production. *American Journal of Clinical Nutrition, 76,* 454-459.

Harland, J. (2005). Nutrition and genetics: Mapping individual health. *ILSI Europe Concise Monograph Series.*

Hegele, R.A., Jugenberg, M., Connelly, P.W., and Jenkins, D.J.A. (1997). Evidence for gene-diet interaction in the response of blood pressure to dietary fibre. *Nutrition Research, 17,* 1229-1238.

Holick, M.F. (2006). Vitamin D: Its role in cancer prevention and treatment. *Progress in Biophysics and Molecular Biology, 52,* 49-59.

Jenab, M., Ferrari, P., Mazuir, M., Tjonneland, A., Clavel-Chapelon, F., Linseisen, J., Trichopoulou, A., Tumino, R., Bueno-de-Mesquita, H.B., Lund, E., Gonzalez, C.A., Johansson, G., Key, T.J., and Riboli, E. (2005). European prospective investigation into cancer and nutrition (EPIC) study. Variations in lycopene blood levels and tomato consumption across European countries based on the European prospective investigation into cancer and nutrition (EPIC) study. *Journal of Nutrition, 135*(8), 2032S-2036S.

Kaput, J. Ordovas, J.M., Ferguson, L., van Ommen, B., Rodriguez, R.L., Allen, L., Ames, B.N., Dawson, K., German, B., Krauss, R., Malyj, W., Archer, M.C., Barnes, S., Bartholomew, A., Birk, R., van Bladeren, P., Bradford, K.J., Brown, K.H., Caetano, R., Castle, D., Chadwick, R., Clarke, S., Clement, K., Cooney, C.A., Corella, D., Manica da Cruz, I.B., Daniel, H., Duster, T., Ebbesson, S.O.E., Elliott, R., Fairweather-Tait, S., Felton, J., Fenech, M., Finley, J.W., Fogg-Johnson, N., Gill-Garrison, R., Gibney, M.J., Gillies, P.J., Gustafsson, J.A., Hartman, J.L. IV, He, L., Hwang, J.-K., Jais, J.-P., Jang, Y., Joost, H., Junien, C., Kanter, M., Kibbe, W.A., Koletzko, B., Korf, B.R., Kornman, K., Krempin, D.W., Langin, D., Lauren, D.R., Lee, J.H., Leveille, G.A., Lin, S.-J., Mathers, J., Mayne, M., McNabb, W., Milner, J.A., Morgan, P., Muller, M., Nikolsky, Y., van der Ouderaa, F., Park, T., Pensel, N., Perez-Jimenez, F., Poutanen, K., Roberts, M., Saris, W.H.M., Schuster, G., Shelling, A.N., Simopoulos, A.P., Southon, S., Tai, E.S., Towne, B., Trayhurn, P., Uauy, R., Visek, W.J., Warden, C., Weiss, R., Wiencke, J., Winkler, J., Wolff, G.L., Zhao-Wilson, X., and Zucker, J.-D. (2005). The case for strategic international alliances to harness nutritional genomics for public and personal health. *British Journal of Nutrition, 94*, 623-632.

Kidd, L.C., Paltoo, D.N., Wang, S., Chen, W., Akereyeni, F., Isaacs, W., Ahaghotu, C., and Kittles, R. (2005). Sequence variation within the 5' regulatory regions of the vitamin D binding protein and receptor genes and prostate cancer risk. *Prostate, 64*(3), 272-282.

Kohlmeier, L. (1995). Future of dietary exposure assessment. *American Journal of Clinical Nutrition, 61*(3 Suppl.), 702S-709S.

Krieger, N. (2005). Stormy weather: Race, gene expression, and the science of health disparities. *American Journal of Public Health, 95*(12), 2155-2160.

Kyrtopoulos, S. (2006). Biomarkers in environmental carcinogenesis research: Striving for a new momentum. *Toxicology Letters, 162*(1), 3-15.

Low, Y.L., Grace, P.B., Dowsett, M., Scollen, S., Dunning, A.M., Mulligan, A.A., Welch, A.A., Luben, R.N., Khaw, K.T., Day, N.E., Wareham, N.J., and Bingham, S.A. (2005). Phytoestrogen exposure correlation with plasma estradiol in postmenopausal women in European Prospective Investigation of Cancer and Nutrition-Norfolk may involve diet-gene interactions. *Cancer Epidemiology Biomarkers & Prevention, 14*(1), 213-220.

Luan, J., Browne, P.O., Harding, A.H., Halsall, D.J., O'Rahilly, S., Chatterjee, V.K., and Wareham, N.J. (2001). Evidence for gene-nutrient interaction at the PPARgamma locus. *Diabetes, 50*(3), 686-689.

Manolio, T.A., Bailey-Wilson, J.E., and Collins, F.S. (2006). Genes, environment, and the value of prospective cohort studies. *Nature Reviews Genetics, 7*(10), 812-820.

Marks, L.S., Kojima, M., Demarzo, A., Heber, D., Bostwick, D.G., Qian, J., Dorey, F.J., Veltri, R.W., Mohler, J.L., and Partin, A.W. (2004). Prostate cancer in native Japanese and Japanese-American men: Effects of dietary differences on prostate tissue. *Urology, 64*(4), 765-771.

Masuda, M., Suzui, M., Lim, J.T., and Weinstein, I.B. (2003). Epigallocatechin-3-gallate inhibits activation of HER-2/neu and downstream signaling pathways in human head and neck and breast carcinoma cells. *Clinical Cancer Research, 9*(9), 3486-3491.

Mathers, J. (2004). Chairman's introduction: What can we expect to learn from genomics? *Proceedings of the Nutrition Society, 63*, 1-4.

Menendez, J.A., Lupu, R., and Colomer, R. (2005). Exogenous supplementation with omega-3 polyunsaturated fatty acid docosahexaenoic acid (DHA; 22:6n-3) synergistically enhances taxane cytotoxicity and downregulates Her-2/neu (c-erbB-2) oncogene expression in human breast cancer cells. *European Journal of Cancer Prevention*, 14(3), 263-270.

Milner, J.A. (2003). Incorporating basic nutrition science into health interventions for cancer prevention. *Journal of Nutrition*, 133(11 Suppl. 1), 3820S-3826S.

Mitchell, B.L., Yasui, Y., Lampe, J.W., Gafken, P.R., and Lampe, P.D. (2005). Evaluation of matrix-assisted laser desorption/ionization-time of flight mass spectrometry proteomic profiling: Identification of alpha 2-HS glycoprotein B-chain as a biomarker of diet. *Proteomics*, 5(8), 2238-2246.

Motulsky, A.G. (1999). If I had a gene test, what would I have and who would I tell? *Lancet*, 354(Suppl. 1), Si35-Si37.

National Institutes of Health Consensus Development Panel. (2001). National Institutes of Health Consensus Development Conference Statement: Phenylketonuria, screening and management, October 16-18, 2000. *Pediatrics*, 108(4), 972-982.

Nishida, C., Uauy, R., Kumanyika, S., and Shetty, P. (2004). The joint WHO/FAO expert consultation on diet, nutrition, and the prevention of chronic diseases: Process, product, and policy implications. *Public Health Nutrition*, 7(1A), 245-250.

Organisation for Economic Co-operation and Development (OECD). (2005). *OECD Health Data: How does the United States compare?* Available: http://www.oecd.org/health/healthdata [accessed May 8, 2006].

Ordovas, J.M., Lopez-Miranda, J., Mata, P., Perez-Jimenez, F., Lichtenstein, A.H., and Schaefer, E.J. (1995). Gene-diet interaction in determining plasma lipid response to dietary intervention. *Atherosclerosis*, 118, S11-S27.

Ordovas, J.M. (2006). Genetic interactions with diet influence the risk of cardiovascular disease. *American Journal of Clinical Nutrition*, 83(2), 443S-446S.

Paeratakul, S., Popkin, B.M., Kohlmeier, L., Hertz-Picciotto, I., Guo, X., and Edwards, L.J. (1998). Measurement error in dietary data: Implications for the epidemiologic study of the diet-disease relationship. *European Journal of Clinical Nutrition*, 52(10), 722-727.

Perera, F.P. (1994). Biomarkers and molecular epidemiology in mutation/cancer research. *Mutation Research*, 313, 117-129.

Prentice, R.L., Caan, B., Chlebowski, R.T., Patterson, R., Kuller, L.H., Ockene, J.K., Margolis, K.L., Limacher, M.C., Manson, J.E., Parker, L.M., Paskett, E., Phillips, L., Robbins, J., Rossouw, J.E., Sarto, G.E., Shikany, J.M., Stefanick, M.L., Thomson, C.A., Van Horn, L., Vitolins, M.Z., Wactawski-Wende, J., Wallace, R.B., Wassertheil-Smoller, S., Whitlock, E., Yano, K., Adams-Campbell, L., Anderson, G.L., Assaf, A.R., Beresford, S.A., Black, H.R., Brunner, R.L., Brzyski, R.G., Ford, L., Gass, M., Hays, J., Heber, D., Heiss, G., Hendrix, S.L., Hsia, J., Hubbell, F.A., Jackson, R.D., Johnson, K.C., Kotchen, J.M., LaCroix, A.Z., Lane, D.S., Langer, R.D., Lasser, N.L., and Henderson, M.M. (2006). Low-fat dietary pattern and risk of invasive breast cancer: The Women's Health Initiative Randomized Controlled Dietary Modification Trial. *Journal of the American Medical Association*, 295(6), 629-642.

Rankinen, T., and Bouchard, C. (2006). Genetics of food intake and eating behavior phenotypes in humans. *Annual Review of Nutrition*, 26, 16.1-16.23.

Rapuri, P.B., Gallagher, J.C., Kinyamu, H.K., and Ryschon, K.L. (2001). Caffeine intake increases the rate of bone loss in elderly women and interacts with vitamin D receptor genotypes. *American Journal of Clinical Nutrition*, 74, 694-700.

Redon, R., Ishikawa, S., Fitch, K.R., Feuk, L., Perry, G.H., Andrews, T.D., Fiegler, H., Shapero, M.H., Carson, A.R., Chen, W., Cho, E.K., Dallaire, S., Freeman, J.L., González, J.R., Gratacòs, M., Huang, J., Kalaitzopoulos, D., Komura, D., MacDonald, J.R., Marshall, C.R., Mei, R., Montgomery, L., Nishimura, K., Okamura, K., Shen, F., Somerville, M.J., Tchinda, J., Valsesia, A., Woodwark, C., Yang, F., Zhang, J., Zerjal, T., Zhang, J., Armengol, L., Conrad, D.F., Estivill, X., Tyler-Smith, C., Carter, N.P., Aburatani, H., Lee, C., Jones, K.W., Scherer, S.W., and Hurles, M.E. (2006). Global variation in copy number in the human genome. *Nature, 444,* 444-454.

Reinwald, S., and Weaver, C.M. (2006). Soy isoflavones and bone health: A double-edged sword? *Journal of Natural Products, 69*(3), 450-459.

Ross, S.A. (2003). Diet and DNA methylation interactions in cancer prevention. *Annals of the New York Academy of Sciences, 983,* 197-207.

Schatzkin, A., and Gail, M. (2002). The promise and peril of surrogate end points in cancer research. *Nature Reviews Genetics, 2,* 1-9.

Scheuner, M.T. (2001). Genetic predisposition to coronary artery disease. *Current Opinion in Cardiology, 16*(4), 251-260.

Singer, P.A., and Daar, D.A. (2001). Harnessing genomics and biotechnology to improve global health equity. *Science, 294,* 87-89.

Slattery, M.L., Neuahusen, S.L., Hoffman, M., Caan, B., Curtin, K., Ma, K.N., and Samowitz W. (2004). Dietary calcium, vitamin D, VDR genotypes, and colorectal cancer. *International Journal of Cancer, 111,* 750-756.

Stover, P. (2006). Influence of human genetic variation on nutritional requirements. *American Journal of Nutrition, 83*(Suppl.), 436S-442S.

Thomas, J.C., Irwin, D.E., Shaugnessy Zuiker, E., and Millikan, R.C. (2005). Genomics and the public health code of ethics. *American Journal of Public Health, 95*(12), 2139-2143.

Thomas, T., Gallo, M.A., and Thomas, T.J. (2004). Estrogen receptors as targets for drug development for breast cancer, osteoporosis, and cardiovascular diseases. *Current Cancer Drug Targets, 4*(6), 483-499.

Trujillo, E., Davis, C., and Milner, J. (2006). Nutrigenomics, proteomics, metabolomics, and the practice of dietetics. *Journal of the American Dietetic Association, 106,* 403-413.

Uusitupa, M.I., Ruuskanen, E., Makinen, E., Laitinen, J., Toskala, E., Kervinen, K., and Kesaniemi, Y.A. (1992). A controlled study on the effect of beta-glucan-rich oat bran on serum lipids in hypercholesterolemic subjects: Relation to apolipoprotein E phenotype. *Journal of the American College of Nutrition, 11*(6), 651-659.

van Erk, M.J., Blom, W.A.M., van Ommen, B., and Hendriks, H.F.J. (2006). High-protein and high-carbohydrate breakfasts differentially change the transcriptome of human blood cells. *American Journal of Clinical Nutrition, 84,* 1233-1241.

Viguerie, N., Poitou, C., Cancello, R., Stich, V., Clement, K., and Langin, D. (2005, January). Transcriptomics applied to obesity and caloric restriction. *La biochimie, 87*(1), 117-123.

Warren, A.J., and Shields, P.G. (1997). Molecular epidemiology: Carcinogen-DNA adducts and genetic susceptibility. *Proceedings of the Society for Experimental Biology and Medicine, 216,* 172-180.

Way, T.D., Kao, M.C., and Lin, J.K. (2004). Apigenin induces apoptosis through proteasomal degradation of HER2/neu in HER2/neu-overexpressing breast cancer cells via the phosphatidylinositol 3-kinase/Akt-dependent pathway. *Journal of Biological Chemistry, 279*(6), 4479-4489.

Wiedmann, M.W., and Caca, K. (2005). Molecularly targeted therapy for gastrointestinal cancer. *Current Cancer Drug Targets, 5*(3), 171-193.

Wong, H.L., Seow, A., Arakawa, K., Lee, H.P., Yu, M.C., and Ingles, S.A. (2003). Vitamin D receptor start codon polymorphism and colorectal cancer risk: Effect modification by dietary calcium and fat in Singapore Chinese. *Carcinogenesis, 24,* 1091-1095.

Wooding, S., Kim, U.K., Bamshad, M.J., Larsen, J., Jorde, L.B., and Drayna, D. (2004). Natural selection and molecular evolution in PTC, a bitter-taste receptor gene. *American Journal of Human Genetics, 74*(4), 637-646.

World Cancer Research Fund and American Institute for Cancer Research. (1997). *Food, nutrition, and the prevention of cancer: A global perspective.* Washington, DC: American Institute for Cancer Research.

World Health Organization. (2002). *Genomics and world health.* Geneva, Switzerland: Author.

World Health Organization. (2003). Diet, nutrition, and the prevention of chronic diseases: Report of a joint WHO/FAO consultation. *WHO Technical Report Series,* 916. Available: http://whqlibdoc.who.int/trs/WHO_TRS_916.pdf [accessed May 5, 2006].

World Health Organization. (2004). *Global strategy on diet, physical activity, and health.* Available: http://www.who.int/dietphysicalactivity/strategy/eb11344/strategy_english_web.pdf [accessed April 28, 2006].

World Health Organization. (2005). *Preventing chronic diseases: A vital investment.* Available: http://www.who.int/chp/chronic_disease_report/en/index.html [accessed July 22, 2005].

Yang, C.S., Lee, M.J., and Chen, L. (1999). Human salivary tea catechin levels and catechin esterase activities: Implication in human cancer prevention studies. *Cancer Epidemiology Biomarkers & Prevention, 8*(1), 83-89.

Zhou, J.R., Yu, L., Zhong, Y., and Blackburn, G.L. (2003). Soy phytochemicals and tea bioactive components synergistically inhibit androgen-sensitive human prostate tumors in mice. *Journal of Nutrition, 133*(2), 516-521.

15

Genoeconomics

*Daniel J. Benjamin, Christopher F. Chabris, Edward L. Glaeser,
Vilmundur Gudnason, Tamara B. Harris, David I. Laibson,
Lenore J. Launer,* and *Shaun Purcell*

Since the work of Taubman (1976), twin studies have identified a significant degree of heritability for income, education, and many other economic phenotypes (e.g., Behrman, Hrubec, Taubman, and Wales, 1980; Behrman and Taubman, 1989). These studies estimate heritability by contrasting the correlation of economic phenotypes in monozygotic (identical) twin pairs and dizygotic (fraternal) twin pairs. Recent improvements in the technology of studying the human genome will enable social scientists to expand the study of heritability, by incorporating molecular information about variation in individual genes. This essay describes our hopes and concerns about the new research frontier of genomic economics, or genoeconomics.

The core theme of health economics is that individual behavior and social institutions influence health outcomes (Fuchs, 1974). The primary contribution of genoeconomics is likely to be identifying the many ways in which individual behavior and social institutions moderate or amplify genetic differences.

Within genoeconomics, there will be at least three major types of conceptual contributions. First, economics can contribute a theoretical and empirical framework for understanding how market forces and behavioral responses mediate the influence of genetic factors. Second, incorporating genetics into economic analysis can help economists identify and measure important causal pathways (which may or may not be genetic). Finally, economics can aid in analyzing the policy issues raised by genetic information.

Smoking provides one example of economic analysis that can improve the study of how genetic variation influences phenotypic variation. Traditional heritability studies suggest at least some genetic component to lung cancer (Lichtenstein et al., 2000); molecular genetics identifies a locus of lung cancer susceptibility on chromosome 6q23-25 (Bailey-Wilson et al., 2004). The genetic susceptibility to lung cancer is undoubtedly amplified by cigarette smoking, an economic decision affected by advertising, social norms, cigarette prices, consumer income, and tax rates on cigarettes (Cutler and Glaeser, 2005). Economics can explain how social institutions—like the market for cigarettes—interact with genes to jointly generate important health phenotypes like lung cancer.

More generally, economic institutions may either reduce or amplify the inequalities produced by genetic variation. In some situations, social transfers partially offset genetic factors—for example, when individuals with illness receive extra insurance-based resources to treat or manage their illness. The second subfield uses genetic information to identify causal mechanisms. This subfield will recognize a central fact of empirical economics: the ubiquity of mutual causation—for example, health influences wealth and vice versa (Case, Lubotsky, and Paxson, 2002). Genetic measures can help to separate the causal effect in a particular direction.

For example, a robust literature argues that height, even in adolescence, increases earnings (Persico, Postlewaite, and Silverman, 2004). However, this literature is plagued by difficulty in controlling for the fact that height also reflects better health and nutrition in wealthier families. If height-linked alleles were identified, then they could, in principle, be used to measure the causal impact of exogenous variation in height. More formally, such research would analyze allele variation across siblings to identify the causal effect of genetic predispositions for height (controlling for household background characteristics). To take another example, Ding, Lehrer, Rosenquist, and Audrain-McGovern (2005) address the causal effect of health on educational outcomes, using genetic predictors of health to ameliorate confounding by third factors potentially correlated with both health and educational outcomes.

More generally, cognition-linked alleles will contribute to understanding of the cognitive factors that influence income, or the extent to which cognitive factors influence decision making about savings and wealth. Genetic research will also identify biological mechanisms that interact with environmental factors to jointly influence behavior. We anticipate that crude concepts like "risk aversion" (unwillingness to take risks) and "patience" (willingness to delay gratification) that are central to economic analyses will be decomposed into much more useful subcomponents associated with particular neural mechanisms and their environmental and genetic antecedents (Plomin, Corley, Caspi, Fulker, and DeFries, 1998).

Finally, ongoing research will eventually enable researchers to employ new genetic control variables, thereby improving the power of statistical procedures.

Much of the promise of genoeconomics is based in part on economists' long tradition of policy analysis. The economic approach is one in which governments are not seen as infallible custodians of the public good, but rather as separate actors that often have their own objectives (Stigler, 1971). Information economics may also play an important role in the analysis of policy questions. Economists have identified competitive forces that cause individuals to reveal information that is privately beneficial but potentially socially harmful. Economists understand how the public release of certain genetic information can theoretically undermine insurance institutions and thereby inefficiently increase social inequality. Genoeconomics will also identify specific gene-environment interactions with policy implications. For example, imagine that particular genes turn out to be associated with risk factors for poor educational outcomes, poor performance in the labor market, and consequently low levels of income. Imagine too that particular educational interventions are found that mitigate these disadvantages. Then gene-based policies could target disadvantaged at-risk groups with focused interventions. Such interventions will remain purely speculative until the necessary precursor research is implemented and ethical questions are resolved, but focused interventions nevertheless hold considerable long-run potential.

Despite the promise of genoeconomics, there are clearly enormous pitfalls. Even under the best of circumstances—in which a particular genetic pathway has been clearly established—there are concerns about informing individuals of their own risks, especially when there are few interventions to alleviate those risks or when the risks are very small. Providing information to parents about the genome of a fetus or a child creates a different set of dilemmas, including the risk of selective abortion. This has been well discussed with reference to a genetic endowment as straightforward as gender; in many societies economic investment in a daughter is seen as less beneficial than economic investment in a son (e.g., Garg and Morduch, 1998). If the same issues arose in relation to more complex economic traits, a host of ethical and policy questions would arise. Documenting the power of the genome to society at large also creates risks as identifiable social and ethnic groups may face discrimination (or become beneficiaries of positive discrimination) on the basis of their presumed genetic endowments.

These problems are multiplied when genetic research is done carelessly. Historically, there have been many cases of false positives in which early genetic claims have evaporated under subsequent attempts at replication. These false positives can create tremendous mischief. A failure to highlight the small contribution each gene may make to an outcome, as well as the full extent of the interaction between genes and environment, is also likewise dangerous because the public may come to believe falsely in genetic determinism. The responsible path requires statistical care, attention to how genes and environment jointly determine outcomes, and extreme sensitivity to the ethical issues surrounding genetic knowledge.

Despite these dangers, we think that there is potential for productive collaboration between economists, cognitive scientists, epidemiologists, and genetic researchers. In the rest of this essay, we sketch one vision for this field. In the next section, we discuss methodological challenges that confront research in genoeconomics. We then outline a study that is currently under way, which uses a single-nucleotide polymorphism (SNP) panel to analyze associations between candidate cognitive function genes and economic phenotypes.

METHODOLOGICAL CHALLENGES AND PITFALLS

Successful implementation of the research program described above will require careful attention to many methodological issues, some of which we outline in this section. A critical issue is the choice of economic phenotypes to study. Proximal behavioral phenotypes, such as impatience or risk aversion, are probably more directly related to genetic propensities than more distal economic phenotypes, such as wealth accumulation or labor force participation.

Proximal phenotypes have typically been measured with personality tests. Some personality systems are purely conceptually based (e.g., the five factor model) while others are rooted in neurobiology (e.g., Cloninger's three dimensions tied to the dopamine, serotonin, and norepinephrine systems; Cloninger, 1987, 1993; Cloninger, Adolfsson, and Svrakic, 1996). Recently, some personality attributes have been studied with neuroimaging (e.g., Hariri et al., 2006).

Distal phenotypes—for example, wealth accumulated over a lifetime—may also strongly reflect genetic influences, because they represent the cumulative effect of many specific decisions, and may reflect the expression of genes over a long period of time. Given the current state of knowledge (especially the relative lack of definitive findings relating traditional personality traits to specific genetic polymorphisms; see Ebstein, 2006; Munafo et al., 2003), the wisest course is probably to measure both proximal and distal phenotypes and to investigate how the proximal phenotypes mediate the relationship between genes and more distal phenotypes.

In the rest of this section, we focus on gene-environment interaction studies in the context of quantitative genetic designs and modern association analysis. In that setting we consider issues under three general headings: the nonindependence of genes and environments, the measurement of genetic variation, and problems searching for small, complex effects.

Correlated Genes and Environments

Genes and environments are, for various reasons, often not independent factors. This has implications for statistical designs attempting to uncover genetic influences, environmental influences, and the interactions of genes and environments.

Gene-environment interaction (G×E) can be conceptualized as the genetic control of *sensitivity* to different environments. In contrast, a correlation between genes and environment (GE correlation, rGE) can represent genetic control of *exposure* to different environments (Kendler and Eaves, 1986; Plomin and Bergeman, 1991). For example, Jang, Vernon, and Livesley (2000) show that genetic influences on alcohol and drug misuse are correlated with various aspects of the family and school environment.

We might expect correlations between genes and environments to arise for a number of reasons. For example, individuals do, to some extent, implicitly select their own environments on the basis of innate, genetically influenced characteristics.

One important form of gene-environment correlation arises due to population stratification. A stratified sample is one that contains individuals from two or more subpopulations that may differ in allele frequencies at many sites across the genome. This will induce a correlation in the sample between all allelic variants that differ in frequency between the subpopulations and any environmental factors, diseases, or other measures that also happen to differ (possibly for entirely nongenetic reasons) between the subpopulations. As such, population stratification is an important source of potential confounding in population-based genetic studies. For example, if cases and controls are not matched for ethnic background, population stratification effects can lead to spurious association, or false-positive errors. To address concerns over possible hidden stratification effects, a series of family-based tests of association have been developed. Because related family members necessarily belong to the same population stratum, using relatives as controls automatically ensures protection against the effects of stratification (Spielman, McGinnis, and Ewens, 1993). Recently, a different approach—called genomic control or structured association—has emerged, using DNA markers from across the genome to directly infer ancestry for individuals in the sample or to look for signs of stratification (Devlin and Roeder, 1999; Pritchard, Stephens, and Donnelly, 2000).

An association between an environment and an outcome may arise due to a third variable, namely common genetic inheritance (e.g., DiLalla and Gottesman, 1991). For example, if a gene X is inherited, it might cause phenotypes Y and Z, respectively, in a parent and in a child. Researchers will observe a correlation between the parental phenotype Y and the child's phenotype Z. Researchers may mistakenly infer a causal relationship between Y and Z if they do not control for the real (unobserved) causal mechanism: gene X.

Measuring Genetic Variation

The typical "gene by environment" association study should really be called an "allele by environment" study because, very often, only a single variant within a gene is studied. In the context of standard candidate gene association studies, many researchers are realizing that failure to comprehensively measure all common variation in a gene or region can lead to inconsistent results and makes the interpretation of negative results particularly troublesome. (If you have not adequately measured "G," then it is hard to evaluate its relationship to the phenotype.) With emerging genomic technologies, it will soon be easy to measure myriad single nucleotide polymorphisms or microsatellite markers, even if only one SNP is known to be functional.

The same issue applies to G×E analysis. The question will be how to adapt G×E methods to this new "gene-based" paradigm, in which the gene rather than the specific allele, genotype, or haplotype becomes the central unit of analysis. In addition, if a researcher measures multiple genes (for example, all genes in a pathway, each with multiple markers), then new analytic approaches will be needed to simultaneously model the joint action of the pathway, as well as how the individual genes influence the phenotype or interact with the environment.

Naturally, more comprehensively measuring all common variation in a gene costs more both financially (more genotyping) and statistically (more tests are performed). How to best combine information from multiple markers in a given region is an ongoing issue in statistical genetics. One option is to simply test each variant individually and then adjust the significance levels to account for this multiple testing. Standard procedures, such as the Bonferroni, are typically too conservative because they assume the tests are independent. Instead, it is often better to use permutation procedures to control the family-wise error rate or to control the false discovery rate. A second option is to combine the single variants together, either in a multilocus test (such as Hotelling's T^2 or a set-based test using sum-statistics) or in a haplotype-based test. As mentioned above, this is currently a very active area of research (e.g., Brookes, Chen, Xu, Taylor, and Asherson, 2006).

All these approaches rely on the variation being common. Even for large samples, this means that variants with a population frequency of less than 1 percent are unlikely to be detected. If a gene is important for a given outcome but contains multiple, different rare variants, then many current approaches will fail.

Searching for Small Effects and Interactions

Increasingly, researchers are appreciating the central importance of large sample sizes in genetics to afford sufficient statistical power to detect small effects. For complex, multifactorial traits, many researchers expect the effects of individual variants to be as low as < 1 percent of the total phenotypic variance for quantitative outcomes. For case-control designs, allelic odds ratios of 1.2 and lower are often considered. Such small effects require very large samples—typically thousands of individuals, if more than one variant is to be tested and proper controls for multiple testing are in place. The consequences of chronic low statistical power are sobering. If power is on average only marginally greater than the type I error rate, then a large number of published studies may well be type I errors. Average power around the 50 percent level yields a pattern of inconsistent replication. A great deal of time and money has been spent on poorly designed experiments that, at best, stand little chance of doing what they are supposed to and, at worst, are advancing type I errors in the literature.

Although the individual effects of any one variant may be very small, it is of course a possibility that this is because they represent the marginal effect of an interaction, for example with some environmental factor. In other words, by looking only at a single variant and essentially averaging over all other interacting environmental factors, one would see only an attenuated signal and perhaps miss the link between the gene, environment, and outcome. This is one reason for explicitly considering G×E when searching for genetic variants.

In humans, G×E has been found in monogenic diseases; in plant and animal genetics, there is strong evidence for G×E in complex phenotypes. For example, phenylketonuria is a Mendelian human disorder, but the gene acts to produce the severe symptoms of mental retardation only in the presence of dietary phenylalanine. Research in Drosophila melanogaster has found evidence for G×E in quantitative traits including bristle number, longevity, and wing shape (Mackay, 2001; Clare and Luckinbill, 1985). The detection of G×E in model organisms suggests that it will play an equally important role in complex human phenotypes. Indeed, promising results are emerging (e.g., Caspi et al., 2002, 2003; Dick et al., 2006a; MacDonald, Perkins, Jodouin, and Walker, 2002; Mucci, Wedren, Tamini, Trichopoulos, and Adami, 2001). However, human studies suffer from a crucial methodological difference: the inability to inexpensively manipulate genes and environments experimentally. Epidemiological designs will therefore tend to be less powerful, as well as prone to confounding. Despite these greater challenges, consideration of G×E in human molecular genetic studies potentially offers a number of rewards, including increased power to map genes, to identify high-risk individuals, and to elucidate biological pathways.

Many commentators have noted the general difficulties faced in uncovering interactions of any kind (e.g., Clayton and McKeigue, 2001; Cooper, 2003). Indeed, general epidemiology has struggled for decades to adequately define and test interaction. The central problem, as stated by Fisher and Mackenzie in 1923 when first describing the factorial design and analysis of variance (ANOVA), is that, in statistical terms, "interaction" is simply whatever is left over after the main effects are removed. It follows that the presence or absence of interaction can depend on how the main effects are defined. For dichotomous phenotypes, the presence of a measured interaction effect will depend on the modeling assumption that is used in the empirical analysis (see Campbell, Gatto, and Schwartz, 2005, for another example). For example, if the risk genotype G+ has (likelihood ratio) effect g and the risk environment E+ has (likelihood ratio) effect e, the question is how to specify the joint effect *in the absence of an interaction*. Assuming an additive model implies that the joint effect (without an interaction effect) is $g + e - 1$, whereas a multiplicative model implies that the joint effect (without an interaction effect) is ge. Hence, the absence of an interaction effect in the additive model generically implies the existence of an interaction effect in the multiplicative model (and vice versa). Mathematically, as long as neither g nor e is equal to one, then, $g + e - 1 \neq ge$.

Analogously, for quantitative phenotypes, transformation of scale can induce or remove interaction effects. To see this, imagine a G×E study of amygdala morphology (i.e., measures of the anatomical size of the amygdala based on magnetic resonance images). For illustrative purposes, assume that the amygdala is a sphere with radius given by an additive sum of a gene effect—1 mm—and an environment effect—also 1 mm. Assume too that the radius exhibits no gene-environment interaction.

If the measured phenotype were cross-sectional area (a function of radius squared), however, gene and environment are no longer additive in their effects. There is now G×E, as G+ increases area by 3 units under E- and 5 units under E+. If the phenotype were based on volume, the apparent measurement of G×E is stronger. However, these interaction effects are purely "statistical" and not "biological": that is, G and E do not interact on any causal level. The interactions are effectively a consequence of misspecifying the main effects model (see Table 15-1).

Consider now that a "downstream" phenotype is measured, such as some aspect of the serotonergic system that is influenced by the amygdala. There can be no guarantee that the effects of G and E should necessarily display an additive relationship at this level, considering the various neurochemical cascades and reciprocal feedback loops that are presumably involved in a system as complex as the human brain. Or the measured phenotype may be even further downstream—a clinical diagnosis based on behavioral symptoms, or a 25-item self-report questionnaire measure, log-transformed to approximate normality. Finding G×E at these levels may well be strikingly irrelevant with respect to the presence of interaction at the causal level.

The point of this example is not to claim that the only appropriate causal level is the neurological one. Rather, for complex phenotypes, the level at which genes and environment operate (which need not be the same level) might often be quite distal compared with the level of measured phenotype. Consequently, the distinction between statistical and biological interaction always should be borne in mind. Purely statistical interactions are still useful if one's only goal is prediction, for example, early diagnosis or identification of high-risk individuals. But to help understand mechanisms and pathways, an interaction detected by statistical methods must have some causal, biological, or behavioral counterpart to be of significant interest.

False negatives are also a major concern in the study of G×E. Tests of interaction generally suffer from relatively low power (Wahlsten, 1990). In this case, it is not clear that efforts to detect genes will benefit from

TABLE 15-1 Measurement of G×E Depends on the Modality of Measurement

Radius (mm)		E-	E+
	G-	1	2
	G+	2	3
Area/π (mm^2)		E-	E+
	G-	1	4
	G+	4	9
Volume/$(4\pi/3)$ (mm^3)		E-	E+
	G-	1	8
	G+	8	27

more complex models that allow for potential G×E effects, even if G×E effects are large.

Nature is undoubtedly complex. How complex our statistical models need to be is less clear. Combining the definitional problems of interaction with the low power to detect G×E with the new avenues for multiple-testing abuses brought about by extra E variables, attempting to incorporate G×E could make an already difficult endeavor nearly impossible (Cooper, 2003). However, we see these obstacles as important but not insurmountable: with proper experimental design and better developed statistical tools, G×E will be able to be robustly detected, with relevance to biology, public health, and eventually economics.

Although larger data sets—more individuals, more phenotypic measures, more genetic variants assayed—are desirable for many reasons (some of which have already been mentioned), they also pose a further methodological challenge for detecting G×E. A new wave of whole genome scale studies has already begun, in which as many as half a million SNPs are assayed. Issues of multiple testing and statistical power are already paramount in such studies. Efforts to detect G×E magnify these concerns.

AGES-REYKJAVIK STUDY COLLABORATION

Currently, the main obstacle to bringing genetic research into economics is the fact that few data sets combine economic measures with biosamples that can be genotyped. An exception is the Age, Gene/Environment Susceptibility-Reykjavik Study (AGES-RS). On the basis of the AGES-RS, we are currently exploring associations between candidate genes involved in decision making and economic phenotypes and how these relationships are mediated by the environment. We think our project illustrates one possible direction for research in economic genomics, as well as some of the benefits of multidisciplinary collaboration—including team members with training in economics, cognitive science, epidemiology, medicine, genetics, and statistics.

Administered by the Icelandic Heart Association, the original Reykjavik Study (RS) surveyed 30,795 men and women born between 1907 and 1935 who lived in Reykjavik as of 1967. While the majority of participants were surveyed once between 1967 and 1991, about 5,700 were surveyed twice and as many as 6,000 people were surveyed up to six times over this period. The Older Persons Examination, which contained many components of the RS questionnaire as well as additional health measures, was administered between 1991 and 1997 to a subset of the Reykjavik Study that was ages 70 and older as of 1991. The Laboratory of Epidemiology, Demography, and Biometry of the National Institute on Aging initiated the AGES-RS in 2002 in collaboration with the Icelandic Heart Association to collect genotypic as well as additional phenotypic data from surviving participants of the Reykjavik Study. The AGES-RS includes 5,764 of the 11,549 surviving participants. Currently, 2,300 participants have been genotyped. For more detailed information about the AGES-RS, see Harris et al. (in press).

Although primarily used to study health, the AGES-RS data already contain a number of measures of economic interest, summarized in Table 15-2. Distal economic phenotypes we plan to study include labor supply and wealth accumulation. For example, Figure 15-1 shows the percentage of respondents who have a second job. Figure 15-2 shows the distribution of working hours in the sample. Notice that there is a substantial amount of variation in these phenotypes. The RS questionnaire asks about attributes of participants' house or apartment, from which it is possible to construct a proxy measure of housing wealth. We are currently investigating the feasibility of collecting more extensive measures of wealth and income.

In addition to these distal phenotypes, we plan to study proximal phenotypes—such as impulsiveness, risk aversion, and cognition—that may be more closely related to underlying genetic propensities. A measure of late-life general cognitive function can be constructed from existing data on memory, speed of processing, and working memory. Various questionnaires ask about health-related decisions, such as smoking, drinking, eating habits, and conscientious health behaviors (e.g., getting regular check-ups). Each of these decisions reflects a trade-off between the present and the future, and economic theory postulates that some individuals are more impulsive, or "impatient" in economics jargon. From these decisions, we will construct an index of impulsive behaviors.

We also plan to add standard experimental measures of impulsive and risk-averse preferences to the next wave of the AGES-Reykjavik Study. These protocols ask participants to choose between immediate and delayed monetary rewards or to choose between certain and risky monetary rewards. These choices are played out with real monetary stakes. Such measures correlate with real-world impulsive and risky decisions across a range of contexts (e.g., for discounting: Fuchs, 1982; Bickel, Odum, and Madden, 1999; Petry and Casarella, 1999; Kirby, Petry, and Bickel, 1999; Kirby and Petry, 2004; Ashraf, Karlan, and Yin, 2004; Shapiro, 2005; for risk aversion: Barsky, Juster, Kimball, and Shapiro, 1997; Dohmen et al., 2005; Kimball, Sahm, and Shapiro, 2006). Other experimental decision-making measures yield similar distributions of responses whether they are administered to neurologically healthy older adults or to college-age subjects (Kovalchik, Camerer, Grether, Plott, and Allman, 2003).

Existing research in economics implies that distal phenotypes, such

TABLE 15-2 Measured Phenotypes in the Icelandic AGES-RS Data

Measured Phenotypes	Reykjavik Study 1967-1991	Older Persons Exam 1991-1996	AGES-Reykjavik 2002-2006
Distal economic phenotypes			
Number of jobs and hours worked (labor supply)	X	X	
Attributes of house/ apartment (housing wealth)	X	X	
Occupational history (human capital accumulation)	X	X	X
Years of education (human capital accumulation)	X	X	X
Social networks (social capital accumulation)		X	X
Proximal decision-making phenotypes			
Smoking frequency (impulsivity)	X	X	X
Drinking frequency (impulsivity)		X	X
Exercise frequency (impulsivity)	X	X	X
Eating habits (impulsivity)		X	X
Health conscientiousness (impulsivity)	X	X	
Long-term memory (general cognitive ability)			X
Speed of processing (general cognitive ability)		X	X
Working memory (general cognitive ability)			X
MRI of the brain (general cognitive ability)			X

NOTES: This table displays phenotypic data already collected. For the next wave of the AGES-Reykjavik study, we plan to add additional distal phenotypes (wealth and income) and proximal phenotypes (experimental measures of impulsivity and risk aversion). The cognitive SNP panel will be administered to participants in the AGES-Reykjavik study. In addition to the AGES-Reykjavik questionnaire, participants in the AGES-Reykjavik study have answered the Reykjavik study questionnaire once, twice, or six times during 1967-1991. The Older Persons Exam was administered to those ages 70 and older as of 1991.

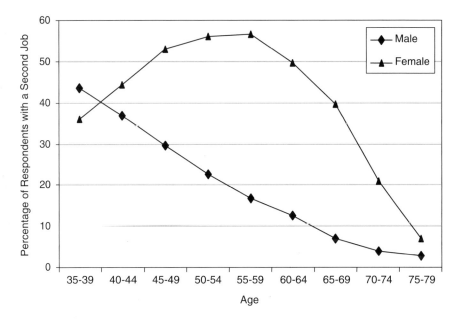

FIGURE 15-1 Percentage of respondents in the Icelandic AGES-RS data who have a second job, by gender and age.
SOURCE: Author's calculations.

as labor supply and wealth accumulation, will be related to proximal phenotypes that matter for decision making, such as impulsiveness, risk aversion, and cognitive function (Barsky et al., 1997; Benjamin, Brown, and Shapiro, 2006; Dohmen et al., 2005). These proximal phenotypes are more likely to be directly associated with underlying genetic propensities and to mediate the relationship between genetic polymorphisms and the distal phenotypes.

Three key empirical findings have motivated our choice of candidate genes for decision making:

1. Research in the new field of neuroeconomics (Glimcher and Rustichini, 2004; Glimcher, Dorris, and Bayer, 2005) has begun to explore the neuroscientific foundations of economic behavior.[1] McClure, Laibson, Loewenstein, and Cohen (2004) found that impulsive behavior, when measured with laboratory tasks, appears to be governed by the interaction between the brain's

---

[1]There is also a related, older literature that explores the relationship between personality and neuropharmacological interventions—for example, see Nelson and Cloninger (1997).

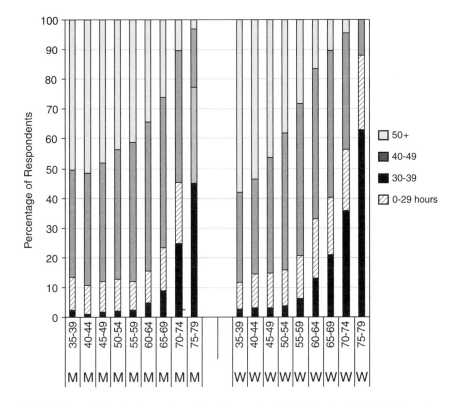

FIGURE 15-2 Distribution of working hours in the Icelandic AGES-RS data, by gender and age.
SOURCE: Author's calculations.

impatient "limbic system" (more accurately, mesolimbic dopa-minergic reward-related regions) and a patient "cortical system," which includes elements of the prefrontal cortex and the parietal cortex. McClure et al. (2004) show that the limbic system is active only when individuals are confronted with choices between immediate and future rewards. By contrast, the cortical system is active for all decisions (whether or not immediate rewards are among the choices), and its activity increases in trials when subjects choose more delayed rewards.

2. Individual differences in the tendency to make impulsive, present-oriented decisions may in part be associated with cognitive function. In both laboratory situations and real-world measures, a correlation has been found between high function and less impulsivity and being more risk-neutral across a variety of decision-

making domains (Benjamin, Brown, and Shapiro, 2006; see also Frederick, 2005), including financial choices, health behaviors, capital accumulation, and the like. Critically, this holds true even when controls for income are included.

3. Differences in cognitive function, in turn, may be mediated predominantly by structural and functional differences in prefrontal and parietal brain regions—the same network of cortical regions that operates to counter the impulsive tendencies of the limbic/reward system (Gray, Chabris, and Braver, 2003; Chabris, 2007).

These results lead us to the working hypothesis that prefrontal/parietal and limbic networks are the neural substrates of the psychological constructs of impulsiveness and cognitive function (that are in turn related to economic decision making). We therefore hypothesize that genes implicated in these traits and brain systems may be associated with economic behavior and outcomes in the AGES-RS data. We have developed a list of these genes and their known or likely functional SNPs (Table 15-3). An SNP panel will be created to rapidly genotype the 2,300 subjects who have already been genotyped with an extensive set of SNPs in the AGES-RS data. These new SNPs will include both functional alleles and SNPs to tag haplotypes of the genes, based on the HapMap.

To select genes for this SNP panel, we focused on specific phenotypes and biological pathways of relevance to the model sketched above. First, we selected genes in two critical neurotransmission pathways, the serotonin and dopamine systems, because both of these pathways have been associated with impulsive behavior. (It is true that these systems are not *exclusively* involved in impulsiveness or decision making in general—all genetic or neurobiological systems, including the putative "language gene" FOXP2, are involved in multiple cognitive and behavioral domains—but these provide useful starting points given the current state of knowledge about the neurobiology of decision making.) Serotonin function has been associated with several aspects of impulsivity, including reward sensitivity and inhibitory cognitive control (e.g., Cools et al., 2005; Walderhaug et al., 2002), as well as prefrontal cortex activity (Rubia et al., 2005), while several dopamine-related genes have been associated with attention deficit hyperactivity disorder (ADHD; see Faraone et al., 2005, for a meta-analysis of association studies) and with limbic/reward system functioning.

Second, we selected genes that have been associated or implicated in phenotypes related to cognitive functions: memory (e.g., de Quervain and Papassotiropoulos, 2006); schizophrenia, which involves neurocognitive dysfunction (Hallmayer et al., 2005); Alzheimer disease; and brain size

TABLE 15-3 Genes That Are Candidates for Inclusion in a Panel of SNPs for Association Studies with Cognitive, Neural, and Economic Phenotypes

Gene	Position	Description and References
Dopamine (DA) System		
TH	11p15.5	Tyrosine hydroxylase
DDC	7p12.2	Dopa decarboxylase
VMAT1	8p21.3	Vesicular monoamine transporter 1
VMAT2	10q25.3	Vesicular monoamine transporter 2
DRD1	5q35.1	Dopamine receptor 1
		ADHD (Bobb et al., 2005)
DRD2	11q23	Dopamine receptor 2
		Neural activation during working memory (Jacobsen et al., 2006)
		DRD2 binding in striatum (Hirvonen et al., 2004)
DRD3	3q13.3	Dopamine receptor 3
DRD4*	11p15.5	Dopamine receptor 4
		ADHD (Faraone et al., 2005)
DRD5	4p16.1	Dopamine receptor 5
		ADHD (Faraone et al., 2005)
CALCYON	10q26.3	Calcyon (DRD1 interacting protein)
		ADHD (Laurin et al., 2005)
DAT1*	5p15.3	Dopamine transporter
		ADHD (Faraone et al., 2005)
COMT	22q11.2	Catechol-o-methyltransferase
		Frontal lobe, executive function (Egan et al., 2001; Meyer-Lindberg et al., 2006)
MAOA*	Xp11.23	Monoamine oxidase A
		NEO personality traits (Rosenberg et al., 2006); aggression G×E interaction (Caspi et al., 2002)
MAOB	Xp11.23	Monoamine oxidase B
DBH	9q34.2	Dopamine beta hydroxylase
		ADHD (Faraone et al., 2005)
Serotonin (5-HT) System		
TPH1	11p15.3	Tryptophan hydroxylase 1
TPH2	12q21.1	Tryptophan hydroxylase 2
HTR1A		Serotonin receptor 1A
HTR1B	6q14.1	Serotonin receptor 1B
		ADHD (Faraone et al., 2005)
HTR2A	13q14.2	Serotonin receptor 2A
		Explicit memory (de Quervain et al., 2003; Papassotiropoulos et al., 2005a; Reynolds et al., 2006)

Continued

TABLE 15-3 Continued

Gene	Position	Description and References
HTR3A	11q23.1	Serotonin receptor 3A Amygdala and frontal lobe function (Iidaka et al., 2005)
HTT*	17q11.1	Serotonin transporter Amygdala function (Hariri et al., 2002) ADHD (Faraone et al., 2005) Cognitive aging (Payton et al., 2005) Under selection in CEU and ASN populations (Voight et al., 2006)

Genes Reported to Be Associated with General Cognitive Ability
(reviewed by Payton, 2006; Plomin et al., in press)

Gene	Position	Description and References
CBS	21q22.3	Cystathionine beta-synthase IQ (Barbaux et al., 2000)
CCKAR	4p15.2	Cholecystokinin A receptor IQ (Shimokata et al., 2005)
CHRM2	7q33	Muscarinic cholinergic receptor 2 IQ (Comings et al., 2003; Gosso et al., 2006) Performance IQ (Dick et al., 2006c)
CTSD	11p15.5	Cathepsin D Mental retardation and microcephaly caused by mutation (Siintola et al., 2006) IQ (Payton et al., 2003, 2006)
IGF2R	6q25.3	Insulin-like growth factor 2 receptor IQ (Chorney et al., 1998; Jirtle, 2005)
KLOTHO	13q13.1	Klotho IQ (Deary et al., 2005b)
MSX1	4p16.2	Muscle segment homeobox, drosophila, homolog of, 1 IQ (Fisher et al., 1999)
NCSTN	1q23.2	Nicastrin IQ (Deary et al., 2005a) AD (Bertram et al., 2007)
PLXNB3	Xq28	Plexin B3 Vocabulary, white matter (Rujescu et al., 2006)
PRNP	20p13	Prion protein IQ (Rujescu et al., 2003; Kachiwala et al., 2005) Brain structure (Rujescu et al., 2002) Long-term memory (Papassotiropoulos et al., 2005b) AD (Bertram et al., 2007)
RECQL2	8p12	RECQ protein-like 2 Cognitive composite in LSADT (Bendixen et al., 2004)

TABLE 15-3 Continued

Gene	Position	Description and References
SSADH	6p22.2	Succinate semi-aldehyde dehydrogenase IQ (Plomin et al., 2004) IQ linkage peak on chr6 is near this gene (Posthuma et al., 2005) Recent positive selection (Blasi et al., 2006)

Candidate Genes Near Linkage Peaks in Studies of IQ
(Posthuma et al., 2005; Luciano et al., 2006; Hallmayer et al., 2005; Dick et al., 2006b)

NR4A2	2q24.1	Nuclear receptor subfamily 4, group A, member 2
SLC25A12	2q31.1	Solute carrier family 25, member 12
SCN1A	2q24.3	Sodium channel, neuronal type 1, alpha subunit
SCN2A	2q24.3	Sodium channel, neuronal type 2, alpha subunit
TBR1	2q24.2	T-box, brain, 1
SCN3A	2q24.3	Sodium channel, neuronal type 3, alpha subunit
KCNH7	2q24.2	Potassium channel, voltage-gated, subfamily H, member 7
GAD1	2q31.1	Gluatamate decarboxylase 1
HOXD1	2q31.1	Homeobox D1
CHN1	2q31.1	Chimerin 1
RAPGEF4	2q31.1	RAP guanine nucleotide exchange factor
NOSTRIN	2q24.3	Nitric oxide synthase trafficker
BBS5	2q31.1	BBS5 gene
DLX1	2q31.1	Distal-less homeobox 1
DLX2	2q31.1	Distal-less homeobox 2
KIF13A	6p22.3	Kinesin family member 13A
NQO2	6p25.2	NAD(P)H dehydrogenase, quinone 2
RANBP9	6p23	RAN-binding protein 9
PNR	6q23.2	Trace amine-associated receptor 5 ("putative neurotransmitter receptor")
NRN1	6p25.1	Neuritin 1
S100B	21q22.3	S100 calcium-binding protein, beta

Genes Associated with Memory Ability
(de Quervain and Papassotiropoulos, 2006)

ADCY8	8q24.2	Adenylate cyclase 8
CAMK2G	10q22	Calcium/calmodulin-dependent protein kinase 2 gamma
GRIN2A	16p13	Ionotropic glutamate receptor, NMDA subunit 2A
GRIN2B	12p12	Ionotropic glutamate receptor, NMDA subunit 2B

Continued

TABLE 15-3 Continued

Gene	Position	Description and References
GRM3	7q21.1	Metabotropic glutamate receptor 3 Frontal and hippocampal function (Egan et al., 2004)
PRKCA	17q22–23.2	Protein kinase C, alpha
PRKACG	9q13	Protein kinase, cAMP-dependent, catalytic, gamma
(Papassotiropoulos et al., 2006)		
KIBRA	5q35.1	Kidney and brain expressed protein
CLSTN2	3q23	Calsyntenin 2
(Kravitz et al., 2006)		
ESR1	6q25.1	Estrogen receptor 1 AD (Bertram et al., 2007)
HSD17B1	17q21.31	Hydroxysteroid (17-beta) dehydrogenase 1

Genes Associated with Schizophrenia (SZ)
(reviewed by Norton et al., 2006; Owen et al., 2005)

Gene	Position	Description
AKT1	14q32.3	V-AKT murine thymoma viral oncogene homolog 1
DAOA	13q34	D-amino acid oxidase activator
DISC1	1q42.1	Disrupted in schizophrenia 1 Hippocampal structure and function (Callicott et al., 2005) Cognitive aging in women (Thomson et al., 2005) Cognitive performance in SZ (Burdick et al., 2005; reviewed by Porteous et al., 2006)
DTNBP1	6p22.3	Dystrobrevin-binding protein 1 g in SZ and controls (Burdick et al., 2006) IQ (Posthuma et al., 2005): linkage peak on chr6 contains this gene PFC function (Fallgatter et al., 2006) Under selection in Europeans (Voight et al., 2006)
NRG1	8p22	Neuregulin 1 Premorbid IQ in high-risk SZ subjects (Hall et al., 2006)
RGS4	1q23.3	Regulator of G-protein signaling 4 (Talkowski et al., 2006)

Genes Associated with Alzheimer Disease (AD)
(reviewed by Bertram et al., 2007; Bertram and Tanzi, 2004)

Gene	Position	Description
ACE	17q23	Angiotensin I-converting enzyme
APOE	19q13.2	Apolipoprotein E Risk factor for AD, general cognitive function (Small et al., 2004)

TABLE 15-3 Continued

Gene	Position	Description and References
BACE1	11q23.3	Beta-site amyloid beta A4 precursor protein-cleaving enzyme 1
		Interacts w/ APOE (Bertram and Tanzi, 2004)
		Modulates myelination in mice (Hu et al., 2006)
CHRNB2	1q21	Cholinergic receptor, neural nicotinic, beta polypeptide 2
CST3	20p11.2	Cystatin 3
GAPDHS	19q13.1	Clyceraldehyde-3 phosphate dehydrogenase, spermatogenic
IDE	10q23.33	Insulin-degrading enzyme 2
		Interacts w/ APOE (Bertram and Tanzi, 2004)
MTHFR	1p36.3	Methylenetetrahydrofolate reductase
PSEN1	14q24.3	Presenilin 1
TF	3q21	Transferrin
TFAM	10q21	Transcription factor A, mitochondrial
TNF	6p21.3	Tumor necrosis factor

Genes Associated with Brain/Head Size
(except for VDR, all have mutations causing microcephaly)

Gene	Position	Description and References
ASPM	1q31.3	Abnormal spindle-like, microcephaly-associated
		Under selection in humans (Mekel-Bobrov et al., 2005)
		Small effect on IQ subtests (Luciano et al., 2006)
		No significant effect on normal-range brain size (Woods et al., 2006)
CDK5RAP2	9q33.2	CDK5 regulatory subunit associated protein 2
		Brain size (Woods et al., 2005; Evans et al., 2006)
		Reverse association w/ verbal IQ (Luciano et al., 2006)
CENPJ	13q12.12	Centromeric protein J
		Brain size; under selection in CEU sample (Voight et al., 2006; cf. Evans et al., 2006)
MCPH1	8p23.1	Microcephalin
		Under selection in humans (Evans et al., 2005)
		No significant effects on IQ subtests (Luciano et al., 2006), normal-range brain size (Woods et al., 2006)

Continued

TABLE 15-3 Continued

Gene	Position	Description and References
VDR	12q13.11	Vitamin D receptor Head size (Handoko et al., 2006), not associated with schizophrenia

Genes Associated with Miscellaneous Brain and Cognitive Functions

Gene	Position	Description and References
BDNF	11p14.1	Brain-derived neurotrophic factor Memory, hippocampus (Egan et al., 2003; Dempster et al., 2005) Age-related cognitive decline (Harris et al., 2006) Not associated with working memory performance (Hansell et al., 2006)
CHRNA4	20q13.2	Neuronal nicotinic cholinergic receptor alpha polypeptide 4 Attentional function (Greenwood et al., 2005; Parasuraman et al., 2005)
CHRNA7	15q13.3	Neuronal nicotinic cholinergic receptor alpha polypeptide 7 Schizophrenia and auditory processing (Leonard et al., 2002)
NET1	16q12.2	Norepinephrine transporter ADHD (Bobb et al., 2005)
OXTR	3p26.2	Oxytocin receptor Trust; autism (Wu et al., 2005; Ylisaukko-Oja et al., 2005)
PAX6	11p13	Paired box gene 6 Development of executive function networks (Ellison-Wright et al., 2004)
SNAP25	20p12.2	Synaptosomal-associated protein, 25-KD ADHD (Faraone et al., 2005) Performance IQ (Gosso et al., 2006)
FADS2	11q12–q13	Fatty-acid desaturase 2 ADHD (Brookes et al., 2006)
NOS1	12q24	Neuronal nitric oxide synthase PFC function, schizophrenia (Reif et al., 2006)
CETP	16q21	Cholesterol ester transfer protein Better MMSE performance in centenarians (Barzilai et al., 2006)

NOTE: Table indicates possible mechanisms mediating genetic influences on these phenotypes (or other reasons for including the gene). Both known or suspected functional SNPs in these genes, as well as tagging SNPs from the HapMap, would be used. Names and genomic positions are taken from OMIM or the UCSC Genome Browser. Genes marked with an asterisk (*) have known or probable functional alleles that are *not* SNPs. Citations given for each gene are meant to be representative of the suggestive evidence in the literature (through 2006), not exhaustive lists of relevant publications on the gene.

(for a meta-analysis, see McDaniel, 2005; for candidate genes, see Gilbert, Dobyns, and Lahn, 2005; Woods, Bond, and Enard, 2005).

Finally, we added several genes associated with specific cognitive abilities, such as memory and attention, or that are linked to cognition via other mechanisms (Goldberg and Weinberger, 2004). Naturally, there is overlap among these categories; for example, COMT (catechol-O-methyltransferase) is part of the dopamine pathway, and it also has a common SNP that is associated with measures of executive function and frontal lobe activation (Egan et al., 2001); HTR2A (serotonin receptor 2A) is a serotonin receptor gene that has been associated with long-term memory ability (de Quervain et al., 2003); and while HTT (serotonin transporter) is a part of the serotonin system, it has also been associated with ADHD and cognitive processes. Table 15-3 is therefore not meant to be an exhaustive or final list of possible candidate genes for economic behavior, but rather our estimate of the best starting points for study, given the literature published through the end of 2006.

In addition to the considerable behavioral and medical phenotypes, the AGES-RS data includes several measures of cognitive function: speed of processing, working memory, and long-term memory, as well as educational achievement, the mini-mental state exam, and a clinical dementia evaluation. An index of general cognitive function (g) can be inferred from a principal components analysis of the individual cognitive tests; indeed, working memory and processing speed are prominent components of g (Chabris, 2007). It should be emphasized that AGES-RS participants are 67 years and older and current cognitive functions reflect important contributions of diseases of old age. Each subject in the AGES follow-up also received structural magnetic resonance imaging (MRI) of the brain with evaluations of atrophy, infarcts, white matter lesions, and high-resolution T1-weighted images for voxel-based morphometric analysis.

We plan to examine direct associations between the genes in our SNP panel and the distal economic outcomes measured in the AGES-RS data—for instance, labor force participation and housing wealth. We will also investigate whether these associations are mediated by proximal variables like cognitive function, brain morphology, and impatience.

To implement these analyses, we will construct composite phenotypic measures. Such composites will reduce measurement error, increase power, and reduce the number of statistical tests. Moreover, rather than simply testing each SNP genotype individually, we will construct composite "SNP sets" that index the "load" of sets of SNPs that individually may have small effects but collectively explain more variance in an outcome measure (for examples of this methodology, see Harlaar et al., 2005, for general cognitive ability; de Quervain and Papassotiropoulos, 2006, for memory; and Comings et al., 2001, for pathological gambling behavior).

CONCLUSION

This essay reviews our hopes and concerns about the joint study of genetic variation and variation in economic phenotypes. The new field of genoeconomics will study the ways in which genetic variation interacts with social institutions and individual behavior to jointly influence economic outcomes.

Genetic research and economic research will have three major points of contact. First, economics can contribute a theoretical and empirical framework for understanding how individual behavior and economic markets mediate the influence of genetic factors. Second, incorporating (exogenous) genetic variation into empirical analysis can help economists identify and measure causal pathways and mechanisms that produce individual differences. Finally, economics can aid in analyzing the policy issues raised by the existence of genetic knowledge and its potential societal diffusion.

Despite the promise of genoeconomics, there are numerous pitfalls. Ethical issues crop up at every juncture, both during the research process and once the research results are disseminated. The problems are even greater when genetic research is done carelessly or reported misleadingly. Historically, there have been many cases of false positives in which preliminary genetic claims have subsequently collapsed as a result of unsuccessful replications. Communication about research results must also highlight the fact that genes alone do not determine outcomes. A highly complex set of gene effects, environment effects, and gene-environment interactions jointly cause phenotypic variation.

The way forward requires statistical care, attention to how the environment mediates genes, and sensitivity to the ethical issues surrounding genetic knowledge. We think that there is potential for productive collaboration between economists, cognitive scientists, epidemiologists, and genetic researchers. Indeed, we end by summarizing a study that is currently under way, which uses a SNP panel to analyze associations between candidate cognitive genes and economic phenotypes.

REFERENCES

Ashraf, N., Karlan, D.S., and Yin, W. (2004). Tying odysseus to the mast: Evidence from a commitment savings product in the Philippines. *Quarterly Journal of Economics, 121*(2), 635-672.

Bailey-Wilson, J.E, Amos, C.I, Pinney, S.M., Petersen, G.M., de Andrade, M., Wiest, J.S, Fain, P., Schwartz, A.G., You, M., Franklin, W., Klein, C., Gazdar, A., Rothschild, H., Mandal, D., Coons, T., Slusser, J., Lee, J., Gaba, C., Kupert, E., Perez, A., Zhou, X., Zeng, D., Liu, Q., Zhang, Q., Seminara, D., Minna, J., and Anderson, M. (2004). A major lung cancer susceptibility locus maps to chromosome 6q23-25. *American Journal of Human Genetics, 75*, 460-474.

Barbaux, S., Plomin, R., and Whitehead, A.S. (2000). Polymorphisms of genes controlling homocysteine/folate metabolism and cognitive function. *Neuroreport, 11*(5), 1133-1136.

Barsky, R.B., Juster, F.T., Kimball, M.S., and Shapiro, M.D. (1997). Preference parameters and behavioral heterogeneity: An experimental approach in the Health and Retirement Study. *Quarterly Journal of Economics, 112*(2), 537-579.

Barzilai, N., Atzmon, G., Derby, C.A., Bauman, J.M., and Lipton, R.B. (2006). A genotype of exceptional longevity is associated with preservation of cognitive function. *Neurology, 67*(12), 2170-2175.

Behrman, J.R., Hrubec, Z., Taubman, P., and Wales, T.J. (1980). *Socioeconomic success: A study of the effects of genetic endowments, family environment, and schooling.* New York: North-Holland.

Behrman, J.R., and Taubman, P. (1989). Is schooling mostly in the genes? Nature-nurture decomposition using data on relatives. *Journal of Political Economy, 97,* 1425-1446.

Bendixen, M.H., Nexo, B.A., Bohr, V.A., Frederiksen, H., McGue, M., Kolvraa, S., and Christensen, K. (2004). A polymorphic marker in the first intron of the Werner gene associates with cognitive function in aged Danish twins. *Experimental Gerontology,* 1101-1107.

Benjamin, D.J., Brown, S.A., and Shapiro, J.M. (2006). Who is behavioral? Cognitive ability and anomalous preferences. Submitted for publication.

Bertram, L., McQueen, M.B., Mullin, K., Blacker, D., and Tanzi, R.E. (2007). Systematic meta-analyses of Alzheimer disease genetic association studies: The AlzGene database. *National Genetology, 39*(1), 17-23.

Bertram, L, and Tanzi, R.E. (2004). Alzheimer's disease: One disorder, too many genes? *Human Molecular Genetics, 13*(1), R135-R141.

Bickel, W.K., Odum, A.L., and Madden, G.J. (1999). Impulsivity and cigarette smoking: Delay discounting in current, never, and ex-smokers. *Psychopharm, 146*(4), 447-454.

Blasi, P., Palmerio, F., Aiello, A., Rocchi, M., Malaspina, P., and Novelletto, A. (2006, July). SSADH variation in primates: intra- and interspecific data on a gene with a potential role in human cognitive functions. *Journal of Molecular Evolution, 63*(1), 54-68.

Bobb, A.J., Addington, A.M., Sidransky, E., Gornick, M.C., Lerch, J.P., Greenstein, D.K., Clasen, L.S., Sharp, W.S., Inoff-Germain, G., Wavrant-De Vrieze, F., Arcos-Burgos, M., Straub, R.E., Hardy, J.A., Castellanos, F.X., and Rapoport, J.L. (2005). Support for association between ADHD and two candidate genes: NET1 and DRD1. *American Journal of Medical Genetics Part B: Neuropsychiatric Genetics, 134*(1), 67-72.

Brookes, K.J., Chen, W., Xu, X., Taylor, E., and Asherson, P. (2006, August). Association of fatty acid desaturase genes with attention-deficit/hyperactivity disorder. *Biological Psychiatry, 60*(10), 1053-1061.

Burdick, K.E., Hodgkinson, C.A., Szeszko, P.R., Lencz, T., Ekholm, J.M., Kane, J.M., Goldman, D., and Malhotra, A.K. (2005). DISC1 and neurocognitive function in schizophrenia. *Neuroreport, 16*(12), 1399-1402.

Burdick, K.E., Lencz, T., Funke, B., Finn, C.T., Szeszko, P.R., Kane, J.M., Kucherlapati, R., and Malhotra, A.K. (2006). Genetic variation in DTNBP1 influences general cognitive ability. *Human Molecular Genetics, 15*(10), 1563-1568.

Callicott, J.H., Straub, R.E., Pezawas, L., Egan, M.F., Mattay, V.S., Hariri, A.R., Verchinski, B.A., Meyer-Lindenberg, A., Balkissoon, R., Kolachana, B., Goldberg, T.E., and Weinberger, D.R. (2005). Variation in DISC1 affects hippocampal structure and function and increases risk for schizophrenia. *Proceedings of the National Academy of Sciences, 102*(24), 8627-8632.

Campbell, U.B., Gatto, N.M., and Schwartz, S. (2005). Distributional interaction: Interpretational problems when using incidence odds ratios to assess interaction. *Epidemiologic Perspectives in Innovation, 2*(1), 1.

Case, A., Lubotsky, D., and Paxson, C. (2002). Economic status and health in childhood: The origins of the gradient. *American Economic Review, 92*(5), 1308-1334.

Caspi, A., McClay, J., Moffitt, T.E., Mill, J., Martin, J., Craig, I.W., Taylor, A., and Poulton, R. (2002). Role of genotype in the cycle of violence in maltreated children. *Science, 297*(5582), 851-854.

Caspi, A., Sugden, K., Moffitt, T.E., Taylor, A., Craig, I.W., Harrington, H., McClay, J., Mill, J., Martin, J., Braithwaite, A., and Poulton, R. (2003). Influence of life stress on depression: Moderation by a polymorphism in the 5-HTT gene. *Science, 301*(5631), 386-389.

Chabris, C.F. (2007). Cognitive and neurobiological mechanisms of the Law of General Intelligence. In M.J. Roberts (Ed.), *Integrating the mind.* Hove, UK: Psychology Press.

Chorney, M.J., Chorney, K., Seese, N., Owen, M.J., Daniels, J., McGuffin, P., Thompson, L.A., Detterman, D.K., Benbow, C.P., Lubinski, D., Eley, T.C., and Plomin, R. (1998). A quantitative trait locus (QTL) associated with cognitive ability in children. *Psychological Science, 9,* 159-166.

Clare, M.J., and Luckinbill, L.S. (1985). The effects of gene-environment interaction on the expression of longevity. *Heredity, 55*(1), 19-26.

Clayton, D., and McKeigue, P.M. (2001). Epidemiological methods for studying genes and environmental factors in complex diseases. *Lancet, 358,* 1356-1360.

Cloninger, C.R. (1987). Neurogenetic adaptive mechanisms in alcoholism. *Science, 236*(4800), 410-416.

Cloninger, C.R. (1993). Unraveling the causal pathway to major depression. *American Journal of Psychiatry, 150*(8), 1137-1138.

Cloninger, C.R., Adolfsson, R., Svrakic, N.M. (1996, January). Mapping genes for human personality. *Nature Genetics, 12*(1), 3-4.

Comings, D.E., Gade-Andavolu, R., Gonzalez, N., Wu, S., Muhleman, D., Chen, C., Koh, P., Farwell, K., Blake, H., Dietz, G., MacMurray, J.P., Lesieur, H.R., Rugle, L.J., and Rosenthal, R.J. (2001). The additive effect of neurotransmitter genes in pathological gambling. *Clinical Genetics, 60,* 107-116.

Comings, D.E., Wu, S., Rostamkhani, M., McGue, M., Lacono, W.G., Cheng, L.S., and MacMurray, J.P. (2003). Role of the cholinergic muscarinic 2 receptor (CHRM2) gene in cognition. *Molecular Psychiatry, 8*(1), 10-11.

Cools, R., Blackwell, A., Clark, L., Menzies, L., Cox, S., and Robbins, T.W. (2005). Tryptophan depletion disrupts the motivational guidance of goal-directed behavior as a function of trait impulsivity. *Neuropsychopharmacology, 30,* 1362-1373.

Cooper, R.S. (2003). Gene-environment interactions and etiology of common complex disease. *Annals of Internal Medicine, 139,* 437-440.

Cutler, D.M., and Glaeser, E. (2005). What explains differences in smoking, drinking, and other health-related behaviors? *American Economic Review Papers and Proceedings, 95,* 238-242.

de Quervain, D.J., Henke, K., Aerni, A., Coluccia, D., Wollmer, M.A., Hock, C., Nitsch, R.M., and Papassotiropoulos, A. (2003). A functional genetic variation of the 5-HT2a receptor affects human memory. *Nature Neuroscience, 6,* 1141-1142.

de Quervain, D.J., and Papassotiropoulos, A. (2006). Identification of a genetic cluster influencing memory performance and hippocampal activity in humans. *Proceedings of the National Academy of Sciences, 103,* 4270-4274.

Deary, I.J., Hamilton, G., Hayward, C., Whalley, L.J., Powell, J., Starr, J.M., and Lovestone, S. (2005a). Nicastrin gene polymorphisms, cognitive ability level and cognitive ageing. *Neuroscience Letters, 373*(2), 110-114.

Deary, I.J., Harris, S.E., Fox, H.C., Hayward, C., Wright, A.F., Starr, J.M., and Whalley, L.J. (2005b). KLOTHO genotype and cognitive ability in childhood and old age in the same individuals. *Neuroscience Letters, 378*(1), 22-27.

Dempster, E., Toulopoulou, T., McDonald, C., Bramon, E., Walshe, M., Filbey, F., Wickham, H., Sham, P.C., Murray, R.M., and Collier, D.A. (2005). Association between BDNF val66 met genotype and episodic memory. *American Journal of Medical Genetics Part B: Neuropsychiatric Genetics, 134*(1), 73-75.

Devlin, B., and Roeder, K. (1999). Genomic Control for Association Studies. *Biometrics, 55*, 997-1004.

Dick, D.M., Agrawal, A., Schuckit, M.A., Bierut, L., Hinrichs, A., Fox, L., Mullaney, J., Cloninger, C.R., Hesselbrock, V., Nurnberger, J.I., Jr., Almasy, L., Foroud, T., Porjesz, B., Edenberg, H., and Begleiter, H. (2006a, March). Marital status, alcohol dependence, and GABRA2: Evidence for gene-environment correlation and interaction. *Journal of Studies on Alcohol, 67*(2), 185-194.

Dick, D.M., Aliev, F., Bierut, L., Goate, A., Rice, J., Hinrichs, A., Bertelsen, S., Wang, J.C., Dunn, G., Kuperman, S., Schuckit, M., Nurnberger, J., Jr., Porjesz, B., Beglieter, H., Kramer, J., and Hesselbrock, V. (2006b, January). Linkage analyses of IQ in the Collaborative Study on the Genetics of Alcoholism (COGA) Sample. *Behavior Genetics, 36*(1), 77-86.

Dick, D.M., Aliev, F., Kramer, J., Wang, J.C., Hinrichs, A., Bertelsen, S., Kuperman, S., Schuckit, M., Nurnberger, J., Jr., Edenberg, H.J., Porjesz, B., Begleiter, H., Hesselbrock, V., Goate, A., and Bierut, L. (2006c). Association of CHRM2 with IQ: Converging evidence for a gene influencing intelligence. *Behavior Genetitcs, 37*(2), 265-272.

DiLalla, L.F., and Gottesman, I.I. (1991). Biological and genetic contributors to violence: Widom's untold tale. *Psychological Bulletin, 109*, 125-129.

Ding, W., Lehrer, S.F., Rosenquist, J.N., and Audrain-McGovern, J. (2005). The impact of health on academic performance: New evidence using genetic markers. University of Pennsylvania, mimeo.

Dohmen, T., Falk, A., Huffman, A., Sunde, U., Schupp, J., Wagner, G.G. (2005, September). Individual risk attitudes: Evidence from a large, representative, experimentally-validated survey. IZA Discussion Paper No. 1730.

Ebstein, R.P. (2006, May). The molecular genetic architecture of human personality: Beyond self-report questionnaires. *Molecular Psychiatry, 11*(5), 427-445.

Egan, M.F., Goldberg, T.E., Kolachana, B.S., Callicott, J.H., Mazzanti, C.M., Straub, R.E., Goldman, D., Weinberger, D.R. (2001). Effect of COMT Val108/158 Met genotype on frontal lobe function and risk for schizophrenia. *Proceedings of the National Academy of Sciences, 98*, 6917-6922.

Egan, M.F., Kojima, M., Callicott, J.H., Goldberg, T.E., Kolachana, B.S., Bertolino, A., Zaitsev, E., Gold, B., Goldman, D., Dean, M., Lu, B., and Weinberger, D.R. (2003). The BDNF val66met polymorphism affects activity-dependent secretion of BDNF and human memory and hippocampal function. *Cell, 112*(2), 257-269.

Egan, M.F., Straub, R.E., Goldberg, T.E., Yakub, I., Callicott, J.H., Hariri, A.R., Mattay, V.S., Bertolino, A., Hyde, T.M., Shannon-Weickert, C., Akil, M., Crook, J., Vakkalanka, R.K., Balkissoon, R., Gibbs, R.A., Kleinman, J.E., and Weinberger, D.R. (2004). Variation in GRM3 affects cognition, prefrontal glutamate, and risk for schizophrenia. *Proceedings of the National Academy of Sciences, 101*(34), 12604-12609.

Ellison-Wright, Z., Heyman, I., Frampton, I., Rubia, K., Chitnis, X., Ellison-Wright, I., Williams, S.C., Suckling, J., Simmons, A., and Bullmore, E. (2004). Heterozygous PAX6 mutation, adult brain structure, and fronto-striato-thalamic function in a human family. *European Journal of Neuroscience, 19*(6), 1505-1512.

Evans, P.D., Gilbert, S.L., Mekel-Bobrov, N., Vallender, E.J., Anderson, J.R., Vaez-Azizi, L.M., Tishkoff, S.A., Hudson, R.R., and Lahn, B.T. (2005). Microcephalin, a gene regulating brain size, continues to evolve adaptively in humans. *Science, 309*(5741), 1717-1720.

Evans, P.D., Vallender, E.J., and Lahn, B.T. (2006, June). Molecular evolution of the brain size regulator genes CDK5RAP2 and CENPJ. *Gene, 375, 75-79.*

Fallgatter, A.J., Herrmann, M.J., Hohoff, C., Ehlis, A.C., Jarczok, T.A., Freitag, C.M., and Deckert, J. (2006, September). DTNBP1 (dysbindin) gene variants modulate prefrontal brain function in healthy individuals. *Neuropsychopharmacology, 31*(9), 2002-2010.

Faraone, S.V., Perlis, R.H., Doyle, A.E., Smoller, J.W., Goralnick, J.J., Holmgren, M.A., and Sklar, P. (2005). Molecular genetics of attention-deficit/hyperactivity disorder. *Biological Psychiatry, 57,* 1313-1323.

Fisher, P.J., Turic, D., Williams, N.M., McGuffin, P., Asherson, P., Ball, D., Craig, I., Eley, T., Hill, L., Chorney, K., Chorney, M.J., Benbow, C.P., Lubinski, D., Plomin, R., and Owen, M.J. (1999). DNA pooling identifies QTLs on chromosome 4 for general cognitive ability in children. *Human Molecular Genetics, 8*(5), 915-922.

Fisher, R.A., and Mackenzie, W.A. (1923). Studies in crop variation. II. The manurial responses of different potato varieties. *Journal of Agricultural Science, 13,* 311-320.

Frederick, S. (2005). Cognitive reflection and decision making. *Journal of Economic Perspectives, 19,* 24-42.

Fuchs, V. (1974). *Who shall live? Health, Economics, and Social Choice.* New York: Basic Books.

Fuchs, V. (1982). Time preference and health: An exploratory study. In V. Fuchs (ed.), *Economic aspects of health* (pp. 93-120). Chicago, IL: University of Chicago Press.

Garg, A., and Morduch, J. (1998). Sibling rivalry and the gender gap: Evidence from child health outcomes in Ghana. *Journal of Population Economics, 11,* 471-493.

Gilbert, S.L., Dobyns, W.B., and Lahn, B.T. (2005). Genetic links between brain development and brain evolution. *Nature Reviews Genetics, 6*(7), 581-590.

Glimcher, P.W., Dorris, M.C., and Bayer, H.M. (2005). Physiological utility theory and the neuroeconomics of choice. *Games and Economic Behavior 52*(2), 213-256.

Glimcher, P.W., and Rustichini, A. (2004). Neuroeconomics: The consilience of brain and decision. *Science, 306*(5695), 447-452.

Goldberg, T.E., and Weinberger, D.R. (2004). Genes and the parsing of cognitive processes. *Trends in Cognitive Sciences, 8*(7), 325-335.

Gosso, M.F., de Geus, E.J., van Belzen, M.J., Polderman, T.J., Heutink, P., Boomsma, D.I., and Posthuma, D. (2006a, September). The SNAP-25 gene is associated with cognitive ability: evidence from a family-based study in two independent Dutch cohorts. *Molecular Psychiatry, 11*(9), 878-886.

Gosso, M.F., van Belzen, M., de Geus, E.J., Polderman, J.C., Heutink, P., Boomsma, D.I., and Posthuma, D. (2006b, November). Association between the CHRM2 gene and intelligence in a sample of 304 Dutch families. *Genes, Brain, and Behavior, 5*(8), 577-584.

Gray, J.R., Chabris, C.F., and Braver, T.S. (2003). Neural mechanisms of general fluid intelligence. *Nature Neuroscience, 6,* 316-322.

Greenwood, P.M., Fossella, J.A., and Parasuraman, R. (2005). Specificity of the effect of a nicotinic receptor polymorphism on individual differences in visuospatial attention. *Journal of Cognitive Neuroscience, 17*(10), 1611-1620.

Hallmayer, J.F., Kalaydjieva, L., Badcock, J., Dragovic, M., Howell, S., Michie, P.T., Rock, D., Vile, D., Williams, R., Corder, E.H., Hollingsworth, K., and Jablensky, A. (2005). Genetic evidence for a distinct subtype of schizophrenia characterized by pervasive cognitive deficit. *American Journal of Human Genetics, 77,* 468-476.

Handoko, H.Y., Nancarrow, D.J., Mowry, B.J., and McGrath, J.J. (2006). Polymorphisms in the vitamin D receptor and their associations with risk of schizophrenia and selected anthropometric measures. *American Journal of Human Biology, 18*(3), 415-417.

Hansell, N.K., James, M.R., Duffy, D.L., Birley, A.J., Luciano, M., Geffen, G.M., Wright, M.J., Montgomery, G.W., and Martin, N.G. (2006). Effect of the BDNF V166M polymorphism on working memory in healthy adolescents. *Genes, Brain, and Behavior, 6*(3), 260-269.

Hariri, A.R., Brown, S.M., Williamson, D.E., Flory, J.D., de Wit, H., and Manuck, S.B. (2006). Preference for immediate over delayed rewards is associated with magnitude of ventral striatal activity. *Journal of Neuroscience, 26,* 13213-13217.

Hariri, A.R., Mattay, V.S., Tessitore, A., Kolachana, B., Fera, F., Goldman, D., Egan, M.F., and Weinberger, D.R. (2003). Serotonin transporter genetic variation and the response of the human amygdala. *Science, 297*(5580), 400-403.

Harris, S.E., Fox, H., Wright, A.F., Hayward, C., Starr, J.M., Whalley, L.J., and Deary, I.J. (2006, May). The brain-derived neurotrophic factor Val66Met polymorphism is associated with age-related change in reasoning skills. *Molecular Psychiatry, 11*(5), 505-513.

Harris, T.B., Launer, L.J., Eiriksdottir, G., Kjartansson, O., Jonsson, P.V., Sigurdsson, G., Thorgeirsson, G., Aspelund, T., Garcia, M.E., Cotch, M.F., Hoffman, H.J., and Gudnason, V., for the Age, Gene/Environment Susceptibility-Reykjavik Study Investigators. (in press). Age, gene/environment susceptibility—Reykjavik Study: Multidisciplinary applied phenomics. *American Journal of Epidemiology.*

Hu, X., Hicks, C.W., He, W., Wong, P., Macklin, W.B., Trapp, B.D., and Yan, R. (2006, December). Bace1 modulates myelination in the central and peripheral nervous system. *Nature Neuroscience, 9*(12),1520-1525.

Iidaka, T., Ozaki, N., Matsumoto, A., Nogawa, J., Kinoshita, Y., Suzuki, T., Iwata, N., Yamamoto, Y., Okada, T., and Sadato, N. (2005). A variant C178T in the regulatory region of the serotonin receptor gene HTR3A modulates neural activation in the human amygdala. *Journal of Neuroscience, 25*(27), 6460-6466.

Jacobsen, L.K., Pugh, K.R., Mencl, W.E., and Gelernter, J. (2006). C957T polymorphism of the dopamine D2 receptor gene modulates the effect of nicotine on working memory performance and cortical processing efficiency. *Psychopharmacology, 188*(4), 530-540.

Jang, K.L., Vernon, P.A., and Livesley, W.J. (2000). Personality disorder traits, family environment, and alcohol misuse: A multivariate behavioural genetic analysis. *Addiction, 95,* 873-888.

Jirtle, R.L. (2005). Biological consequences of divergent evolution of M6P/IGF2R imprinting. Presented at the Environmental Epigenomics, Imprinting and Disease Susceptibility conference, Durham, N.C., November, 2-4.

Kachiwala, S.J., Harris, S.E., Wright, A.F., Hayward, C., Starr, J.M., Whalley, L.J., and Deary, I.J. (2005). Genetic influences on oxidative stress and their association with normal cognitive ageing. *Neuroscience Letters, 386*(2), 116-120.

Kendler, K., and Eaves, L. (1986). Models for the joint effect of genotype and environment on liability to psychiatric diseases. *American Journal of Psychiatry, 143,* 279-289.

Kimball, M.S., Sahm, C., and Shapiro, M.D. (2006). Using survey-based risk tolerance. Submitted for publication.

Kirby, K.N., and Petry, N.M. (2004). Heroin and cocaine abusers have higher discount rates for delayed rewards than alcoholics or non-drug-using controls. *Addiction, 99,* 461-471.

Kirby, K.N., Petry, N.M., and Bickel, W.K. (1999). Heroin addicts have higher discount rates for delayed rewards than non-drug-using controls. *Journal of Experimental Psychology, 128*(1), 78-87.

Kovalchik, S., Camerer, C.F., Grether, D.M., Plott, C.R., and Allman, J.M. (2003). Aging and decision making: A broad comparative study of decision behavior in neurologically healthy elderly and young individuals. Caltech manuscript.

Kravitz, H.M., Meyer, P.M., Seeman, T.E., Greendale, G.A., and Sowers, M.R. (2006, September). Cognitive functioning and sex steroid hormone gene polymorphisms in women at midlife. *American Journal of Medicine, 119*(9 Suppl. 1), S94-S102.

Laurin, N., Misener, V.L., Crosbie, J., Ickowicz, A., Pathare, T., Roberts, W., Malone, M., Tannock, R., Schachar, R., Kennedy, J.L., and Barr, C.L. (2005). Association of the calcyon gene (DRD1IP) with attention deficit/hyperactivity disorder. *Molecular Psychiatry, 10*, 1117-1125.

Leonard, S., Gault, J., Hopkins, J., Logel, J., Vianzon, R., Short, M., Drebing, C., Berger, R., Venn, D., Sirota, P., Zerbe, G., Olincy, A., Ross, R.G., Adler, L.E., and Freedman, R. (2002). Association of promoter variants in the alpha7 nicotinic acetylcholine receptor subunit gene with an inhibitory deficit found in schizophrenia. *Archives of General Psychiatry, 59*(12), 1085 1096.

Lichtenstein, P., Holm, N.V., Verkasalo, P.K., Iliadou, A., Kaprio, J., Koskenvuo, M., Pukkala, E., Skytthe, A., and Hemminki, K. (2000). Environmental and heritable factors in the causation of cancer: Analyses of cohorts of twins from Sweden, Denmark, and Finland. *New England Journal of Medicine 343*, 78-85.

Luciano, M., Lind, P., Wright, M., Duffy, D., Wainwright, M., Montgomery, G., and Martin, N. (2006, August 6-10). Candidate genes for human brain evolution: Do they relate to head size and cognitive ability? Presented at the International Congress of Human Genetics, Brisbane, Australia.

Luciano, M., Wright, M.J., Duffy, D.L., Wainwright, M.A., Zhu, G., Evans, D.M., Geffen, G.M., Montgomery, G.W., and Martin, N.G. (2006). Genome-wide scan of IQ finds significant linkage to a quantitative trait locus on 2q. *Behavior Genetics, 36*(1), 45-55.

MacDonald, S.D., Perkins, S.L., Jodouin, M.L.T., and Walker, M.C. (2002). Folate levels in pregnant women who smoke: An important gene/environment interaction. *American Journal of Obstetrics and Gynecology, 187*, 620-625.

MacKay, T.F. (2001). The genetic architecture of quantitative traits. *Annual Review of Genetics, 35*, 303-339.

McClure, S.M., Laibson, D.I., Loewenstein, G., and Cohen, J.D. (2004). Separate neural systems value immediate and delayed monetary rewards. *Science, 306*, 503-507.

McDaniel, M.A. (2005). Big-brained people are smarter: A meta-analysis of the relationship between in vivo brain volume and intelligence. *Intelligence, 33*, 337-346.

Mekel-Bobrov, N., Gilbert, S.L., Evans, P.D., Vallender, E.J., Anderson, J.R., Hudson, R.R., Tishkoff, S.A., and Lahn, B.T. (2005). Ongoing adaptive evolution of ASPM, a brain size determinant in Homo sapiens. *Science, 309*(5741), 1720-1722.

Meyer-Lindenberg, A., Nichols, T., Callicott, J.H., Ding, J., Kolachana, B., Buckholtz, J., Mattay, V.S., Egan, M., and Weinberger, D.R. (2006, September). Impact of complex genetic variation in COMT on human brain function. *Molecular Psychiatry, 11*(9), 867-877.

Mucci, L.A., Wedren, S., Tamimi, R.M., Trichopoulos, D., and Adami, H-O. (2001). The role of gene-environment interaction in the aetiology of human cancer: Examples from cancers of the large bowel, lung and breast. *Journal of Internal Medicine, 249*(6), 519-524.

Munafo, M.R., Clark, T.G., Moore, L.R., Payne, E., Walton, R., and Flint, J. (2003, May). Genetic polymorphisms and personality in healthy adults: A systematic review and meta-analysis. *Molecular Psychiatry, 8*(5), 471-484.

Nelson, E., and Cloninger, C.R. (1997). Exploring the TPQ as a possible predictor of antidepressant response to nefazodone in a large multi-site study. *Journal of Affective Disorder, 44*(2-3), 197-200.

Norton, N., Williams, H.J., and Owen, M.J. (2006). An update on the genetics of schizophrenia. *Current Opinion in Psychiatry, 19*(2), 158-164.

Owen, M.J., Craddock, N., and O'Donovan, M.C. (2005). Schizophrenia: Genes at last? *Trends in Genetics, 21*(9), 518-525.

Papassotiropoulos, A., Henke, K., Aerni, A., Coluccia, D., Garcia, E., Wollmer, M.A., Huynh, K.D., Monsch, A.U., Stahelin, H.B., Hock, C., Nitsch, R.M., and de Quervain, D.J. (2005a). Age-dependent effects of the 5-hydroxytryptamine-2a-receptor polymorphism (His452Tyr) on human memory. *Neuroreport, 16*(8), 839-842.

Papassotiropoulos, A., Wollmer, M.A., Aguzzi, A., Hock, C., Nitsch, R.M., and de Quervain, D.J. (2005b). The prion gene is associated with human long-term memory. *Human Molecular Genetics, 14*(15), 2241-2246.

Papassotiropoulos, A., et al. (2006). Common Kibra alleles are associated with human memory performance. *Science, 314,* 475-478.

Parasuraman, R., Greenwood, P.M., Kumar, R., and Fossella, J. (2005). Beyond heritability: Neurotransmitter genes differentially modulate visuospatial attention and working memory. *Psychological Science, 16,* 200-207.

Payton, A. (2006). Investigating cognitive genetics and its implications for the treatment of cognitive deficit. *Genes, Brain, and Behavior, 5*(Suppl. 1), 44-53.

Payton, A., Holland, F., Diggle, P., Rabbitt, P., Horan, M., Davidson, Y., Gibbons, L., Worthington, J., Ollier, W.E., and Pendleton, N. (2003). Cathepsin D exon 2 polymorphism associated with general intelligence in a healthy older population. *Molecular Psychiatry, 8*(1), 14-18.

Payton, A., Gibbons, L., Davidson, Y., Ollier, W., Rabbitt, P., Worthington, J., Pickles, A., Pendleton, N., and Horan, M. (2005). Influence of serotonin transporter gene polymorphisms on cognitive decline and cognitive abilities in a nondemented elderly population. *Molecular Psychiatry, 10*(12), 1133-1139.

Payton, A., van den Boogerd, E., Davidson, Y., Gibbons, L., Ollier, W., Rabbitt, P., Worthington, J., Horan, M., and Pendleton, N. (2006). Influence and interactions of cathepsin D, HLA-DRB1 and APOE on cognitive abilities in an older non-demented population. *Genes, Brain, and Behavior, 5*(Suppl. 1), 23-31.

Persico, N., Postlewaite, A., and Silverman, D. (2004) The effect of adolescent experience on labor market outcomes: The case of height. *Journal of Political Economy, 112,* 1019-1053.

Petry, N.M., and Casarella, T. (1999). Excessive discounting of delayed rewards in substance abusers with gambling problems. *Drug and Alcohol Dependence, 56*(1-2), 25-32.

Plomin, R., and Bergeman, C.S. (1991). The nature of nurture: genetic influence on environmental measures. *Behavioral and Brain Sciences, 14*(3), 373-386.

Plomin, R., Kennedy, J.K.J., and Craig, I.W. (in press). The quest for quantitative trait loci associated with intelligence. *Intelligence.*

Plomin, R., Turic, D.M., Hill, L., Turic, D.E., Stephens, M., Williams, J., Owen, M.J., and O'Donovan, M.C. (2004). A functional polymorphism in the succinate-semialdehyde dehydrogenase (aldehyde dehydrogenase 5 family, member A1) gene is associated with cognitive ability. *Molecular Psychiatry, 9*(6), 582-586.

Porteous, D.J., Thomson, P., Brandon, N.J., and Millar, J.K. (2006, July). The genetics and biology of DISC1: An emerging role in psychosis and cognition. *Biological Psychiatry, 60*(2), 123-131.

Posthuma, D., Luciano, M., Geus, E.J., Wright, M.J., Slagboom, P.E., Montgomery, G.W., Boomsma, D.I., Martin, N.G. (2005). A genomewide scan for intelligence identifies quantitative trait loci on 2q and 6p. *American Journal of Human Genetics, 77*(2), 318-326.

Pritchard, J.K., Stephens, M., and Donnelly, P.J. (2000). Inference of population structure using multilocus genotype data. *Genetics, 155,* 945-959.

Reif, A., Herterich, S., Strobel, A., Ehlis, A.C., Saur, D., Jacob, C.P., Wienker, T., Topner, T., Fritzen, S., Walter, U., Schmitt, A., Fallgatter, A.J., and Lesch, K.P. (2006, March). A neuronal nitric oxide synthase (NOS-I) haplotype associated with schizophrenia modifies prefrontal cortex function. *Molecular Psychiatry, 11*(3), 286-300.

Rosenberg, S., Templeton, A.R., Feigin, P.D., Lancet, D., Beckmann, J.S., Selig, S., Hamer, D.H., and Skorecki, K. (2006, November). The association of DNA sequence variation at the MAOA genetic locus with quantitative behavioural traits in normal males. *Human Genetics, 120*(4), 447-459.

Rubia, K., Lee, F., Cleare, A.J., Tunstall, N., Fu, C.H., Brammer, M., and McGuire, P. (2005). Tryptophan depletion reduces right inferior prefrontal activation during response inhibition in fast, event-related fMRI. *Psychopharmacology, 179*, 791-803.

Rujescu, D., Meisenzahl, E.M., Giegling, I., Kirner, A., Leinsinger, G., Hegerl, U., Hahn, K., and Moller, H.J. (2002). Methionine homozygosity at codon 129 in the prion protein is associated with white matter reduction and enlargement of CSF compartments in healthy volunteers and schizophrenic patients. *Neuroimage, 15*(1), 200-206.

Rujescu, D., Hartmann, A.M., Gonnermann, C., Moller, H.J., and Giegling, I. (2003). M129V variation in the prion protein may influence cognitive performance. *Molecular Psychiatry, 8*(11), 937-941.

Rujescu, D., Meisenzahl, E.M., Krejcova, S., Giegling, I., Zetzsche, T., Reiser, M., Born, C.M., Moller, H.J., Veske, A., Gal, A., and Finckh, U. (2006). Plexin B3 is genetically associated with verbal performance and white matter volume in human brain. *Molecular Psychiatry, 12*, 190-194.

Shapiro, J.M. (2005). Is there a daily discount rate? Evidence from the food stamp nutrition cycle. *Journal of Public Economics, 89*(2), 303-325.

Shimokata, H., Ando, F., Niino, N., Miyasaka, K., and Funakoshi, A. (2005). Cholecystokinin A receptor gene promoter polymorphism and intelligence. *Annals of Epidemiology, 15*(3), 196-201.

Siintola, E., Partanen, S., Stromme, P., Haapanen, A., Haltia, M., Maehlen, J., Lehesjoki, A.E., and Tyynela, J. (2006, June). Cathepsin D deficiency underlies congenital human neuronal ceroid-lipofuscinosis. *Brain, 129*(Pt 6), 1438-1445.

Small, B.J., Rosnick, C.B., Fratiglioni, L., and Backman, L. (2004). Apolipoprotein E and cognitive performance: A meta-analysis. *Psychology and Aging, 19*(4), 592-600.

Smith, A. (1776). *An inquiry into the nature and causes of the wealth of nations, Volumes I and II.* R.H. Campbell and A.S. Skinner (Eds.). Indianapolis, IN: Liberty Fund.

Spielman, R.S., McGinnis, R.E., and Ewens, W.J. (1993). Transmission test for linkage disequilibrium: The insulin gene region and insulin-dependent diabetes mellitus (IDDM). *American Journal of Human Genetics, 52*(3), 506-516.

Stigler, G.J. (1971). The theory of economic regulation. *Bell Journal of Economics, 2*(1), 3-21.

Taubman, P. (1976). The determinants of earnings. Genetics, family, and other environments: A study of white male twins. *American Economic Review, 66*, 858-870.

Thomson, P.A., Harris, S.E., Starr, J.M., Whalley, L.J., Porteous, D.J., and Deary, I.J. (2005). Association between genotype at an exonic SNP in DISC1 and normal cognitive aging. *Neuroscience Letters, 389*(1), 41-45.

Voight, B.F., Kudaravalli, S., Wen, X., and Pritchard, J.K. (2006). A map of recent positive selection in the human genome. *Public Library of Science Biology, 4*(3), e72.

Wahlsten, D. (1990). Insensitivity of the analysis of variance to heredity-environment interaction. *Behavioral and Brain Sciences, 13*, 109-161.

Walderhaug, E., Lunde, H., Nordvik, J.E., Landro, N.I., Refsum, H., and Magnusson, A. (2002). Lowering of serotonin by rapid tryptophan depletion increases impulsiveness in normal individuals. *Psychopharmacology, 164*, 385-391.

Woods, C.G., Bond, J., and Enard, W. (2005). Autosomal recessive primary microcephaly (MCPH): A review of clinical, molecular, and evolutionary findings. *American Journal of Human Genetics, 76*, 717-728.

Woods, R.P., Freimer, N.B., De Young, J.A., Fears, S.C., Sicotte, N.L., Service, S.K., Valentino, D.J., Toga, A.W., and Mazziotta, J.C. (2006, June). Normal variants of Microcephalin and ASPM do not account for brain size variability. *Human Molecular Genetics, 15*(12), 2025-2029.

Wu, S., Jia, M., Ruan, Y., Liu, J., Guo, Y., Shuang, M., Gong, X., Zhang, Y., Yang, X., and Zhang D. (2005). Positive association of the oxytocin receptor gene (OXTR) with autism in the Chinese Han population. *Biological Psychiatry, 58*(1), 74-77.

Ylisaukko-Oja, T., Alarcon, M., Cantor, R.M., Auranen, M., Vanhala, R., Kempas, E., von Wendt, L., Jarvela, I., Geschwind, D.H., and Peltonen, L. (2005). Search for autism loci by combined analysis of Autism Genetic Resource Exchange and Finnish families. *Annals of Neurology, 59*(1), 145-155.

16

Mendelian Randomization: Genetic Variants as Instruments for Strengthening Causal Inference in Observational Studies

George Davey Smith and *Shah Ebrahim*

T he incorporation of biomarkers into population-based health surveys is generally intended to improve categorization of exposures or health outcome measures (National Research Council, 2001). An unintended consequence of the growing use of biomarkers—for example, in assessing nutritional status—is that investigators are less aware of the continued threats to validity of their findings caused by measurement error, confounding, and reverse causality, which affect biomarkers in the same way as exposures and outcomes measured using less precise methods. This chapter briefly outlines when and why conventional observational approaches have been misleading and then introduces the Mendelian randomization approach, a form of the use of genes as instrumental variables as briefly discussed by Douglas Ewbank in the earlier *Cells and Surveys* volume (Ewbank, 2001). The variety of inferences that can be drawn from this approach is illustrated and then potential limitations and ways to address these limitations are outlined. The chapter concludes by summarizing the ways in which Mendelian randomization approaches differ from other methodologies that depend on the use of genetic markers in population-based research.

LIMITS OF OBSERVATIONAL EPIDEMIOLOGY

To investigators interested in the health consequences of a modifiable environmental exposure—say, a particular aspect of diet—the obvious approach would be to directly study dietary intake and how this relates to the risk of disease. Why, then, should an alternative approach be advanced? The impetus for thinking of new approaches is that conventional observational study designs have yielded findings that have failed to be confirmed by randomized controlled trials (Davey Smith and Ebrahim, 2002). Observational studies demonstrated that beta carotene intake was associated with a lower risk of lung cancer mortality, and this stimulated an already active market for vitamin supplements that was based on the notion that they substantially influence chronic disease risk (Figure 16-1). The scientists involved in conducting the observational studies advocated taking supplements in material intended for the public (Willett, 2001) and also, relying on observational data, concluded "Available data thus strongly support the hypothesis that dietary carotenoids reduce the risk of lung cancer" (Willett, 1990). However large-scale randomized controlled trials reported disappointing findings: beta carotene supplementation produced no reduction in risk of lung cancer (Alpha-Tocopherol and Beta-Carotene Cancer Prevention Study Group, 1994).

With respect to cardiovascular disease, observational studies suggesting that beta carotene (Manson et al., 1991), vitamin E supplements (Rimm et al., 1993; Stampfer et al., 1993), vitamin C supplements (Osganian et al., 2003), and hormone replacement therapy (Stampfer and Colditz, 1991) were protective were followed by large trials showing no such protection (Omenn et al., 1996; Alpha-Tocopherol and Beta-Carotene Cancer Prevention Study Group, 1994; Lancet, 1999; Heart Protection Study Collaborative Group, 2002; Beral, Banks, and Reeves, 2002). In each case special pleading was advanced to explain the discrepancy: Were the doses of vitamins given in the trials too high or too low to be comparable to the observational studies? Did hormone replacement therapy use start too late in the trials? Were differences explained by the duration of follow-up or other design aspects? Were interactions with other factors, such as smoking or alcohol consumption, key? Rather than such particular explanations being true (with the happy consequence that both the observational studies and the trials had got the right answers, but to different questions), it is likely that a general problem of confounding—by lifestyle and socioeconomic factors, or by baseline health status and prescription policies—is responsible. Indeed, in the vitamin E supplements example, the observational studies and the trials tested precisely the same thing. Figures 16-2a and 16-2b show the findings from observational studies of taking vitamin E supplements (Rimm et al., 1993; Stampfer et al., 1993) and a meta-analysis of trials of supplements (Eidelman, Hollar, Hebert, Lamas, and Hennekens, 2004). The point here is that the observational studies specifically investigated the effect of taking supplements for a short period (2-5 years) and found an apparent, robust, and large protective effect, even after adjustment for confounders. The trials tested ran-

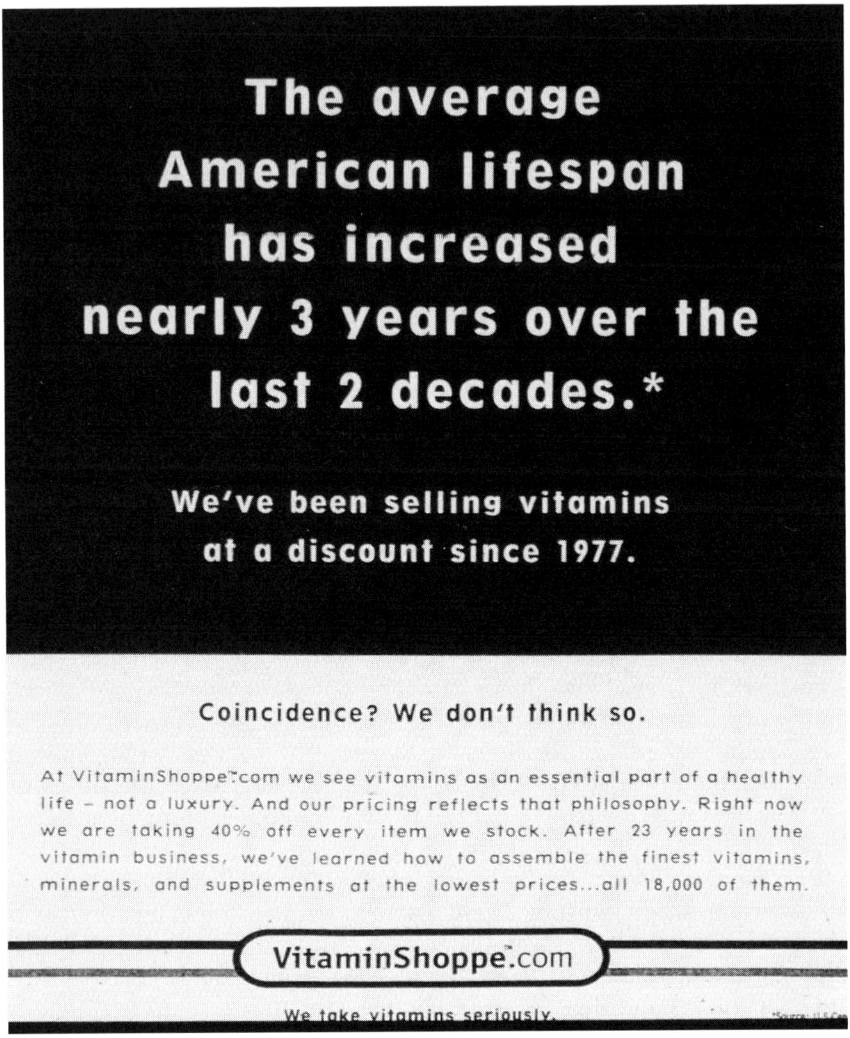

FIGURE 16-1 Advertisement from the *Boston Globe*.

domization to essentially the same supplements for the same period and found no protective effect. Importantly, the trial findings cannot be attributed to confounding or self-selection of healthier people into a vitamin-taking group, as taking or not taking vitamin E was determined randomly, which (providing it is done properly) avoids these sources of bias.

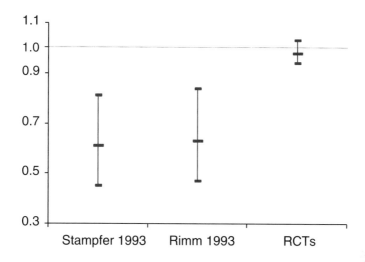

FIGURE 16-2a Vitamin E supplement use and risk of coronary heart disease in two observational studies (Rimm et al., 1993; Stampfer et al., 1993) and in a meta-analysis of randomized controlled trials (RCTs) (Eidelman et al., 2004).

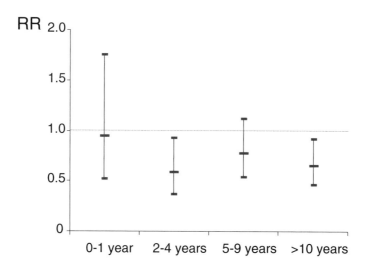

FIGURE 16-2b Health Professional Follow-up Study (Rimm et al., 1993). NOTE: Observed effect of duration of vitamin E use compared with no use on coronary heart disease events in the Health Professional Follow-up Study.

Other processes in addition to confounding can generate robust, but noncausal, associations in observational studies. Reverse causation—in which the disease influences the apparent exposure, rather than vice versa—may generate strong and replicable associations. For example, many studies have found that people with low circulating cholesterol levels are at increased risk of several cancers, including colon cancer. If causal, this is an important association, as it might mean that efforts to lower cholesterol levels would increase the risk of cancer. However, it is possible that the early stages of cancer may, many years before diagnosis or death, lead to a lowering in cholesterol levels, rather than low cholesterol levels increasing the risk of cancer. Similarly in studies of inflammatory markers, such as C-reactive protein and cardiovascular disease risk, it is possible that early stages of atherosclerosis—which is an inflammatory processes—elevate circulating inflammatory markers, and since people with atherosclerosis are more likely to experience cardiovascular events, a robust but noncausal association between levels of inflammatory markers and incident cardiovascular disease is generated. A form of reverse causation can also occur through reporting bias, with the presence of disease influencing reporting disposition. In case-control studies, people with the disease under investigation may report on their prior exposure history in a different way than do controls—perhaps because the former will think harder about potential reasons that account for why they have developed the disease.

These problems of confounding and bias produce associations in observational studies that are not reliable indicators of the true causation. Furthermore, the strength of associations between truly causal risk factors and disease in observational studies is underestimated due to random measurement imprecision in indexing the exposure. A century ago, Charles Spearman demonstrated mathematically how such measurement imprecision would lead to what he termed the "attenuation by errors" of associations (Spearman, 1904; Davey Smith and Phillips, 1996), later renamed "regression dilution bias."

Observational studies can and do produce findings that either spuriously enhance or downgrade estimates of causal associations between modifiable exposures and disease. For these reasons, alternative approaches—including those within the Mendelian randomization framework—need to be applied.

MENDELIAN RANDOMIZATION

The basic principle utilized in Mendelian randomization is that genetic variants that either alter the level of, or mirror the biological effects of, a modifiable environmental exposure that itself alters disease risk should be related to disease risk to the extent predicted by their influence on exposure to the environmental risk factor. Common genetic polymorphisms that have a well-characterized biological function (or are markers for such variants) can therefore be utilized to study the effect of a suspected environmental exposure on disease risk (Davey Smith and Ebrahim, 2003, 2004, 2005; Davey Smith, 2006). The exploitation of situations in which genotypic differences produce effects similar to environmental factors (and vice versa) clearly resonates with the concepts of phenocopy (Goldschmidt, 1938) and genocopy (Schmalhausen, 1938, cited by Gause, 1942) in developmental genetics. Phenocopy refers to the situation in which an environmental effect produces the same effect as that produced by a genetic mutation. Genocopy, the reverse of phenocopy, is when genetic variation generates an outcome that could be produced by an environmental stimulus (Jablonka-Tavory, 1982).

Why use genetic variants as proxies for environmental exposures rather than measure the exposures themselves? First, unlike environmental exposures, genetic variants are not generally associated with the wide range of behavioral, social, or physiological factors that, for example, confound the association between vitamin C and coronary heart disease. This means that if a genetic variant is used to proxy for an environmentally modifiable exposure, it is unlikely to be confounded in the way that direct measures of the exposure will be. Furthermore, aside from the effects of population structure (see Palmer and Cardon, 2005, for a discussion of the likely impact of this), such variants will not be associated with other genetic variants, except those with which they are in linkage disequilibrium. This latter assumption follows from the law of independent assortment (sometimes referred to as Mendel's second law), hence the term "Mendelian randomization." We illustrate this powerful aspect of Mendelian randomization in Tables 16-1a and 16-b, showing the strong associations between a wide range of variables and blood C-reactive protein (CRP) levels, but no association of the same factors with genetic variants in the CRP gene. The only factor related to genotype is the expected, biological influence of the genetic variant on CRP levels.

Second, we have seen how inferences drawn from observational studies may be subject to bias due to reverse causation. Disease processes may influence exposure levels, such as alcohol intake, or measures of intermediate phenotypes, such as cholesterol levels and C-reactive protein. However germ line genetic variants associated with average alcohol intake or circulating levels of intermediate phenotypes will not be influenced by the onset of disease. This will be equally true with respect to reporting bias generated by knowledge of disease status in case-control studies or of differential reporting bias in any study design.

Third, associative selection bias in which participants' entry to a study

TABLE 16-1a Means or Proportions of Blood Pressure, Pulse Pressure, Hypertension, and Potential Confounders by Quarters of C-Reactive Protein (CRP) N = 3,529

	Means or Proportions by Quarters of C-Reactive Protein (Range mg/L)				P Trend Across Categories
	1 (0.16-0.85)	2 (0.86-1.71)	3 (1.72-3.88)	4 (3.89-112.0)	
Hypertension (%)	45.8	49.7	57.5	60.0	< 0.001
Body mass index (kg/m2)	25.2	27.0	28.5	29.7	< 0.001
High-density lipoprotein cholesterol (mmol/l)	1.80	1.69	1.0	1.53	< 0.001
Lifecourse socioeconomic position score	4.08	4.37	4.46	4.75	< 0.001
Doctor diagnosis of diabetes (%)	3.5	2.8	4.1	8.4	< 0.001
Current smoker (%)	7.9	9.6	10.9	15.4	< 0.001
Physically inactive (%)	11.3	14.9	20.1	29.6	< 0.001
Moderate alcohol consumption (%)	22.2	19.6	18.8	14.0	< 0.001

SOURCE: Davey Smith et al. (2005).

TABLE 16-1b Means or Proportions of CRP Systolic Blood Pressure, Hypertension, and Potential Confounders by 1059G/C Genotype

	Means or Proportions by Genotype		
	GG	GC or CC	P
CRP (mg/L log scale)[a]	1.81	1.39	< 0.001
Hypertension (%)	53.3	53.1	0.95
Body mass index (kg/m^2)	27.5	27.8	0.29
High-density lipoprotein cholesterol (mmol/l)	1.67	1.65	0.38
Lifecourse socioeconomic position score	4.35	4.42	0.53
Doctor diagnosed diabetes (%)	4.7	4.5	0.80
Current smoker (%)	11.2	9.3	0.24
Physically inactive (%)	18.9	18.9	1.0
Moderate alcohol consumption (%)	18.6	19.8	0.56

[a]Geometric means and proportionate (%) change for a doubling of CRP.
CRP = C-reactive protein.
SOURCE: Davey Smith et al. (2005).

is related to both their exposure level and disease risk can generate spurious associations. This is unlikely to occur with respect to genetic variants. There is empirical evidence that a wide range of genetic variants and participation rates in etiological studies are not associated. Odds ratios for differences in the prevalence of genetic variants between those willing and less willing to participate in studies are generally null, showing no strong evidence to support associative selection bias in genetic studies (Bhatti et al., 2005). As these investigators noted, it is important that researchers test this assumption in their own data, as it is possible that other genotypes, particularly those associated with health-relevant behaviors (e.g., alcohol consumption), may show associations.

Finally, a genetic variant will indicate long-term levels of exposure, and if the variant is taken as a proxy for such exposure, it will not suffer from the measurement error inherent in phenotypes that have high levels of variability and are poorly estimated by a single measure. For example, groups defined by cholesterol level–related genotype will, over a long period, experience the cholesterol difference seen between the groups. Indeed, use of the Mendelian randomization approach predicts a strength of association that is in line with randomized controlled trial findings of effects of cholesterol lowering when the increasing benefits seen over the relatively short trial period are projected to the expectation for differences over a lifetime (Davey Smith and Ebrahim, 2004).

Categories of Mendelian Randomization

Several categories of inference can be drawn from studies utilizing the Mendelian randomization paradigm. In the most direct forms, genetic variants can be related to the probability or level of exposure ("exposure propensity") or to intermediate phenotypes believed to influence disease risk. Less direct evidence can come from genetic variant–disease associations that indicate that a particular biological pathway may be of importance, perhaps because the variants modify the effects of environmental exposures. Several examples of these categories have been given elsewhere (Davey Smith and Ebrahim, 2003, 2004; Davey Smith, 2006); here illustrative cases of the first two categories are briefly outlined.

Exposure Propensity: Alcohol Intake and Health

The possible protective effect of moderate alcohol consumption on risk of coronary heart disease (CHD) remains controversial (Marmot, 2001; Bovet and Paccaud, 2001; Klatsky, 2001). Nondrinkers may be at a higher risk of coronary heart disease because health problems (perhaps induced by previous alcohol abuse) dissuade them from drinking (Shaper,

1993). As well as this form of reverse causation, confounding could play a role, with nondrinkers being more likely to display an adverse profile of socioeconomic or other behavioral risk factors for coronary heart disease (Hart, Davey Smith, Hole, and Hawthorne, 1999). Alternatively, alcohol may have a direct biological effect that lessens the risk of coronary heart disease—for example, by increasing the levels of protective high-density lipoprotein (HDL) cholesterol (Rimm, 2001). It is, however, unlikely that a randomized controlled trial of alcohol intake, able to test whether there is a protective effect of alcohol on CHD events, will be carried out.

Alcohol is oxidized to acetaldehyde, which in turn is oxidized by aldehyde dehydrogenases (ALDHs) to acetate. Half of Japanese people are heterozygotes or homozygotes for a null variant of ALDH2, and peak blood acetaldehyde concentrations post alcohol challenge are 18 times and 5 times higher, respectively, among homozygous null variant and heterozygous individuals compared with homozygous wild type individuals (Enomoto, Takase, Yasuhara, and Takada, 1991). This renders the consumption of alcohol unpleasant through inducing facial flushing, palpitations, drowsiness, and other symptoms. As Figure 16-3a shows,

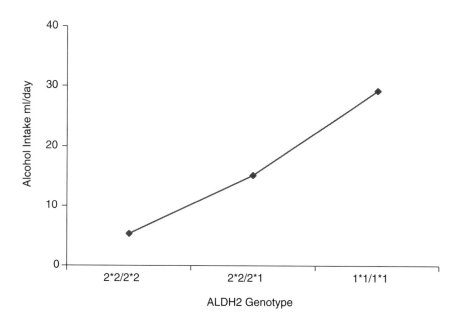

FIGURE 16-3a Relationship between characteristics and alcohol consumption.
SOURCE: Takagi et al. (2002).

there are very considerable differences in alcohol consumption according to genotype (Takagi et al., 2002). The principles of Mendelian randomization are seen to apply: two factors that would be expected to be associated with alcohol consumption, age and cigarette smoking, which would confound conventional observational associations between alcohol and disease, are not related to genotype despite the strong association of genotype with alcohol consumption (Figure 16-3b).

It would be expected that the ALDH2 genotype influences diseases known to be related to alcohol consumption, and as proof of principle it has been shown that ALDH2 null variant homozygosity—associated with low alcohol consumption—is indeed related to a lower risk of liver cirrhosis (Chao et al., 1994). Considerable evidence, including data from randomized controlled trials, suggests that alcohol increases HDL cholesterol levels (Haskell et al., 1984; Burr, Fehily, Butland, Bolton, and Eastham, 1986), which should protect against coronary heart disease. In line with this, ALDL2 genotype is strongly associated with HDL cholesterol in the expected direction (Figure 16-3c). Given the apparent protective effect of alcohol against CHD risk seen in observational studies, possession of the null ALDH2 allele—associated with lower alcohol consumption—should be associated with a greater risk of myocardial infarction, and this is what was seen in a case-control study (Takagi et al., 2002). Men either homozygous or heterozygous for null ALDH2 were at twice the risk of myocardial infarction. Statistical adjustment for HDL cholesterol greatly attenuated the association between ALDH2 genotype and coronary heart disease, indicating that the cardio-protective effect of alcohol is mediated by increased levels of HDL cholesterol.

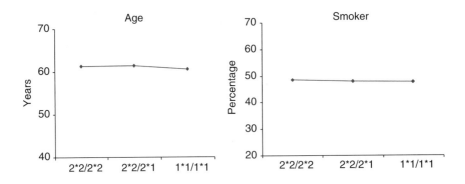

FIGURE 16-3b Relationship between characteristics and ALDH2 genotype.
SOURCE: Takagi et al. (2002).

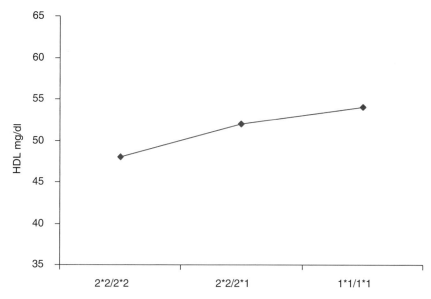

FIGURE 16-3c Relationship between HDL cholesterol and ALDH2 genotype.
SOURCE: Tagaki et al. (2002).

Intermediate Phenotypes

Genetic variants can influence such circulating biochemical factors as cholesterol, homocysteine, or fibrinogen levels. This provides a method for assessing causality in associations observed between these measures (*intermediate phenotypes*) and disease, and thus whether interventions to modify the intermediate phenotype could be expected to influence disease risk. Proof of principle for this approach is provided by familial hypercholesterolemia genetic variants that are associated with higher circulating cholesterol levels, which increase risk of coronary heart disease. These observational data are in line with the randomized controlled trial evidence confirming that lowering cholesterol reduces the risk of coronary heart disease (Davey Smith and Ebrahim, 2004).

C-Reactive Protein and Coronary Heart Disease

Strong associations of CRP, an acute phase inflammatory marker, with hypertension, insulin resistance, and coronary heart disease have been repeatedly observed (Danesh et al., 2004; Wu, Dorn, Donahue, Sempos, and Trevisan, 2002; Pradhan, Manson, Rifai, Buring, and Ridker, 2001; Han et al., 2002; Sesso et al., 2003; Hirschfield and Pepys, 2003; Hu, Meigs, Li, Rifai, and Manson, 2004), with the obvious inference that CRP is a cause of these conditions (Ridker et al., 2005; Sjöholm and Nyström, 2005; Verma, Szmitko, and Ridker, 2005). A Mendelian randomization study has examined the association between polymorphisms of the CRP gene and demonstrated that, although serum CRP differences were highly predictive of blood pressure and hypertension, the CRP variants—which are related to sizeable serum CRP differences—were not associated with these same outcomes (Davey Smith et al., 2005b). It is likely that these divergent findings are explained by the extensive confounding between serum CRP and outcomes (as shown in Table 16-1). Current evidence on this issue, although statistically underpowered, also suggests that CRP levels do not lead to elevated risk of insulin resistance (Timpson et al., 2005) or coronary heart disease (Casas et al., 2006). Again, confounding, and reverse causation—in which existing coronary disease or insulin resistance may influence CRP levels—could account for this discrepancy. Similar findings have been reported for serum fibrinogen, variants in the beta fibrinogen gene, and coronary heart disease (Davey Smith et al., 2005a; Keavney et al., 2006). The CRP and fibrinogen examples demonstrate that Mendelian randomization can increase evidence for a causal effect of an environmentally modifiable factor (as in the cases of alcohol and cholesterol levels discussed earlier) as well as provide evidence against causal effects, which can help direct efforts away from targets of no preventative or therapeutic relevance.

Implications of Mendelian Randomization Study Findings

Establishing the causal influence of environmentally modifiable risk factors from Mendelian randomization designs informs policies for improving population health through population-level interventions. They do not imply that the appropriate strategy is genetic screening to identify those at high risk and application of selective exposure reduction policies. For example, establishing the association between genetic variants (such as familial defective ApoB) associated with elevated cholesterol level and CHD risk strengthens causal evidence that elevated cholesterol is a modifiable risk factor for coronary heart disease for the whole population. Thus even though the population attributable risk for coronary heart disease of this variant is small, it usefully informs public health approaches to improving population health. It is this aspect of Mendelian randomization that illustrates its distinction from conventional risk identification and genetic screening purposes of genetic epidemiology.

Mendelian Randomization and Randomized Controlled Trials

Randomized controlled trials are clearly the definitive means of obtaining evidence on the effects of modifying disease risk processes. There are similarities in the logical structure of randomized controlled trials and Mendelian randomization, however. Figure 16-4 illustrates this, drawing attention to the unconfounded nature of exposures proxied for by genetic variants (analogous to the unconfounded nature of a randomized intervention), the lack of possibility of reverse causation as an influence on exposure-outcome associations in both Mendelian randomization and randomized controlled trial settings, and the importance of intention to treat analyses—that is, comparisons of groups defined by genetic variant, irrespective of associations between the genetic variant and the proxied for exposure within any particular individual.

The analogy with randomized controlled trials is also useful in understanding why an objection to Mendelian randomization—that the environmentally modifiable exposure proxied for by the genetic variants (such as alcohol intake or circulating CRP levels) are influenced by many other factors in addition to the genetic variants (Jousilahti and Salomaa, 2004)—while true, is of no consequence. Consider a randomized con-

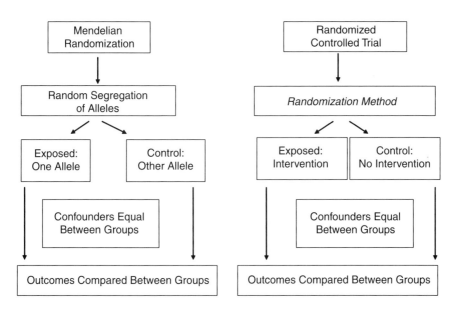

FIGURE 16-4 Mendelian randomization and randomized controlled trial designs compared.

trolled trial of blood pressure–lowering medication. Blood pressure is influenced mainly by factors other than taking blood pressure–lowering medication. Obesity, alcohol intake, salt consumption and other dietary factors, smoking, exercise, physical fitness, genetic factors, and early life developmental influences are all of importance. However, the randomization that occurs in trials ensures that these factors are balanced between the group that receives the blood pressure–lowering medication and the control group that does not. Thus the fact that many other factors are related to the modifiable exposure does not vitiate the power of randomized controlled trials; neither does it vitiate the strength of Mendelian randomization designs.

A related objection is that genetic variants often explain only a trivial proportion of the variance in the environmentally modifiable risk factor that is being proxied for (Glynn, 2006). Again, consider a randomized controlled trial of blood pressure–lowering medication, in which 50 percent of participants receive the medication and 50 percent receive a placebo. If the antihypertensive therapy reduced blood pressure by a quarter of a standard deviation, which is approximately the situation for such pharmacotherapy, then within the whole study group the treatment assignment (i.e., antihypertensive use versus placebo) will explain 1.5 percent of the variance in blood pressure. In the example of CRP haplotypes used as markers for CRP levels, these haplotypes explain 1.7 percent of the variance in CRP levels in the population (Lawlor, Harbord, Sterne, and Davey Smith, 2007). As can be seen, the quantitative association of genetic variants as proxies can be similar to that of randomized treatments with respect to biological processes that such treatments modify. Both logic and quantification fail to support criticisms of the Mendelian randomization approach on the basis of either the obvious fact that many factors influence most phenotypes of interest or that particular genetic variants account for only a small proportion of variance in the phenotype.

Mendelian Randomization and Instrumental Variable Approaches

As well as the analogy with randomized controlled trials, Mendelian randomization can also be likened to instrumental variable approaches that have been heavily utilized in econometrics and social science. In this approach, the instrument is a variable that is related to the outcome only through its association with the modifiable exposure of interest. The instrument is not related to confounding factors, nor is its assessment biased in a manner that would generate a spurious association with the outcome. Furthermore the instrument will not be influenced by the development of the outcome (i.e., there will be no reverse causation). Figure 16-5 presents this basic schema, where the dotted line between genotype

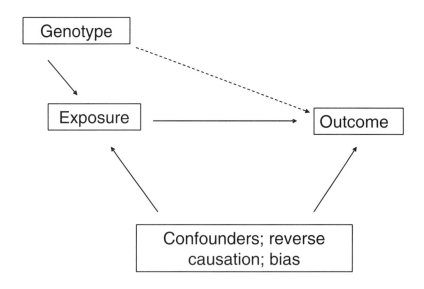

FIGURE 16-5 Mendelian randomization as an instrumental variables approach.

and the outcome provides an unconfounded and unbiased estimate of the causal association between the exposure that the genotype is proxying for and the outcome. The development of instrumental variable methods in econometrics, in particular, has led to a sophisticated range of statistical methods for estimating causal effects, and these have now been applied in Mendelian randomization studies (e.g., Davey Smith et al., 2005a, 2005b; Timpson et al., 2005). The parallels between Mendelian randomization and instrumental variable approaches are discussed in more detail elsewhere (Thomas and Conti, 2004; Didelez and Sheehan, 2007; Lawlor et al., 2007).

Mendelian Randomization and Gene-Environment Interaction

Mendelian randomization is one way in which genetic epidemiology can inform understanding about environmental determinants of disease. A more conventional approach has been to study interactions between environmental exposures and genotype (Perera, 1997; Mucci, Wedren, Tamimi, Trichopoulos, and Adami, 2001). From epidemiological and Mendelian randomization perspectives, several issues arise with gene-environment interactions.

The most reliable findings in genetic association studies relate to the main effects of polymorphisms on disease risk (Clayton and McKeigue, 2001). The power to detect meaningful gene-environment interaction is low (Wright, Carothers, and Campbell, 2002), resulting in a large number of reports of spurious gene-environment interactions in the medical literature (Colhoun, McKeigue, and Davey Smith, 2003). The presence or absence of statistical interactions depends on the scale used (i.e., linear or logarithmic association between the exposure and disease outcome) and the difficulty in defining whether deviation from either an additive or multiplicative model exists, given the imprecision of estimation. Measurement error—particularly if differential with respect to other factors influencing disease risk—makes interactions both difficult to detect and often misleading when they are apparently found (Clayton and McKeigue, 2001). Furthermore, the biological implications of interactions (however defined) are generally uncertain (Thompson, 1991).

The situation may be different with exposures that differ qualitatively rather than quantitatively between individuals. Consider the possible influence of smoking tobacco on bladder cancer risk. Observational studies suggest an association, but clearly confounding, and a variety of biases could generate such an association. The potential carcinogens in tobacco smoke of relevance to bladder cancer include aromatic and heterocyclic amines, which are detoxified by N-acetyl transferase 2 (NAT2). Genetic variation in NAT2 enzyme levels leads to slower or faster acetylation states. If the carcinogens in tobacco smoke do increase the risk of bladder cancer, then it would be expected that slow acetylators, who have a reduced rate of detoxification of these carcinogens, would be at an increased risk of bladder cancer if they were smokers, whereas if they were not exposed to these carcinogens (and the major exposure route for those outside of particular industries is through tobacco smoke) then an association of genotype with bladder cancer risk would not be anticipated. Table 16-2 tabulates findings from the largest study to date reported in a way that allows analysis of this simple hypothesis (Garcia-Closas et al., 2005). As can be seen, the influence of the NAT2 slow acetylation genotype is appreciable only among those exposed to heavy smoking. Since the geno-

TABLE 16-2 Association of NAT2 Slow Acetylation Genotype with Bladder Cancer in Never and Ever Smokers and Overall. Odds Ratio (95% confidence intervals)

Overall	Never Smokers	Ever Smokers
1.4 (1.2-1.7)	0.9 (0.6-1.3)	1.6 (1.3-1.9)

SOURCE: Garcia-Closas et al. (2005).

type will be unrelated to confounders, it is difficult to reason why this situation should arise, unless smoking is a causal factor with respect to bladder cancer. Thus the presence of a sizable effect of genotype in heavy smokers but not nonsmokers provides evidence of the causal nature of an environmentally modifiable risk factor, in this example, smoking.

However, gene-environment interactions interpreted within the Mendelian randomization framework are not protected from confounding in the same way as the main genetic effects are.

PROBLEMS AND LIMITATIONS OF MENDELIAN RANDOMIZATION

We consider Mendelian randomization to be one of the brightest current prospects for improving causal understanding in population-based studies. There are, however, several potential limitations to the application of this methodology (Davey Smith and Ebrahim, 2003; Little and Khoury, 2003), which we discuss below.

Failure to Establish Reliable Genotype-Intermediate Phenotype or Genotype-Disease Associations

If the associations between genotype and a potential intermediate phenotype, or between genotype and disease outcome, are not reliably estimated, then interpreting these associations in terms of their implications for potential environmental causes of disease will clearly be inappropriate. This is not an issue peculiar to Mendelian randomization; instead, the nonreplicable nature of perhaps most apparent findings in genetic association studies is a serious limitation to the whole enterprise. In Box 16-1 we summarize possible reasons for the nonreplication of findings (Cardon and Bell, 2001; Colhoun et al., 2003). Population stratification—that is, the confounding of genotype-disease associations by factors related to subpopulation group membership in the overall population in a study—is unlikely to be a major problem in most situations (Wacholder, Rothman, and Caporaso, 2000; Wacholder et al., 2002; Palmer and Cardon, 2005). Genotyping errors can, of course, lead to failures of replication of genotype-disease associations. When intermediate phenotypes can be measured, as in the case of CRP, a demonstration of the expected relationship between genotype and intermediate phenotype in such studies indicates that genotyping errors are not to blame.

Regarding failure to replicate results in genetic epidemiology, true variation between studies is clearly possible—for example, people heterozygous for familial hypercholesterolemia seem to experience increased mortality only in populations with substantial dietary fat intake and the presence of other CHD risk factors (Sijbrands et al., 2001; Pimstone et al.,

---

BOX 16-1
Reasons for Inconsistent Genotype-Phenotype Associations

True variation
Variation of allelic association between subpopulations: (1) disease causing allele in linkage disequilibrium with different marker alleles in different populations; or (2) different variants within the same geno contribute to disease risk in different populations

Effect modification by other genetic or environmental factors that vary between populations

Spurious variation
Genotyping errors
Misclassification of phenotype
Confounding by population structure
Lack of power
Chance
Publication bias

SOURCES: Cardon and Bell (2001); Colhoun et al. (2003).

1998). Nevertheless, the major factor for nonreplication is probably inadequate statistical power (generally reflecting limited sample size), coupled with publication bias (Colhoun et al., 2003).

Confounding of Genotype: Environmentally Modifiable Risk Factor–Disease Associations

The power of Mendelian randomization lies in its ability to avoid the often substantial confounding seen in conventional observational epidemiology. However, confounding can be reintroduced into Mendelian randomization studies and needs to be considered when interpreting the results.

Linkage Disequlibrium

It is possible that the locus under study is in linkage disequilibrium with another polymorphic locus. Confounding will result if both the study locus and that with which it is in linkage disequilibrium are both associated with the outcome of interest. It may seem unlikely—given the relatively short distances over which linkage disequilibrium is seen in the human genome—that a polymorphism influencing, say, CHD risk would be associated with another polymorphism influencing CHD risk (and thus producing confounding). There are, nevertheless, cases of different genes influencing the same metabolic pathway being in physical proximity. For example, different polymorphisms influencing alcohol metabolism appear to be in linkage disequilibrium (Osier et al., 2002).

Pleiotropy and the Multifunction of Genes

Mendelian randomization is most useful when it can be used to relate a single intermediate phenotype to a disease outcome. However, polymorphisms may (and probably often will) influence more that one intermediate phenotype, and this may mean that they proxy for more than one environmentally modifiable risk factor. This can be the case through multiple effects mediated by their RNA expression or immediate protein coding, through alternative splicing, in which one polymorphic region contributes to alternative forms of more than one protein (Glebart, 1998), or through other mechanisms. The most robust interpretations will be possible when the functional polymorphism appears to directly influence the level of the intermediate phenotype of interest (as in the CRP example), but such examples are probably going to be less common in Mendelian randomization than in cases in which the polymorphism can influence several systems, with different potential interpretations of how the effect on outcome is generated.

The association of possession of the ApoE-2 allele with cholesterol levels and coronary heart disease might be an example of pleiotropic effects, since carriers of this allele have lower cholesterol levels but do not have the degree of protection against coronary heart disease that would be anticipated from this (Keavney et al., 2004; Song, Stampfer, and Liu, 2004). In addition to lower cholesterol levels, the ApoE-2 allele is associated with less efficient transfer of very low density lipoproteins and chylomicrons from the blood to the liver, greater postprandial lipemia, and an increased risk of type III hyperlipidemia (Smith, 2002; Eichner et al., 2002). These differences will accompany the lower cholesterol levels and may counterbalance the predicted benefits.

Multiple Instruments as an Approach to Confounding in Mendelian Randomization

Linkage disequilibrium and pleiotropy can reintroduce confounding and vitiate the power of the Mendelian randomization approach. Genomic knowledge may help in estimating the degree to which these are likely to be problems in any particular Mendelian randomization study, through, for instance, explication of genetic variants that may be in linkage disequilibrium with the variant under study, or the function of a particular variant and its known pleiotropic effects. Furthermore, genetic variation can be related to measures of potential confounding factors in each study, and the magnitude of such confounding estimated. Empirical studies to date suggest that common genetic variants are largely unrelated to the behavioral and socioeconomic factors considered to be important confounders in conventional observational studies (Smits et al., 2004; Bhatti et al., 2005; Davey Smith et al., 2005b, 2007; Chatterjee, Kalaylioglu, and Carroll, 2005; Umbach and Weinberg, 1997). However, relying on measurement of confounders does, of course, remove the central purpose of Mendelian randomization, which is to balance unmeasured as well as measured confounders (as randomization does in randomized controlled trials).

It may be possible to identify two separate genetic variants, which are not in linkage disequilibrium with each other, but which both serve as proxies for the environmentally modifiable risk factor of interest. If both variants are related to the outcome of interest and point to the same underlying association, then it becomes much less plausible that reintroduced confounding explains the association, since it would have to be acting in the same way for these two unlinked variants. This can be likened to randomized controlled trials of different blood pressure–lowering agents, which work through different mechanisms and have different potential side effects but lower blood pressure to the same degree. If the different agents produce the same reductions in cardiovascular disease risk, then it is unlikely that this is through agent-specific effects of the drugs; instead, it points to blood pressure lowering as being key. The use of multiple genetic variants working through different pathways has not been applied in Mendelian randomization to date, but it represents an important potential development in the methodology.

Canalization and Developmental Stability

Perhaps a greater potential problem for Mendelian randomization than reintroduced confounding arises from the developmental compensation that may occur through a polymorphic genotype being expressed during fetal or early postnatal development, and thus influencing development in such a way as to buffer against the effect of the polymorphism. Such compensatory processes have been discussed since C.H. Waddington introduced the notion of canalization in the 1940s (Waddington, 1942). Canalization refers to the buffering of the effects of either environmental or genetic forces attempting to perturb development, and Waddingtion's ideas have been well developed both empirically and theoretically (Wilkins, 1997; Rutherford, 2000; Gibson and Wagner, 2000; Hartman, Garrik, and Hartwell, 2001; Debat and David, 2001; Kitami and Nadeau, 2002; Gu et al., 2003; Hornstein and Shomron, 2006). Such buffering can be achieved either through genetic redundancy (more than one gene having the same or similar function) or through alternative metabolic routes, in which the complexity of metabolic pathways allows recruitment of different pathways to reach the same phenotypic end point. In effect, a functional polymorphism expressed during fetal development or postnatal growth may influence the expression of a wide range of other genes, leading to changes that may compensate for the influence of the polymorphism.

In the field of animal genetic engineering studies, such as knockout preparations or transgenic animals manipulated so as to overexpress foreign DNA, the interpretive problem created by developmental compensation is well recognized (Morange, 2001; Shastry, 1998; Gerlai, 2001; Williams and Wagner, 2000). Conditional preparations, in which the level of transgene expression can be induced or suppressed through the application of external agents, are now being utilized to investigate the influence of such altered gene expression after the developmental stages during which compensation can occur (Bolon and Galbreath, 2002). Thus further evidence on the issue of genetic buffering should emerge to inform interpretations of both animal and human studies.

Most examples of developmental compensation relate to dramatic genetic or environmental insults, so it is unclear whether the generally small phenotypic differences induced by common functional polymorphisms will be sufficient to induce compensatory responses. The fact that the large gene-environment interactions that have been observed often relate to novel exposures (e.g., drug interactions) that have not been present during the evolution of a species (Wright et al., 2002) may indicate that homogenization of response to exposures that are widely experienced—as would be the case with the products of functional polymorphisms or common mutations—has occurred; canalizing mechanisms could be particularly relevant in these cases. Further work on the basic mechanisms of developmental stability and how this relates to relatively small exposure differences during development will allow these considerations to be understood. Knowledge of the stage of development at which a genetic variant has functional effects will also allow the potential of developmental compensation to buffer the response to the variant to be assessed.

In some Mendelian randomization designs, developmental compensation is not an issue. For example, when maternal genotype is utilized as an indicator of the intrauterine environment, then the response of the fetus will not differ whether the effect is induced by maternal genotype or by environmental perturbation, and the effect on the fetus can be taken to indicate the effect of environmental influences during the intra-uterine period. Also, in cases in which a variant influences an adulthood environmental exposure—for example, ALDH2 variation and alcohol intake—developmental compensation to genotype will not be an issue. In many cases of gene-environment interaction interpreted with respect to causality of the environmental factor, the same applies. However, in some situations there remains the somewhat unsatisfactory position of Mendelian randomization facing a potential problem that cannot currently be adequately assessed.

Lack of Suitable Genetic Variants to Proxy for the Exposure of Interest

An obvious limitation of Mendelian randomization is that it can only examine areas for which there are functional polymorphisms (or genetic markers linked to such functional polymorphisms) that are relevant to the modifiable exposure of interest. In the context of genetic association studies more generally, it has been pointed out that, in many cases, even if a locus is involved in a disease-related metabolic process, there may be no suitable marker or functional polymorphism to allow study of this process (Weiss and Terwillger, 2000). Since one of our examples, used in an earlier paper (Davey Smith and Ebrahim, 2003), of how observational epidemiology appeared to have got the wrong answer related to vitamin C, we considered whether the association between vitamin C and coronary heart disease could have been studied utilizing the principles of Mendelian randomization. We stated that polymorphisms exist that are related to lower circulating vitamin C levels—for example, the haptoglobin polymorphism (Langlois, Delanghe, DeBuyzere, Bernard, and Ouyang, 1997; Delanghe, Langlois, Duprez, DuBuyzere, and Clement, 1999)—but in this case the effect on vitamin C is at some distance from the polymorphic protein and, as in the apolipoprotein E example, other phenotypic differences could have an influence on CHD risk that would distort examination of the influence of vitamin C levels through relating genotype to disease. SLC23A1—a gene encoding for the vitamin C transporter SVCT1, vitamin C transport by intestinal cells— would be an attractive candidate for Mendelian randomization studies (Erichsen, Eck, Levine, and Chanock, 2001). However, by 2003 (the date of our earlier paper) a search for variants had failed to find any common single-nucleotide polymorphism that could be used in such a way. However, since then, functional variation in SLC23A1 that is related to circulating vitamin C levels has been identified (Timpson et al., personal communication). Rapidly developing knowledge of human genomics will identify more variants that can serve as instruments for Mendelian randomization studies.

CONCLUSIONS: MENDELIAN RANDOMIZATION, WHAT IT IS, AND WHAT IT IS NOT

Mendelian randomization is not predicated on the presumption that genetic variants are major determinants of health and disease in populations. There are many cogent critiques of genetic reductionism and the overselling of "discoveries" in genetics that reiterate obvious truths so clearly (albeit somewhat repetitively) that there is no need to repeat them here (e.g., Berkowitz, 1996; Baird, 2000; Holtzman, 2001; Strohman, 1993; Rose, 1995). Mendelian randomization does not depend on there being genes "for" particular traits, and certainly not in the strict sense of a gene "for" a trait being one that is maintained by selection because of its causal association with that trait (Kaplan and Pigliucci, 2001). The association of genotype and the environmentally modifiable factor that it proxies for will be, like most genotype-phenotype associations, one that is contingent and cannot be reduced to individual-level prediction, but within environmental limits will pertain at a group level (Wolf, 1995). This is analogous to a randomized controlled trial of antihypertensive agents, in which at a collective level the group randomized to active medication will have lower mean blood pressure than the group randomized to placebo, but at an individual level many participants randomized to active treatment will have higher blood pressure than many individuals randomized to placebo. These group-level differences are what create the analogy between Mendelian randomization and randomized controlled trials, outlined in Figure 16-4.

Finally, the associations that Mendelian randomization depend on do need to pertain to a definable group at a particular time, but they do not need to be immutable. Thus ALDH2 variation will not be related to alcohol consumption in a society in which alcohol is not consumed, and the association will vary by gender and by cultural group and may change over time (Higuchi et al., 1994; Hasin et al., 2002). Within the setting of a study of a well-defined group, however, the genotype will be associated with group-level differences in alcohol consumption, and group assignment will not be associated with confounding variables.

Mendelian Randomization and Genetic Epidemiology

Critiques of contemporary genetic epidemiology often focus on two features of findings from genetic association studies: that the population-attributable risk of the genetic variants is low, and that in any case the influence of genetic factors is not reversible (Terwilliger and Weiss, 2003). These evaluations of the role of genetic epidemiology are not relevant when considering the potential contributions of Mendelian randomization. This approach is not concerned with the population attributable risk of any particular genetic variant, but the degree to which associations between the genetic variant and disease outcomes can demonstrate the importance of environmentally modifiable factors as causes of disease. Consider, for example, the case of familial hypercholesterolemia or familial defective ApoB. The genetic mutations associated with these conditions will account for only a trivial percentage of cases of coronary heart disease in the population—that is, the population attributable risk will be low, despite a high relative risk of coronary heart disease (Tybjaerg H et al., 1998). However, by identifying blood cholesterol levels as a causal factor for coronary heart disease, the triangulation between genotype, blood cholesterol, and CHD risk identifies an environmentally modifiable factor with a very high population attributable risk—assuming that 50 percent of the population have raised blood cholesterol above 6.0 mmol/l, and this is associated with a relative risk of two-fold, a population attributable risk of 33 percent is obtained. The same reasoning applies to the nonmodifiable nature of genotype-disease associations. The point of Mendelian randomization approaches is to utilize these associations to strengthen inferences regarding modifiable *environmental* risks for disease and then reduce disease risk in the population through applying this knowledge.

Mendelian randomization differs from other contemporary approaches to genetic epidemiology in that its central concern is not with the magnitude of genetic variant influences on disease, but rather on what the genetic associations tell us about environmentally modifiable causes of disease. As David B. Abrams, director of the Office of Behavioral and Social Sciences Research at the U.S. National Institutes of Health has said, "The more we learn about genes, the more we see how important environment and lifestyle really are." Many years earlier, the pioneering geneticist Thomas Hunt Morgan articulated a similar sentiment in his Nobel prize acceptance speech, when he contrasted his views with the then popular genetic approach to disease—eugenics. He thought that "through public hygiene and protective measures of various kinds we can more successfully cope with some of the evils that human flesh is heir to. Medical science will here take the lead—but I hope that genetics can at times offer a helping hand" (Morgan, 1935). More than seven decades later, it might now be time that genetic research can directly strengthen the knowledge base of public health.

REFERENCES

Alpha-Tocopherol, Beta Carotene Cancer Prevention Study Group. (1994). The effect of vitamin E and beta carotene on the incidence of lung cancer and other cancers in male smokers. *New England Journal of Medicine, 330,* 1029-1035.

Baird, P. (2000). Genetic technologies and achieving health for populations. *International Journal of Health Services, 30,* 407-424.

Beral, V., Banks, E., and Reeves, G. (2002). Evidence from randomized trials of the long-term effects of hormone replacement therapy. *Lancet, 360,* 942-944.

Berkowitz, A. (1996). Our genes, ourselves? *Bioscience, 46,* 42-51.

Bhatti, P., Sigurdson, A.J., Wang, S.S., Chen, J., Rothman, N., Hartge, P., Bergen, A.W., and Landi, M.T. (2005). Genetic variation and willingness to participate in epidemiological research: Data from three studies. *Cancer Epidemiology Biomarkers and Prevention, 14,* 2449-2453.

Bolon, B., and Galbreath, E. (2002). Use of genetically engineered mice in drug discovery and development: Wielding Occam's razor to prune the product portfolio. *International Journal of Toxicology, 21,* 55-64.

Bovet, P., and Paccaud, F. (2001). Alcohol, coronary heart disease, and public health: Which evidence-based policy? *International Journal of Epidemiology, 30,* 734-737.

Burr, M.L., Fehily, A.M., Butland, B.K., Bolton, C.H., and Eastham, R.D. (1986). Alcohol and high-density-lipoprotein cholesterol: A randomised controlled trial. *British Journal of Nutrition, 56,* 81-86.

Cardon, L.R., and Bell, J.I. (2001). Association study designs for complex diseases. *Nature Reviews: Genetics, 2,* 91-99.

Casas, J.P., et al. (2006). Insight into the nature of the CRP–coronary event association using Mendelian randomization. *International Journal of Epidemiology, 35,* 922-931.

Chao, Y.-C., Liou, S.-R., Chung, Y.-Y., Tang, H.-S., Hsu, C.-T., Li, T.-K., and Yin, S.-J. (1994). Polymorphism of alcohol and aldehyde dehydrogenase genes and alcoholic cirrhosis in Chinese patients. *Hepatology, 19,* 360-366.

Chatterjee, N., Kalaylioglu, Z., and Carroll, R.J. (2005). Exploiting gene-environment independence in family-based case-control studies: Increased power for detecting associations, interactions, and joint effects. *Genetic Epidemiology, 28,* 138-156.

Clayton, D., and McKeigue, P.M. (2001). Epidemiological methods for studying genes and environmental factors in complex diseases. *Lancet, 358,* 1356-1360.

Colhoun, H., McKeigue, P.M., and Davey Smith, G. (2003). Problems of reporting genetic associations with complex outcomes. *Lancet, 361,* 865-872.

Danesh, J., Wheeler, J.G., Hirschfield, G.M., Eda, S., Eiriksdottir, G., Rumley, A., Lowe, G.D.O., Pepys, M.B., and Gudnason, V. (2004). C-reactive protein and other circulating markers of inflammation in the prediction of coronary heart disease. *New England Journal of Medicine, 350,* 1387-1397.

Davey Smith, G. (2006). Cochrane lecture: Randomised by (your) god: Robust inference from an observational study design. *Journal of Epidemiology and Community Health, 60,* 382-388.

Davey Smith, G., and Ebrahim, S. (2002). Data dredging, bias, or confounding (editorial). *British Medical Journal, 325,* 1437-1438.

Davey Smith, G., and Ebrahim, S. (2003). Mendelian randomization: Can genetic epidemiology contribute to understanding environmental determinants of disease? *International Journal of Epidemiology, 32,* 1-22.

Davey Smith, G., and Ebrahim, S. (2004). Mendelian randomization: Prospects, potentials, and limitations. *International Journal of Epidemiology, 33,* 30-42.

Davey Smith, G., and Ebrahim, S. (2005). What can Mendelian randomization tell us about modifiable behavioural and environmental exposures? *British Medical Journal, 330,* 1076-1079.

Davey Smith, G., Harbord, R., Milton, J., Ebrahim, S., and Sterne, J.A.C. (2005a). Does elevated plasma fibrinogen increase the risk of coronary heart disease? Evidence from a meta-analysis of genetic association studies. *Arteriosclerosis, Thrombosis, and Vascular Biology, 25,* 2228-2233.

Davey Smith, G., Lawlor, D., Harbord, R., Timpson, N., Rumley, A., Lowe, G., Day, I., and Ebrahim, S. (2005b). Association of C-reactive protein with blood pressure and hypertension: Lifecourse confounding and Mendelian randomization tests of causality. *Arteriosclerosis, Thrombosis, and Vascular Biology, 25,* 1051-1056.

Davey Smith, G., Lawlor, D., Harbord, R., Timpson, N., Day, I., and Ebrahim, S. (2007). Clustered environments and randomized genes: A fundamental distinction between conventional and genetic epidemiology. *Public Library of Science Medicine.*

Davey Smith, G., and Phillips, A.N. (1996). Inflation in epidemiology: The proof and measurement of association between two things revisited. *British Medical Journal, 312,* 1659-1661.

Debat, V., and David, P. (2001). Mapping phenotypes: Canalization, plasticity, and developmental stability. *Trends in Ecology and Evolution, 16,* 555-561.

Delanghe, J., Langlois, M., Duprez, D., De Buyzere, M., and Clement, D. (1999). Haptoglobin polymorphism and peripheral arterial occlusive disease. *Atherosclerosis, 145,* 287-292.

Didelez, V., and Sheehan, N.A. (2007). Mendelian randomization: Why epidemiology needs a formal language for causality. In F. Russo and J. Williamson (Eds.), *Causality and Probability in the Sciences* (pp. 1-30). London: College Publications.

Eichner, J.E., Dunn, S.T., Perveen, G., Thompson, D.M., Stewart, K.E., and Stroehla, B.C. (2002). Apolipoprotein E polymorphism and cardiovascular disease: A HuGE Review. *American Journal of Epidemiology, 155,* 487-495.

Eidelman, R.S., Hollar, D., Hebert, P.R., Lamas, G.A., and Hennekens, C.H. (2004). Randomized trials of vitamin E in the treatment and prevention of cardiovascular disease. *Archives of Internal Medicine, 164,* 1552-1556.

Enomoto, N., Takase, S., Yasuhara, M., and Takada, A. (1991). Acetaldehyde metabolism in different aldehyde dehydrogenase-2 genotypes. *Alcoholism: Clinical and Experimental Research, 15,* 141-144.

Erichsen, H.C., Eck, P., Levine, M., and Chanock, S. (2001). Characterization of the genomic structure of the human vitamin C transporter SVCT1 (*SLC23A2*). *Journal of Nutrition, 131,* 2623-2627.

Ewbank, D. (2001). Demography in the age of genomics: A first look at the prospects. In National Research Council, *Cells and surveys: Should biological measures be included in social science research?* (pp. 64-109), Committee on Population, C.E. Finch, J.E. Vaupel, and K. Kinsella (Eds.). Washington, DC: National Academy Press.

Garcia-Closas, M., et al. (2005). NAT2 slow acetylation, *GSTM1* null genotype, and risk of bladder cancer: Results from the Spanish Bladder Cancer Study and meta-analyses. *Lancet, 366,* 649-659.

Gause, G.F. (1942). The relation of adaptability to adaptation. *Quarterly Review of Biology, 17,* 99-114.

Gerlai, R. (2001). Gene targeting: Technical confounds and potential solutions in behavioural and brain research. *Behavioural Brain Research, 125,* 13-21.

Gibson, G., and Wagner, G. (2000). Canalization in evolutionary genetics: A stabilizing theory? *BioEssays, 22,* 372-380.

Glebart, W.M. (1998). Databases in genomic research. *Science, 282,* 659-661.

Glynn, R.K. (2006). Commentary: Genes as instruments for evaluation of markers and causes. *International Journal of Epidemiology, 35,* 932-934.

Goldschmidt, R.B. (1938). *Physiological genetics.* New York: McGraw Hill.

Gu, Z., Steinmetz, L.M., Gu, X., Scharfe, C., Davis, R.W., and Li, W.-H. (2003). Role of duplicate genes in genetic robustness against null mutations. *Nature, 421,* 63-66.

Han, T.S., Sattar, N., Williams, K., Gonzalez-Villalpando, C., Lean, M.E., and Haffner, S.M. (2002). Prospective study of C-reactive protein in relation to the development of diabetes and metabolic syndrome in the Mexico City diabetes study. *Diabetes Care, 25,* 2016-2021.

Hart, C., Davey Smith, G., Hole, D., and Hawthorne, V. (1999). Alcohol consumption and mortality from all causes, coronary heart disease, and stroke: Results from a prospective cohort study of Scottish men with 21 years of follow up. *British Medical Journal, 318,* 1725-1729.

Hartman, J.L., Garvik, B., and Hartwell, L. (2001). Principles for the buffering of genetic variation. *Science, 291,* 1001-1004.

Hasin, D., Aharonovich, E., Liu, X., Mamman, Z., Matseoane, K., Carr, L., and Li, T.K. (2002). Alcohol and ADH2 in Israel: Ashkenazis, Sephardics, and recent Russian immigrants. *American Journal of Psychiatry, 159*(8), 1432-1434.

Haskell, W.L., Camargo, C., Williams, P.T., Vranizan, K.M., Krauss, R.M., Lindgren, F.T., and Wood, P.D. (1984). The effect of cessation and resumption of moderate alcohol intake on serum high-density-lipoprotein subfractions. *New England Journal of Medicine, 310,* 805-810.

Heart Protection Study Collaborative Group. (2002). MRC/BHF Heart Protection Study of antioxidant vitamin supplementation in 20536 high-risk individuals: A randomised placebo-controlled trial. *Lancet, 360,* 23-33.

Higuchi, S., Matsushita, S., Imazeki, H., Kinoshita, T., Takagi, S., and Kono, H. (1994). Aldehyde dehydrogenase genotypes in Japanese alcoholics. *Lancet, 343,* 741-742.

Hirschfield, G.M., and Pepys, M.B. (2003). C-reactive protein and cardiovascular disease: New insights from an old molecule. *Quarterly Journal of Medicine, 9,* 793-807.

Holtzman, N.A. (2001). Putting the search for genes in perspective. *International Journal of Health Services, 31,* 445.

Hornstein, E., and Shomron, N. (2006). Canalization of development by microRNAs. *Nature Genetics, 38,* S20-S24.

Hu, F.B., Meigs, J.B., Li, T.Y., Rifai, N., and Manson, J.E. (2004). Inflammatory markers and risk of developing type 2 diabetes in women. *Diabetes, 53,* 693-700.

Jablonka-Tavory, E. (1982). Genocopies and the evolution of interdependence. *Evolutionary Theory, 6,* 167-170.

Jousilahti, P., and Salomaa, V. (2004). Fibrinogen, social position, and Mendelian randomisation. *Journal of Epidemiology and Community Health, 58,* 883.

Kaplan, J.M., and Pigliucci, M. (2001). Genes "for" phenotypes: A modern history view. *Biology and Philosophy, 16,* 189-213.

Keavney, B., Danesh, J., Parish, S., Palmer, A., Clark, S., Youngman, L., Delépine, M., Lathrop, M., Peto, R., and Collins, R. (2006). The International Studies of Infarct Survival (ISIS) collaborators. Fibrinogen and coronary heart disease: Test of causality by Mendelian randomization. *International Journal of Epidemiology, 35,* 935-943.

Keavney, B., Palmer, A., Parish, S., Clark, S., Youngman, L., Danesh, J., McKenzie, C., Delépine, M., Lathrop, M., Peto, R., and Collins, R. (2004). Lipid-related genes and myocardial infarction in 4685 cases and 3460 controls: Discrepancies between genotype, blood lipid concentrations, and coronary disease risk. *International Journal of Epidemiology, 33,* 1002-1013.

Kitami, T., and Nadeau, J.H. (2002). Biochemical networking contributes more to genetic buffering in human and mouse metabolic pathways than does gene duplication. *Nature Genetics, 32,* 191-194.

Klatsky, A.L. (2001). Commentary: Could abstinence from alcohol be hazardous to your health? *International Journal of Epidemiology, 30,* 739-742.

Lancet. (1999). Dietary supplementation with n-3 polyunsaturated fatty acids and vitamin E after myocardial infarction: Results of the GISSI-Prevenzione trial. Gruppo Italiano per lo Studio della Sopravvivenza nell'Infarto miocardico. *Lancet, 354*(9177), 447-455.

Langlois, M.R., Delanghe, J.R., De Buyzere, M.L., Bernard, D.R., and Ouyang, J. (1997). Effect of haptoglobin on the metabolism of vitamin C. *American Journal of Clinical Nutrition, 66,* 606-610.

Lawlor, D.A., Harbord, R.M., Sterne, J.A.C., and Davey Smith, G. (2007). Mendelian randomization: Using genes as instruments for making causal inferences in epidemiology. *Statistics in Medicine.*

Little, J., and Khoury, M.J. (2003). Mendelian randomization: A new spin or real progress? *Lancet, 362,* 930-931.

Manson, J., Stampfer, M.J., Willett, W.C., Colditz, G., Rosner, B., Speizer, F.E., and Hennekens, C.H. (1991). A prospective study of antioxidant vitamins and incidence of coronary heart disease in women. *Circulation, 84*(Suppl. II), 546.

Marmot, M. (2001). Reflections on alcohol and coronary heart disease. *International Journal of Epidemiology, 30,* 729-734.

Morange, M. (2001). *The misunderstood gene.* Cambridge, MA: Harvard University Press.

Morgan, T.H. (1935). The relation of genetics to physiology and medicine. *Scientific Monthly, 41,* 5-18.

Mucci, L.A., Wedren, S., Tamimi, R.M., Trichopoulos, D., and Adami, H.O. (2001). The role of gene-environment interaction in the aetiology of human cancer: Examples from cancers of the large bowel, lung, and breast. *Journal of Internal Medicine, 249,* 477-493.

National Research Council (2001). *Cells and surveys. Should biological measures be included in Social Science Research?* Committee on Population, C.E. Finch, J.E. Vaupel, and K. Kinsella (Eds.). Washington, DC: National Academy Press.

Omenn, G.S., Goodman, G.E., Thornquist, M.D., Balmes, J., Cullen, M.R., Glass, A., Keogh, J.P., Meyskens, F.L., Valanis, B., Williams, J.H., Barnhart, S., and Hammar, S. (1996). Effects of a combination of beta carotene and vitamin A on lung cancer and cardiovascular disease. *New England Journal of Medicine, 334,* 1150-1155.

Osganian, S.K., Stampfer, M.J., Rimm, E., Spiegelman, D., Hu, F.B., Manson, J.E., and Willett, W.C. (2003). Vitamin C and risk of coronary heart disease in women. *Journal of the American College of Cardiology, 42,* 246-252.

Osier, M.V., Pakstis, A.J., Soodyall, H., Comas, D., Goldman, D., Odunsi, A., Okonofua, F., Parnas, J., Schulz, L.O., Bertranpetit, J., Bonne-Tamir, B., Lu, R.–B., Kidd, J.R., and Kidd, K.K. (2002). A global perspective on genetic variation at the ADH genes reveals unusual patterns of linkage disequilibrium and diversity. *American Journal of Human Genetics, 71,* 84-99.

Palmer, L., and Cardon, L. (2005). Shaking the tree: Mapping complex disease genes with linkage disequilibrium. *Lancet, 366,* 1223-1234.

Perera, F.P. (1997). Environment and cancer: Who are susceptible? *Science, 278,* 1068-1073.

Pimstone, S.N., Sun, X.-M., du Souich, C., Frohlich, J.J., Hayden, M.R., and Soutar, A.K. (1998). Phenotypic variation in heterozygous familial hypercholesterolaemia. *Arteriosclerosis, Thrombosis, and Vascular Biology, 18,* 309-315.

Pradhan, A.D., Manson, J.E., Rifai, N., Buring, J.E., and Ridker, P.M. (2001). C-reactive protein, interleukin 6, and risk of developing type 2 diabetes mellitus. *Journal of the American Medical Association, 286,* 327-334.

Ridker, P.M., Cannon, C.P., Morrow, D., Rifai, N., Rose, L.M., McCabe, C.H., Pfeffer, M.A., Braunwald, E. (2005). C-reactive protein levels and outcomes after statin therapy. *New England Journal of Medicine, 352,* 20-28.

Rimm, E. (2001). Commentary: Alcohol and coronary heart disease—Laying the foundation for future work. *International Journal of Epidemiology, 30,* 738-739.

Rimm, E.B., Stampfer, M.J., Ascherio, A., Giovannucci, E., Colditz, G.A., and Willett, W.C. (1993). Vitamin E consumption and the risk of coronary heart disease in men. *New England Journal of Medicine, 328,* 1450-1456.

Rose, S. (1995). The rise of neurogenetic determinism. *Nature, 373,* 380-382.

Rutherford, S.L. (2000). From genotype to phenotype: Buffering mechanisms and the storage of genetic information. *BioEssays, 22,* 1095-1105.

Sesso, D., Buring, J.E., Rifai, N., Blake, G.J., Gaziano, J.M., and Ridker, P.M. (2003). C-reactive protein and the risk of developing hypertension. *Journal of the American Medical Association, 290,* 2945-2951

Shaper, A.G. (1993). Editorial: Alcohol, the heart, and health. *American Journal of Public Health, 83,* 799-801.

Shastry, B.S. (1998). Gene disruption in mice: Models of development and disease. *Molecular and Cellular Biochemistry, 181,* 163-179.

Sijbrands, E.J.G., Westengorp, R.G.J., Defesche, J.C., De Meier, P.H.E.M., Smelt, A.H.M., and Kastelein, J.J.P. (2001). Mortality over two centuries in large pedigree with familial hypercholesterolaemia: Family tree mortality study. *British Medical Journal, 322,* 1019-1023.

Sjöholm, A., and Nyström, T. (2005). Endothelial inflammation in insulin resistance. *Lancet, 365,* 610-612.

Smith, J. (2002). Apolipoproteins and aging: Emerging mechanisms. *Ageing Research Reviews, 1,* 345-365.

Smits, K.M., et al. (2004). Association of metabolic gene polymorphisms with tobacco consumption in healthy controls. *International Journal of Cancer, 110*(2), 266-270.

Song, Y., Stampfer, M.J., and Liu, S. (2004). Meta-analysis: Apolipoprotein E genotypes and risk for coronary heart disease. *Annals of Internal Medicine,* 137-147.

Spearman, C. (1904). The proof and measurement of association between two things. *American Journal of Psychology, 15,* 72-101.

Stampfer, M.J., and Colditz, G.A. (1991). Estrogen replacement therapy and coronary heart disease: A quantitative assessment of the epidemiologic evidence. *Preventative Medicine, 20,* 47-63.

Stampfer, M.J., Hennekens, C.H., Manson, J.E., Colditz, G.A., Rosner, B., and Willett, W.C. (1993). Vitamin E consumption and the risk of coronary disease in women. *New England Journal of Medicine, 328,* 1444-1449.

Strohman, R.C. (1993). Ancient genomes, wise bodies, unhealthy people: The limits of a genetic paradigm in biology and medicine. *Perspectives in Biology and Medicine, 37,* 112-145.

Takagi, S., Iwai, N., Yamauchi, R., Kojima, S., Yasuno, S., Baba, T., Terashima, M., Tsutsumi, Y., Suzuki, S., Morii, I., Hanai, S., Ono, K., Baba, S., Tomoike, H., Kawamura, A., Miyazaki, S., Nonogi, H., and Goto, Y. (2002). Aldehyde dehydrogenase 2 gene is a risk factor for myocardial infarction in Japanese men. *Hypertension Research, 25,* 677-681.

Terwilliger, J.D., and Weiss, W.M. (2003). Confounding, ascertainment bias, and the blind quest for a genetic fountain of youth. *Annals of Medicine, 35,* 532-544.

Thomas, D.C., and Conti, D.V. (2004). Commentary on the concept of Mendelian randomization. *International Journal of Epidemiology, 33,* 17-21.

Thompson, W.D. (1991). Effect modification and the limits of biological inference from epidemiological data. *Journal of Clinical Epidemiology, 44,* 221-232.

Timpson, N.J., Lawlor, D.A., Harbord, R.M., Gaunt, T.R., Day, I.N.M., Palmer, L.J., Hattersley, A.T., Ebrahim, S., Lowe, G.D.O., Rumley, A., and Davey Smith, G. (2005). C-reactive protein and its role in metabolic syndrome: Mendelian randomization study. *Lancet, 366,* 1954-1959.

Tybjaerg-Hansen, A., Steffensen, R., Meinertz, H., Schnohr, P., and Nordestgaard, B.G. (1998). Association of mutations in the apolipoprotein B gene with hypercholesterolemia and the risk of ischemic heart disease. *New England Journal of Medicine, 338,* 1577-1584.

Umbach, D.M., and Weinberg, C.R. (1997). Designing and analysing case-control studies to exploit independence of genotype and exposure. *Statistics in Medicine, 16,* 1731-1743.

Verma, S., Szmitko, P.E., and Ridker, P.M. (2005). C-reactive protein comes of age. *Nature Clinical Practice, 2,* 29-36.

Wacholder, S., Rothman, N., and Caporaso, N. (2000). Population stratification in epidemiologic studies of common genetic variants and cancer: Quantification of bias. *Journal of the National Cancer Institute, 92,* 1151-1158.

Wacholder, S., Rothman, N., and Caporaso, N. (2002). Counterpoint: Bias from population stratification is not a major threat to the validity of conclusions from epidemiological studies of common polymorphisms and cancer. *Cancer Epidemiology, Biomarkers and Prevention, 11,* 513-520.

Waddington, C.H. (1942). Canalization of development and the inheritance of acquired characteristics. *Nature, 150,* 563-565.

Weiss, K., and Terwilliger, J. (2000). How many diseases does it take to map a gene with SNPs? *Nature Genetics, 26,* 151-157.

Wilkins, A.S. (1997). Canalization: A molecular genetic perspective. *BioEssays, 19,* 257-262.

Willett, W.C. (1990). Vitamin A and lung cancer. *Nutrition Reviews, 48,* 201-211.

Willett, W.C. (2001). *Eat, drink, and be healthy: The Harvard Medical School guide to healthy eating.* New York: Free Press.

Williams, R.S., and Wagner, P.D. (2000). Transgenic animals in integrative biology: Approaches and interpretations of outcome. *Journal of Applied Physiology, 88,* 1119-1126.

Wright, A.F., Carothers, A.D., and Campbell, H. (2002). Gene-environment interactions: The BioBank UK study. *Pharmacogenomics Journal, 2,* 75-82.

Wolf, U. (1995). The genetic contribution to the phenotype. *Human Genetics, 95,* 127-148.

Wu, T., Dorn, J.P., Donahue, R.P., Sempos, C.T., and Trevisan, M. (2002). Associations of serum C-reactive protein with fasting insulin, glucose, and glycosylated hemoglobin: The Third National Health and Nutrition Examination Survey, 1988-1994. *American Journal of Epidemiology, 155,* 65-71.

17

Multilevel Investigations:
Conceptual Mappings and Perspectives

John T. Cacioppo, Gary G. Berntson, and *Ronald A. Thisted*

A century ago, antibiotics were nonexistent, public health was underdeveloped, leisure time was largely reserved for the wealthy, and germ-based diseases were among the major causes of adult morbidity and mortality. Improvements in living standards, public health efforts, leisure and lifestyles, and medical technology have made population health a notable success story in developed nations. Along with this improvement came an increase in life expectancy and a shift in the kinds of illnesses that cause death. In the early 21st century, most people today will avoid or survive the infections that were the major causes of mortality a century ago and instead die late in life from chronic degenerative conditions, such as cancer and cardiovascular disease. The combination of longer lives and increased prevalence of chronic conditions has raised concerns that people will spend their later years sick, limited by physical disabilities, and saddled with costly health care expenses. The biological and social sciences are rallying to address these issues. An important part of this response rests on data from the introduction of biological indicators and genetic information in social science surveys, which permit the investigation of associations across levels of organization that were inconceivable a century ago.

It is important to recognize that complex health states and outcomes tend to be multiply determined and are subject to contextual (e.g., environmental, cultural, social) as well as biological influences. Statistical models now exist that include stochastic error terms at various hierarchical levels of aggregation, which are applicable to the data matrices that span biological and social levels of organization. Our goal here is not to review these statistical models but to provide a more generic discussion of the conceptual issues that arise in the analysis of social and biological determinants of a healthy life span. We suggest that we must move beyond associations to mechanisms to meet the challenge of identifying the social and behavioral factors that influence the likelihood of remaining healthy and functional for the entire life span. The identification of associations and mechanisms depends on the accurate mapping of biological measures (e.g., biomarkers) to social and behavioral constructs in surveys. Such mappings will be aided by experimental or statistical controls for other factors (e.g., medications, time of day, activity level, body mass index) that influence biomarker expressions; attention to contextual variables (e.g., ethnicity) that may moderate the nature of the mappings; and a careful consideration of the sensitivity, specificity, and generality of the mapping in any given investigation. Before delving into these points, however, we describe briefly the nature of the data sets increasingly available to biological and social scientists.

CONFLUENCE OF DATA MATRICES

The contributions to this volume demonstrate that scientific and technological advances have dramatically altered the data available to study complex behaviors and healthy aging. Estimates among biologists a decade ago were that 100,000 genes were needed for the cellular processes that are responsible for human behavior and aging, but humans have only a quarter that number of genes (Pennisi, 2005). This finding has fostered a recognition that a gene may have multiple small effects (pleiotropy), that many genes may act in additive and configural fashions to produce small effects both on specific abilities and on general abilities, and that genetic expression can be altered by the social as well as the physical environment in which humans live and work. The advent of single-nucleotide polymorphism (SNP) microarrays permits genome-wide association studies that would have been considered impossible less than a decade ago, and microarrays are on the horizon with which to study many if not all functional DNA polymorphisms in the genome (Butcher, Kennedy, and Plomin, 2006).

In addition to the global analysis of genes (genomics), technologies now exist for large-scale analyses of gene transcripts (transcriptomics), proteins (proteomics), and metabolites (metabolomics) in cells, tissues, and organisms. Among the important advances in quantitative analyses of these data is multivariate genetic analysis, which goes beyond analyzing the variance of each phenotype considered separately to analyze the covariance between them (Butcher, Kennedy, and Plomin, 2006). The number of SNPs and the number of various combinations of SNPs can be very large, however, and the complexity of the mapping problem is magnified by the presence of nonisomorphic intervening steps, which, for example, contribute to variation of phenotypic expression as a function of the physical or social context. Bioinformatics tools such as TELiS, which can be used to examine similarities in signaling pathways (transcription factor binding motifs) by genes that are found to differ between groups of interest, may be used to construct an intermediate level of organization, thereby improving the mapping of genotypes to phenotypes.

Developments in tissue and blood assays, ambulatory recording devices, noncontact recording instruments, and powerful and mobile computing devices have also burst onto the scene in recent years. These technologies make it possible to measure a variety of biological parameters in naturalistic as well as laboratory settings and in population-based health research. One such development is the use of drops of whole blood collected on filter paper from a simple finger prick to collect and analyze biological samples that previously required venipuncture (e.g., McDade et al., 2000). McDade, Williams, and Snodgrass (2006) identify over 100 analytes that can now be measured in dried blood spot samples, approximately half of which have particular relevance to population-level health research (e.g., cortisol, CD4+ lymphocytes, C-reactive protein, glycosilated hemoglobin, immunoglobin tumor necrosis factor, Epstein Barr Virus). With the inclusion of these measures in population-based health research, the weak associations that one would predict to exist between multiply determined variables (e.g., stress and C-reactive protein or blood pressure) and the potential influences of moderator variables (e.g., age, ethnicity, socioeconomic status) can be tested. These potential moderator variables may operate through differential reactivity (e.g., certain ethnicities or age cohorts may show salt sensitivity), differential exposure (e.g., certain ethnicities or age cohorts may consume more salt in their diet), or both. Distinguishing between these processes is crucial to moving from the description of associations to the delineation of causal mechanisms.

Recent "epidemics," such as obesity and cardiovascular disease, cannot be fully explained in terms of genes alone, because major shifts in the human genome require much longer periods of time to unfold. These new health challenges require consideration of environmental exposures (e.g., the deployment of soda machines and fast food options in public schools) and individual differences in response to exposures (e.g., individual consumption patterns, salt sensitivity). Importantly, cultural, economic, political, social, psychological, behavioral, and environmental assessments are becoming more detailed, multidimensional, reliable, sensitive, and temporally rich. Early measures of social and behavioral predispositions were once characterized by general indices with poor reliabilities and validity. Although self-report measures often are viewed with suspicion because respondents may not be willing or able to respond accurately, the proper application of psychometric procedures to scale construction and validation and the inclusion of validating behavioral metrics have produced self-report measures with reliability and validity coefficients that rival or exceed the psychometrics of many physiological assessments (e.g., Burleson et al., 2003).

Experience sampling methods (Larson and Csikszentmihalyi, 1983) and day reconstruction methods (Kahneman, Krueger, Schkade, Schwarz, and Stone, 2004) are the sociobehavioral equivalent of ambulatory physiological recordings and make feasible the frequent sampling of social, psychological, behavioral, and biological states. The introduction of multilevel modeling (MLM) with temporal lags (Hawkley, Preacher, and Cacioppo, 2006) further permits frequent random-interval sampling and longitudinal analyses of environments, behaviors, social status, and biological responses. These new analytic techniques permit more powerful tests of mappings between social and biological domains.

Finally, spatially multidimensional electromagnetic, hemodynamic, and optical imaging devices, coupled with temporally precise electrophysiological methodologies, now make it possible to track changes in brain activity with impressive spatial and temporal resolution. The resulting data structures contain millions of elements that can span multiple levels of organization (Herrington, Sutton, and Miller, 2007). Although not yet appropriate for inclusion in population-based health research, these techniques make it possible to test specific hypotheses about the brain mechanisms underlying a variety of psychological processes, and the transduction of these psychological factors into peripheral biological activities and healthy aging. These laboratory techniques are already being used to test a subset of respondents in population-based studies, and ambulatory versions of electroencephalography are currently being developed.

Investigations of orderly associations in these data matrices, and especially of associations and causal connections across levels of representation, create significant challenges as well as opportunities. Knowledge and principles of physiological mechanisms, biometric and psychometric properties of the measures, statistical representation and analysis of multivariate data, and the structure of scientific inference are all important if veridical information is to be extracted from the confluence of these data matrices. In the remainder of this chapter, we outline a simple model to aid the mapping of elements (e.g., constructs, measures) across levels of representation.

MAPPING ACROSS LEVELS OF REPRESENTATION

The simplest method of mapping across levels is the correlative approach. There are notable success stories that have employed this approach, such as the Framingham Heart Study, which the U.S. Public Health Service launched in 1948. The early directors of this study, Roy Dawber and Bill Kannel, began with examinations, detailed medical histories, and blood tests of the more than 5,000 Framingham, Massachusetts, residents biennially. Since the inception of the study, new technologies and measures have been added, and epidemiological and data management methods have been incorporated to improve the scientific yield. Originally envisioned as a 20-year study, these researchers and their successors have now followed the health and lifestyles of the residents of Framingham for 60 years. The Framingham study has resulted in the publication of more than 1,200 peer-reviewed scientific articles (Levy and Brink, 2005) and the advancement of understanding of lifestyle factors in the etiology of cardiovascular disease.

By contemporary standards, the number of sociobehavioral and biological measures queried for possible associations was small. Still, the sample size, frequency of assessment, and duration of the longitudinal study were substantial. As Issa (2005) noted, the Framingham study illustrates how well-designed longitudinal studies that follow population-based cohorts can provide information on the etiology and natural history of the course of disease, generating hypotheses that are testable in laboratory research and clinical trials.

It is important to recognize, however, that not every association identified in the Framingham study proved to be robust or informative, and that for every Framingham study, there are others whose scientific yield is disappointing. The associations uncovered in correlative research, especially atheoretical correlative investigations, run a special risk of yielding false discoveries (i.e., nonreplicable associations), and this possibility increases with the number of possible associations that are examined. False discovery rate techniques have been developed to help mitigate this problem, but these techniques do not eliminate it (see Munafo et al., 2003).

The development and adoption of false discovery rate methods represent an advance in dealing with type I error rates, because the cost of near-zero false discovery rates was a high false negative rate. The ratio of the number of missed small discoveries to false discoveries can be substantially greater than one in studies of complex health outcomes, and small associations can carry large economic ramifications once scaled to the level of the population. Therefore, the cost of missing important but small associations (type II errors) can sometimes be greater than the cost of a type I error.

A strength of a correlative approach is the identification of associations that might be replicable and worthy of further study. An important goal of scientific theory is to describe the causal interrelationships among factors, thereby explicating the mechanism responsible for an association. Moving from the specification of associations to mechanisms is therefore an important objective for future research using biological measures in social science surveys. The correlative approach may generate variables (e.g., genes, neurophysiological circuits, demographic or lifestyle factors) or contextual moderators that are candidates for a causal mechanism.

In addition, the correlative approach also may not indicate the nature of the specificity of the association across levels of representation. For convenience, consider the constructs or measures at each level of representation as elements within a domain or set. The mapping between elements across such sets can take one of the following forms (see Figure 17-1):

1. A one-to-one relation, such that an element in one set or level of representation is associated with one and only one element in another set, and vice versa. An example of a one-to-one relation, discussed below, is the prostate-specific antigen (PSA) assay as a measure of prostate cell activity.

2. A one-to-many relation, meaning that an element of interest in the one set is associated with multiple elements in another set. An example is the orienting response and its mapping into a phasic heart rate deceleration and skin conductance response

3. A many-to-one relation, meaning that two or more elements in one set are associated with one element in another set. (This differs from the preceding only when the order of the mapping across levels of representation—for example, social to biological—is specified.) An example discussed below is psychological stress, exercise, time of day, and other factors that can produce increased cortisol activation.

4. A many-to-many relation, meaning two or more elements in one set are associated with the same (or an overlapping) subset of elements in another set. Not only can psychological stress and exercise, for example, influence cortisol activation but they can also have similar influences on autonomic, catecholaminergic, and immune activity.

5. A null relation, meaning there is no association between the specified element in one set and those observed in another set.

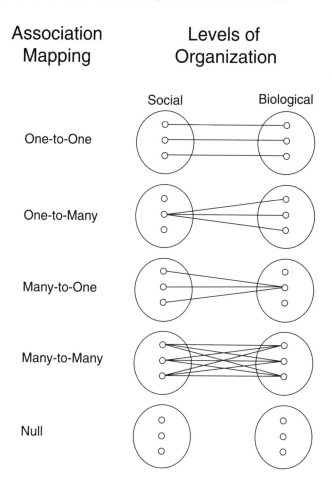

Association Mapping

Levels of Organization

Social   Biological

One-to-One

One-to-Many

Many-to-One

Many-to-Many

Null

FIGURE 17-1 Possible relationships between elements in two adjacent levels of representation (domains). For illustrative purposes, these domains have been labeled "Social" and "Biological."

Association studies involving elements with a one-to-one relation (absent confoundings and measurement error) produce high correlations, whereas association studies involving elements characterized by a null relation yield an essentially zero correlation. The strength of the association between elements across levels of representation can vary a great deal, however, for mappings (2) through (4), and a many-to-many mapping between two elements across levels of representation can produce correlation coefficients that are quite small, making them difficult to distinguish from a null relation, unless the sample size is large. Thus, the initial establishment of an association between elements across levels of representation through a correlative approach is not sufficient to determine the specificity of the mapping.

Why might it be important to go beyond thinking of associations to the interfaces between levels of representation? First, it is important if we are to move efficiently from association to the specification of mechanisms in our investigations. Second, the nature of the mappings between elements at different levels of representation determines the limits of interpretation one can draw about an association (Cacioppo and Tassinary, 1990).

Consider research in which a biological measure (e.g., salivary cortisol level) is known to correlate with a diagnostic category (e.g., a hypothesized state or condition such as "stress"). This established correlation may then be used to justify an interpretation of differences in the biomarker (e.g., salivary cortisol level) as evidence of differences in the diagnostic category (e.g., stress). This form of inference can be problematic, however. Even if we knew that variations in stress were associated with corresponding variations in salivary cortisol, inferring stress based on cortisol represents an error in interpretation because it ignores the possibility that there are other antecedent conditions that could also produce variations in cortisol. That is, it ignores the specificity of the association or mapping to the construct about which one would like to draw the inference.

It is tempting to suggest that these issues do not apply to genetics because there is no doubt that genes play a causal role in the production of complex behaviors and in age-related changes in these behaviors. To say that genes are causal is not equivalent, however, to specifying which gene, or set of genes, is associated with and causal in a particular phenotypic expression or, for that matter, in specifying the mechanism by which associated genes might influence a particular phenotype. Gottesman and Gould (2003) suggested that the number of genes involved in a phenotype is directly related to both the complexity of the phenotype and the difficulty of genetic analyses, and Butcher et al. (2006, p. 6) concluded that:

> Multivariate genetic research consistently points to a single set of generalist genes that accounts for much genetic influence on diverse cognitive abilities . . . . Although each of the many generalist [quantitative trait loci or QTLs] will involve different molecular mechanisms, a QTL set will be useful in tracing the pleiotropic pathways between genes and cognition through the brain to understand how generalist genes have their diffuse effects. These pathways will be complex and determining direct causation will be difficult.

Although difficult, such causal linkages will be more easily resolved if attention is paid to the implications of the many-to-many mapping problem. The mapping between elements across levels of representation may become more complex (e.g., many-to-many) as the number of intervening levels of representation increases.[1] Accordingly, the likelihood of complex and potentially obscure mappings increases as one fails to consider intervening levels of representations. Admittedly, it is not always obvious which of several levels of representation might be "adjacent," except perhaps when levels of representation refer to temporal rather than spatial scope. This caveat that mapping across levels of representation may be fostered by the incremental mapping of elements between proximal levels of organization nevertheless may have heuristic value. For example, endophenotypes such as neurocognitive deficits have proven valuable explanatory constructs between genes and psychiatric diseases (e.g., Gottesman and Gould, 2003; Nuechterlein, Robbins, and Einat, 2005), and in theory the same situation should apply to any mapping that goes from surveys to cells. For this reason, we focus here on the mappings between two adjacent levels of representation. The issues raised about the mappings between adjacent levels of representation can be extended to any number of adjacent levels of organization.

Taxonomy of Mappings

Tests, assays, or measures more generally have two different but related sets of characteristics. Analytic *sensitivity* is the ability to detect very low levels of the target analyte, whereas analytic *specificity* means that detection indicates the presence only of the target analyte. For example, blood sugar levels will vary in a predictable fashion for several hours after ingesting a dosage of glucose. Deviations from the normative values in blood sugar level across time mark a possible problem in metabolism because the blood glucose tolerance test (a procedure for mapping the glucose–blood sugar association) is sensitive and specific as long as the appropriate testing procedures are followed (e.g., fasting prior to the test) to eliminate the other known influences on the observed blood sugar excursions over the course of the test.

The properties of sensitivity and specificity, of course, depend on the elements involved in the mapping.[2] For example, only prostate cells produce PSA, and an assay for PSA has very high sensitivity and speci-

---

[1]The exception to this statement is when mappings among elements across adjacent levels of organization is one-to-one, but such mappings are atypical.

[2]Context here is conceptualized as the constraints limiting the elements that are operating in a given measurement environment.

ficity. But if PSA is mapped to prostate cancer rather than to prostate cell activity, the specificity and sensitivity are quite different. The sensitivity of the assay for PSA for detecting prostate cancer can be low (around 40 percent), as can the specificity (around 60 percent), because high PSA can be produced by active but noncancerous prostate cells. Said differently, the sensitivity and specificity of PSA for prostate cell activity are high, but the sensitivity and specificity of prostate cell activity for prostate cancer are more modest. Distinguishing between the mapping between PSA and prostate cell activity on one hand and prostate cell activity and prostate cancer on the other may make little difference to the physician using a PSA assay to screen for prostate cancer, but it would be important to consider for the researcher who is seeking to understand the *mechanism* for a measured association between PSA and prostate cancer in a large survey.

A third dimension is the *generality* of the mapping. In his influential *Handbook of Experimental Psychology*, S.S. Stevens (1951, p. 20) advised:

> The scientist is usually looking for invariance whether he knows it or not. Whenever he discovers a functional relation between two variables his next question follows naturally: under what conditions does it hold? In other words, under what transformation is the relation invariant? The quest for invariant relations is essentially the aspiration toward generality, and in psychology, as in physics, the principles that have wide application are those we prize.

Is the mapping between two elements across levels of representation universally generalizable, or is it moderated by other factors? If it is generalizable without qualification, then the association requires no attention to characteristics of the context or sample population; that is, the mapping would have external validity. Invariant associations were once assumed, but statistical methods are now well developed to test for potential moderators (e.g., Baron and Kenny, 1986), and increasing attention is being paid to the operation of moderator variables. For instance, we raised the issue of moderators above when discussing differential reactivity and differential exposure.

A taxonomy of associations between elements across levels of representation is summarized in Figure 17-2. The initial step is often to establish that variations in an element in one domain are associated with variations in an element in another, thereby establishing an association. An *outcome* is defined as a mapping in which multiple elements at one level of organization (e.g., biological) are related to an element at another level of organization (e.g., social), and this many-to-one mapping may change across contexts. Initial association studies typically do not address

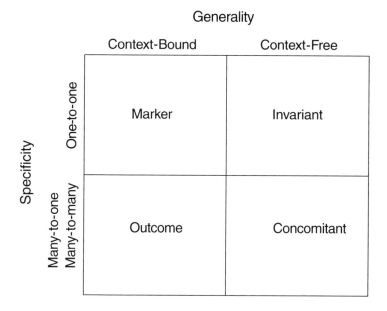

FIGURE 17-2 Taxonomy of mappings among elements between adjacent levels of representation.

issues of specificity or generality, and the treatment of such associations as invariants is premature.

An *invariant* relationship refers to a universal isomorphic (one-to-one) mapping between elements across levels or organization (see Figure 17-2). Invariant mappings permit the inference of an element at one level of organization based on the measurement of its isomorphic element at another. A *marker* is defined as a one-to-one, nonuniversal (e.g., context dependent) relationship between elements across levels of representation (see Figure 17-2). Many medical diagnostic tests, which have sensitivity and specificity only if explicit procedures are followed to eliminate other influences, are examples of markers. As such, inferences based on markers are similar to those for invariants as long as all other elements involved in the mapping are either experimentally or statistically controlled.

Finally, a *concomitant* refers to a many-to-one but universal association between elements across levels of representation and is similar to outcomes, except that the latter is not universal. Outcome and concomitant mappings enable systematic inferences to be drawn about theoretical constructs based only on hypotheticodeductive logic. Specifically, when two theoretical models differ in predictions regarding one or more outcomes or concomitants, then the logic of the experimental design allows theoretical inferences to be drawn about elements at one level of organization based on the measured elements in another. Strong inferences when dealing with outcome or concomitant mappings are limited to hypotheticodeductive reasoning.

When a new effect or association is found not to generalize to specific contexts or individuals, concerns are typically expressed about the methodological differences between the studies. Such a finding raises several important questions, including whether the original association is replicable and, if replicable, whether the diminution in effect size is attributable to measurement issues (e.g., reliability, construct validity) or to the operation of a moderator variable. Careful attention initially to the psychometric properties of all measures, regardless of their level of organization, to ensure their reliability and validity (including construct validity) therefore warrants attention in the design and analysis of studies going from cells to surveys.

In sum, many health states and outcomes can be multiply determined. To the extent that this is the case, investigators who assume rather than establish an invariant relationship between elements in the social and biological domains are at risk for predictably faulty interpretations. Investigators who incorporate cortisol measures in their social science surveys to indicate variations in stress, simply because stress and cortisol are correlated, are unlikely to contribute much to scientific understanding. This is because other factors that influence cortisol (e.g., time of day, time since consuming food) will be unrecognized and uncontrolled (see Adam, 2006). However, if the investigator next asks what is the specificity of this association, other antecedent conditions that influence cortisol are more quickly recognized, and contexts in which these other antecedents can be controlled experimentally or statistically can be developed to allow strong inductive inferences about the state of an individual's stress based on cortisol levels. That is, the sensitivity and specificity of the mapping of biological elements into macro levels of representation may be context dependent, and attention to these issues improves the quality of inductive inferences.

The term "biomarkers" does not distinguish among the various mappings that are possible between biological (e.g., hormonal, genotypic) and social (e.g., individual difference, phenotypic) representations. The taxonomy we have presented—biological outcomes, concomitants, markers, and invariants—may offer greater specification of these mappings and their properties and provide a useful framework within which to view these associations.

CONCLUSION

Interdisciplinary research that crosses biological and social levels of organization raises issues about how might one productively think about concepts, hypotheses, theories, theoretical conflicts, and theoretical tests across levels of organization. Abstract constructs, such as those developed by social scientists, provide a means of understanding highly complex activity without needing to specify each individual action of the simplest components, thereby providing an efficient means of describing the behavior of a complex system (e.g., "healthy aging"). Chemists who work with the periodic table on a daily basis nevertheless use recipes rather than the periodic table to cook, not because food preparation cannot be reduced to chemical expressions but because it is not cognitively efficient to do so. Reductionism, in fact, is one of several approaches to better science based on the value of data derived from distinct levels of organization to constrain and inspire the interpretation of data derived from others levels of organization. In reductionism, the whole is as important to study as are the parts, for only in examining the interplay across levels of organization can the underlying principles and mechanisms be ascertained. The goal of this chapter has been to outline a simple model to aid thinking about elements from different levels of organization.

In sum, the identification of associations and mechanisms in the complex multilevel data sets increasingly available depends on the accurate mapping of biological measures to social and behavioral constructs in surveys. Such mappings will be aided by experimental or statistical controls for other factors (e.g., medications, time of day, activity level, body mass index) that influence biomarker expressions; attention to contextual variables (e.g., ethnicity) that may moderate the nature of the mappings; and a careful consideration of the specificity and generality of the mapping in any given investigation. We hope the proposed formulation aids in this effort.

REFERENCES

Adam, E.K. (2006). Transactions among trait and state emotion and adolescent diurnal and momentary cortisol activity in naturalistic settings. *Psychoneuroendocrinology, 31,* 664-679.

Baron, R.M., and Kenny, D.A. (1986). The moderator-mediator variable distinction in social psychological research: Conceptual, strategic, and statistical considerations. *Journal of Personality and Social Psychology, 51,* 1173-1182.

Burleson, M.H., Poehlmann, K.M., Hawkley, L.C., Ernst, J.M., Berntson, G.G., Malarkey, W.B., Kiecolt-Glaser, J.K., Glaser, R., and Cacioppo, J.T. (2003). Neuroendocrine and cardiovascular reactivity to stress in mid-aged and older women: Long-term temporal consistency of individual differences. *Psychophysiology, 40,* 358-369.

Butcher, L.M., Kennedy, J.K.J., and Plomin, R. (2006). Generalist genes and cognitive neuroscience. *Current Opinion in Neurobiology, 16*, 1-7.

Cacioppo, J.T., and Tassinary, L.G. (1990). Centenary of William James's *Principles of Psychology*: From the chaos of mental life to the science of psychology. *Personality and Social Psychology Bulletin, 16*, 601-611.

Cole, S.W., Hawkley, L.C., Arevalo, J.M., Sung, C.Y., Riose, R.M., and Cacioppo, J.T. (2007). Social regulation of gene expression in human leukocytes. *Genome Biology, 8*(9), R189.

Gottesman, I.I., and Gould, T.D. (2003). The endophenotype concept in psychiatry: Etymology and strategic intentions. *American Journal of Psychiatry, 160*, 636-645.

Hawkley, L.C., Preacher, K.J., and Cacioppo, J.T. (2006). Multilevel modeling of social interactions and mood in lonely and socially connected individuals: The MacArthur social neuroscience studies. In A.D. Ong and M. van Dulmen (Eds.), *Oxford Handbook of Methods in Positive Psychology*. New York: Oxford University Press.

Herrington, J.D., Sutton, B., and Miller, G.A. (2007). Data-file formats in neuroimaging: Background and tutorial. In J.T. Cacioppo, L.G. Tassinary, and G.G. Berntson (Eds.), *Handbook of psychophysiology, third edition*. New York: Cambridge University Press.

Issa, A.M. (2005). Factoring risks to the heart. *Science, 309*, 1679.

Larson, R., and Csikszentmihalyi, M. (1983). The experience sampling method. New *Directions for Methodology of Social and Behavioral Science, 15*, 41-56.

Levy, D., and Brink, S. (2005). *A change of heart: How the Framingham heart study helped unravel the mysteries of cardiovascular disease*. New York: Knopf.

McDade, T.W., Stallings, J.F., Angold, A., Costello, E.J., Burleson, M., Cacioppo, J.T., Glaser, R., and Worthman, C.M. (2000). Epstein-Barr virus antibodies in whole blood spots: A minimally invasive method for assessing cell-mediated immunity. *Psychosomatic Medicine, 62*, 560-568.

McDade, T.W., Williams, S., and Snodgrass, J.J. (2006). What a drop can do: Expanding options for the analysis of blood-based biomarkers in population health research. Paper presented at the annual meeting of Population Association of American, March 31, Los Angeles, CA.

Munafo, M.R., Clark, T.G., Moore, L.R., Payne, E., Walton, R., and Fint, J. (2003). Genetic polymorphisms and personality in healthy adults: A systematic review and meta-analysis. *Molecular Psychiatry, 8*, 471-484.

Nuechterlein, K.H., Robbins, T.W., and Einat, H. (2005). Distinguishing separable domains of cognition in human and animal studies: What separations are optimal for targeting interventions? *Schizophrenia Bulletin, 31*, 870-874.

Pennisi, E. (2005, July 1). Why do humans have so few genes? *Science, 309*, 80.

Platt, J.R. (1964). Strong inference. *Science, 146*, 347-353.

Stevens, S.S. (1951). *Handbook of experimental psychology*. New York: Wiley.

18

Genomics and Beyond: Improving Understanding and Analysis of Human (Social, Economic, and Demographic) Behavior

John Hobcraft

Genes exert their influence by encoding proteins. The level of such gene activity, however, is a regulated process. As molecules, genes are subject to regulation by intracellular factors that, in turn, are a reflection of environmental factors. Neither genes nor environment dominates development; rather there is a continual interaction between genes and the environment. Phenotype emerges as a function of this constant dialogue, and any effort to ascribe percentage values to isolated variables is likely to be biologically meaningless. (National Research Council, 2001b, pp. 63-64)

The genetics of behaviour is much too important a topic to be left to geneticists! (Plomin, 2001, p. 1104)

There is increasing recognition that real progress in understanding human behavior (or health) requires an integrative approach that explores the interplays of pathways within the person and processes whereby the environment influences the person. There is also recognition that developmental progressions through the life course involving these continuous feedbacks and interplays are an essential ingredient. At a minimum, these concerns involve interplays among genes, brains, the phenotype, experience, other persons, and structural contexts. This shift of emphasis is reflected in this volume, and also indicated by a growing number of attempts to provide synthetic frameworks that point toward this integrated approach for a variety of fields, including early child development (Granger and Kivlighan, 2003; National Research Council and Institute of Medicine, 2000); demographic behavior (National Research Council and Institute of Medicine, 2000; National Research Council, 2003; Hobcraft, 2006; Seltzer et al., 2005); social neuroscience (Cacioppo, Berntson, and Adolphs, 2002; Cacioppo, Berntson, Sheridan, and McLintock, 2000; see also Chapter 17 in this volume); health (Johnson and Crow, 2005; National Research Council, 2001b); human bonding (Miller and Rodgers, 2001); resilience (Curtis and Cicchetti, 2003); well-being (Davidson, 2004; Huppert and Bayliss, 2004); and economics (Chapter 15 in this volume).

Thus far, most real progress toward implementing integrative approaches has occurred for health, as reviewed in several chapters of this volume (see also National Research Council, 2001b; Johnson and Crow, 2005), and in psychology, especially psychopathology (see Plomin, DeFries, Craig, and McGuffin, 2003; Rutter, 2006). Bridging the gap between the biological and the social sciences has taken more time, although important developments have occurred at the intersections of economics and psychology in behavioral economics (Brocas and Carillo, 2003, 2004; Camerer, Loewenstein, and Rabin, 2004). However, an increasing number of important prospective national population surveys are collecting (or considering) DNA samples, with the intention of enabling social science researchers, subject to suitable disclosure controls, to access information on a significant range of genetic markers (e.g., in the United States, the National Longitudinal Study of Adolescent Health and the Fragile Families Survey). Although the importance of such large and often nationally representative samples for genetic research is recognized, the longer term potential lies in the ability to explore the interplays between genes and behavior over the life course in response to experiences.

One of the great challenges of an integrative biosocial life-course approach to the study of behavior is the very complexity involved and to find the means of avoiding drowning in the multiple levels and interplays. One of my favorite relevant aphorisms, from the different context of population projections, is John Hajnal's (1955, p. 321) plea for "less computation and more cogitation." At the meeting during the preparation of this volume, Jim Vaupel summed this up differently as "model simple; think complex." My own preference would be to extend this somewhat to "model sufficiently complex, not simplistically; think more, both to complexify and to simplify." The key here is that we need to have more and better theory and conceptualization, including attention to mechanisms and pathways or processes (a theme developed at greater length in Hobcraft, 2006). The problems of data dredging or gene hunting are well recognized, and the need for replication and ever larger samples to avoid false positives is often stressed (e.g., Plomin, 2005).

Yet some of the most influential findings on gene-environment interactions for behavior have emerged from the Dunedin study, which has a total sample of around 1,000 individuals (Caspi et al., 2002, 2003, 2005). Moreover, the best known of these studies, linking the serotonin transporter short allele as a moderator of life stress on depression, has been replicated several times. To give a flavor of these findings, 10 percent of sample members who did not experience stressful events between ages 21 and 26 met diagnostic criteria for depression at age 26, regardless of whether they had two long alleles or one or both alleles short on the 5-HTTLPR gene; however, among those experiencing four or more stressful life events, 17 percent with both alleles long were depressed, compared with 33 percent among those with one or both alleles short. The strategy through which the key investigators approached this research, based on careful evaluation of theory and knowledge, is well laid out in a series of papers (T.E. Moffitt, 2005; Moffitt, Caspi, and Rutter, 2005, 2006; Rutter, Moffitt, and Caspi, 2006). I return to this theme later in this chapter.

A related key topic is the approach to understanding behavior and the challenging issues around causation. Econometricians have devised a range of sophisticated approaches to dealing with unmeasured variables (e.g., Wooldridge, 2002), although when the unmeasured variables are a key contingent part of the process (e.g., as moderators or interactions), these approaches fail. As one of the major contributors to these approaches has put it: "there is no mechanical algorithm for producing a set of 'assumption free' facts or causal estimates based on those facts" (Heckman, 2000, p. 91). Randomized controlled trials have become the gold standard in the health sciences and have also penetrated some other areas, yet "randomized trials can never estimate channels for the effects of treatment" (Moffitt, 2003, p. 453). Once the importance of interplays of genes and experience (and other levels, too) over time is acknowledged, other approaches are needed that emphasize contingent relationships and pathways or progressions. A perceptive account of the wide range of issues involved in gaining understanding from prospective studies is provided by Rutter (1994). Caspi (2004) provides a useful and well-illustrated account of the importance of both self-selection (effects of persons on their environments, whether genetic, from experience, or from choice) and social causation (effects of the environment or contexts on the person, including gene-environment interplays). He argues convincingly that selection is not just a nuisance factor to be controlled away, but that the processes involved are ubiquitous and consequential, pose challenges for policy, have compositional consequences, and the mechanisms must be understood. New and more sophisticated approaches to multilevel life-course model formulation, analysis, and testing are needed, ones that can both achieve some of the requisite rigor in avoiding misattribution of causality to correlation but, perhaps more importantly, also enable serious exploration of pathways, processes, and progressions (Hobcraft, 2006).

SOME ISSUES IN BEHAVIORAL GENOMICS

Until recently most knowledge concerning behavioral genetics came from "quantitative" studies that relied on genetically informative designs (Rutter, Pickles, Murray, and Eaves, 2001), mainly of twins or adoptees. Such studies can provide only indications that there is a genetic component to the variance, which includes any gene-environment interactions and correlations. This black-box approach to partitioning variance relies on a number of contestable assumptions (see Rutter, 2006, pp. 41-54), including reared-together twins experiencing equal environments and twins or adoptees being nonselective of the general population.[1] Moreover, such approaches face additional challenges in dealing with dyads (e.g., both partners in a marriage) or adult environments. Nevertheless, some clues about possible interplays of genes and environment can be obtained, as exemplified by the apparent differential heritability of IQ for groups with different socioeconomic status (SES) (Guo and Stearns, 2002; Rowe, Jacobsen, and Van den Oord, 1999; Turkheimer, Haley, Waldron, D'Onofrio, and Gottesman, 2003), although this may also have arisen from differential variances in experiences by SES group. Moreover, there have been substantial advances in exploring the covariation of multiple characteristics, in order to establish shared components or comorbidities of genetic or environmental variance (Kendler, 2005, refers to this as "advanced genetic epidemiology").

With recent advances in genomics, there has been a major shift toward behavioral genomics, in which specific genetic markers are linked to specific attributes, behaviors, or experiences. Techniques for identifying large numbers of single-nucleotide polymorphisms (SNPs) through gene "chips" or microarrays are becoming cheaper, and much progress has been made in methods of linkage and association analysis (see the recent series of review articles on genetic epidemiology in *Lancet*: Burton, Tobin, and Hopper, 2005; Teare and Barrett, 2005; Cordell and Clayton, 2005; Palmer and Cardon, 2005; Hattersley and McCarthy, 2005; Hopper, Bishop, and Easton, 2005; Davey Smith et al., 2005; see also reviews on genome-wide scans in *Nature Reviews Genetics*: Hirschhorn and Daly, 2005; Wang, Barratt, Clayton, and Todd, 2005).

---

[1]A similar problem of the selectivity of "switchers" bedevils most attempts to control for unobserved variation through fixed-effects models (e.g., restricting analysis to sibling pairs in which one experienced a teenage birth and one did not, or to those who changed their affiliation to a trade union, etc.).

A number of approaches have been developed for "gene hunting," a search for markers on the genome that are associated with a particular life-course outcome. Out of such research a wide range of links to areas of the genome, SNPs, haplotypes, etc., have been identified for many outcomes: attention deficit hyperactivity disorder (ADHD) has received enough attention to warrant a special issue of *Biological Psychiatry* (see Faraone et al., 2005; Sklar, 2005, on haplotype mapping; also Mill et al., 2005). Some progress has also been made in identifying genetic markers for personality traits (Munafò et al., 2003; Van Gestel and Van Broeckhoven, 2003), affective and anxiety disorders (Leonardo and Hen, 2006), gambling (Comings et al., 2001), and intelligence (Harlaar et al., 2005; Plomin, Kennedy, and Craig, 2006). However, many of the identified associations are small and are often not replicated in other studies.

A good example of some of these challenges is the study of general cognitive ability, in which quantitative behavioral genetic studies consistently suggest a high heritability, but few genetic markers have been identified and those that have account for quite small fractions of the variation observed (see Plomin, 2003; Plomin and Spinath, 2004; and Plomin, Kennedy, and Craig, 2006, for good accounts of the methods used and problems encountered). When large numbers of genetic markers are screened (and gene arrays that can identify allelic variation on 900K SNPs are available), there is a huge risk of false positive associations, unless sample sizes are extremely large (Zonderman and Cardon, 2004). However, such studies are one approach to identifying "candidate" genes (see Munafò, 2006) that can be explored in more detail in other studies, if replicable (see also Craig and Plomin, 2006).

A further set of concerns arises in deciding which genetic markers to screen. The human haplotype project is one attempt to systematize such work, by identifying small parts of a chromosome for which there is little or no evidence of recombination (haplotype blocks), such that linkage to areas of the chromosome is relatively robust (e.g., Van den Oord and Neale, 2004; Conrad et al., 2006). Identification of ever increasing numbers of regularly spaced SNPs is another approach to increasing precision in identification of areas of the genome. Both haplotype mapping and SNP identification point to areas on the genome that show a significant association with a characteristic, but neither points to the specific gene that is implicated, and further detailed mapping of identified segments is required. For example, there is increasing evidence for the importance of specific "microsatellite" or "simple sequence repeats" in the context of social behaviors, for which differences in the number of repeats matter (e.g., Bachner-Melman et al., 2005; Hammock and Young, 2005; Kashi and King, 2006).

Even further complexities are introduced as soon as interplays are considered. Gene-environment interactions involve the response to a stimulus varying for different allelic combinations, whereby genes can be seen as moderators of the response (e.g., Caspi et al., 2003; T.E. Moffitt, 2005; Rutter, 2006). Moreover, there is increasing evidence of epistatic effects, in which combinations of genetic markers interact together (including regulatory chains or cascades) rather than singly, and examples have been found in which there is no main effect for single markers but the combination matters (e.g , Brodie, 2000; Grigorenko, 2003; Marchini, Donnelly, and Cardon, 2005; Templeton, 2000). There is also increasing evidence for epigenetic effects, whereby gene expression is altered on a lasting (and possibly heritable) basis through external influences that may operate through DNA methylation (e.g., Cordell, 2002; Eaves, Silberg, and Erklani, 2003; Jaenisch and Bird, 2003; Pastinen et al., 2003; Weaver et al., 2004). A further complexity arises from genes that have many different effects, known as pleiotropy.

All of these complexities have parallels in social science: the relatively rare exploration of pathways or interactions arises in part because of overreliance on statistical models with only main effects. The occurrence of multiple small effects (polygenic for genes) is not at all unusual: for example, associations of differing contexts with early adolescent development are small, but the total variance accounted for by contexts is quite large (Cook, Herman, Phillips, and Settersten, 2002). For several years I have been grappling with the issues involved in the life-course links of experience of disadvantage, in which multiple childhood origins all matter for multiple adult outcomes (e.g., Hobcraft, 2004): such complexities provide major analytic and interpretational challenges, a theme returned to below.

Ultimately, what is required is not just more sophisticated statistical techniques but instead building a much better understanding of pathways and mechanisms involved. For social scientific behavior, many such insights have to come from social scientists. Among other things, we have a large body of evidence relevant to the search for pleiotropic associations, knowing much about how a single stimulus can affect many responses, although the pathways could prove different. But there is also a need for greater interplays among geneticists, neuroscientists, and social scientists to help refine and target what gene-brain-behavior interplays should be prioritized for exploration.

Valuable insights into the processes involved in identifying gene-environment interactions are provided by Moffit et al. (2005), who identify seven steps needed to identify measured gene-environment interactions: (1) consulting quantitative behavioral genetic studies; (2) identifying a candidate environmental factor; (3) optimizing environmental risk measurement; (4) identifying candidate susceptibility genes; (5) testing for an interaction; (6) evaluating whether the interaction extends beyond the gene-environment-outcome triad; and (7) replication and meta-analysis. Moreover, they emphasize the great care required in choosing the candidate environmental and genetic indicators and having strong reasons for expecting the pathway to be plausible.

At a fairly general level, it is hardly surprising that most successful attempts to identify gene-environment interactions relating to behavior have linked to genetic markers for neurotransmitters, since the brain is vital in most behaviors and the ability to cross the blood-brain barrier is likely to be important to responses to external stimuli. For example, Caspi et al. (2003) were able to draw on a body of evidence that suggested that serotonin was involved in pathways to depression. Their choice of stressful events as a likely trigger mechanism was also entirely plausible and evidence-based. Such evidence-based, theoretically informed approaches are more likely to be successful in identifying complex pathways than gene-hunting (although this is one route to identifying candidate genes)— see Rutter et al. (2006). In this area, bringing together social scientists with both geneticists and neuroscientists (and other relevant disciplines) is essential to real progress.

A further example, still being explored, begins in neuroscience. Over a period of several years, the key roles of oxytocin and vasopressin[2] brain receptors in pair bonding have been explored, particularly for voles (Hammock and Young, 2002; Lim, Hammock, and Young, 2004; Young, 2003). Prairie voles bond for life, whereas montane voles do not, and this difference has been linked to a single polymorphism. Other studies of rats, sheep, and hamsters also suggest important roles for oxytocin and vasopressin in maternal imprinting (Insel and Fernald, 2004; Numan and Insel, 2003). For humans, some clues come from functional magnetic resonance imaging (fMRI) scans that suggest that "romantic" and maternal attachments both overlap with sites that are rich in oxytocin and vasopressin receptors (Bartels and Zeki, 2000, 2004). There is further evidence that the dopamine D_2 receptor (part of the reward system) plays a part in such bonding (Insel and Fernald, 2004; Insel and Young, 2001; Young, Wang, and Insel, 2002). The well-established release of oxytocin during breastfeeding is probably also linked to mother-child bonding. Moreover, recent experimental economic research has shown a link between oxytocin (administered as an intranasal spray) and increased levels of trust in humans (Kosfeld, Heinrichs, Zak, Fischbacher, and Fehr, 2005). All of this

---

[2]Oxytocin and vasopressin are closely related neuropeptide hormones and have the property of neurotransmitters that enables them to cross the blood-brain barrier. They have been implicated in a wide range of pro- and antisocial behaviors (for a good overview, see Caldwell and Young, 2006).

body of research points strongly toward genetic markers for oxytocin, vasopressin, and dopamine receptors being candidate genes for human bonding (and possibly partnership breakdown, too?), and exploration has begun (Hammock and Young, 2005).

There is clearly scope for social scientific input into the research on pair bonding and partnership breakdown. We have accumulated a body of knowledge about the personality traits and experiences that are associated with partnership formation and breakdown, and serious work on gene-environment interactions requires identification of the candidate stimuli. Such work could also draw on the integrative synthesis on human bonding of Miller and Rodgers (2001). Do these genetic markers help to explain the greater union fragility of young partnerships? If so, is this linked to late teenage brain development? Several animal studies link the receptors to the olfactory bulb, suggesting that smell (or pheromones) plays an important part, yet the human brain scan studies relied solely on a visual (photographic) stimulus. The neurotransmitters and genes for the receptors involved may be common across species, but the brain areas involved differ, raising unanswered questions about evolution. There are also unanswered questions as to what mechanisms inside the brain result in lasting pair bonding. Is lasting pair bonding a result of an epigenetic effect, involving a lasting change in gene expression? Or is long-term memory involved, and how? And what are the triggers and pathways to generate the oxytocin, vasopressin, and dopamine that the receptors respond to?

COMPLEXITY AND SIMPLIFICATION IN SOCIAL SCIENCE

The balance between sufficient complexity and judicious simplification is needed for real progress to be made in understanding the complexities of human behavior. We have explored some of these issues for behavioral genomics and, fleetingly, neuroscience. But some real progress is also required in the social sciences. Treading the delicate path between mindless empiricism (with parallels to gene hunting) and overblown social theorization (with parallels to some evolutionary theorization) is vital for progress in the science of social science. This requires building (partial) analytic (not grand) theories or frameworks, exploring pathways and mechanisms, finding means of exploring interlinked processes (rather than throwing out any endogenous element), and careful and thorough empirical work, including models that are just complex enough. In this section, some of the complexities that are all too often ignored in oversimplified models are discussed.

A good example of progress toward these goals comes from the recent developments in behavioral economics, in which the realities of departures from a simple rational choice framework are being explored (Brocas and Carillo, 2003, 2004; Camerer et al., 2004). Key to such progress has been the engagement of economics and psychology (e.g., Kahneman, 2003). This section also urges a much greater interaction with psychology, since I think we have much to learn from the underlying approach of some of the best practice there. Moreover, the concerns of this chapter to emphasize the need for an integrated approach to alleles, brains, and contexts and their interplays in understanding human behavior (see also Hobcraft, 2006) also point toward greater connections to or borrowing from psychology, since this is one of the few disciplines that engages across this broad range. Biology is another useful model in this respect— see Lieberson and Lynn (2002) for a thoughtful discussion of the value of some aspects of biology for sociological research.

Development over the life course is complex (e.g., Mortimer and Shanahan, 2003). We have emphasized the role of interplays (and likely inseparability) of pathways within the person, both genetic and neuro-endocrine, and processes, whereby contexts and experiences affect the person, over the life course. This developmental perspective is important: genes, brain, mind, the person, other persons, and structures all interplay over time (see the quotation from the National Research Council, 2001b, at the head of this chapter and Hobcraft, 2006). Multiple dynamic processes thus need to be considered and modeled. We should expect and look for complex feedbacks and interactions, developmental chains involving interlinked sequences, possible key trigger events or experiences, and packages or groups of experiences that together make more than the sum of the parts. Endowments and experiences, both within the person and external, shape the person, who also reacts to and selects contexts and experiences. Refusal to consider intimately interlinked elements of these progressions because of possible biases arising from endogeneity is akin to throwing out the baby with the bathwater.

This perspective raises huge analytic challenges, in addition to the theoretical and conceptual ones. The economist James Duesenbury quipped that "economics is all about how people make choices; sociology is all about how they don't have any choices to make." But the tendency to regard choice (or agency) and structures or constraints (I prefer the broader term contexts) as in opposition should be consigned to the same dustbin of history as the opposition of nature and nurture: both matter a great deal, and they interplay. Neither agency nor structure is a nuisance element that can be discarded from models, and separating the two may not be possible, although disentangling how they interplay matters for understanding.

Models

Several modern approaches to analysis are relevant to the research agenda of trying to improve understanding of human behavior and are increasingly being used. The intrinsically multilevel nature of the gene-brain-person-context interplays virtually demands multilevel models (e.g., Goldstein, 2003), and such models have been adapted for behavioral genetic research (see Guo and Wang, 2002). However, the subtle and likely contingent interplays and feedbacks across these levels are rarely thought through or captured in actual model specification (e.g., Cacioppo et al., 2000; and Chapter 17 in this volume). Relatedly, there have been important developments in multiprocess models that enable the exploration of correlated (endogenous) processes over time (Aassve, Burgess, Propper, and Dickson, 2004; Steele, Joshi, Kallis, and Goldstein, 2006; Steele, Kallis, Goldstein, and Joshi, 2005), although again there is a need to conceptualize and specify the nature of interplays carefully. More use could also be made of structural equation models as used by psychologists, which include such useful features as combining measures in latent constructs and carefully specifying theoretically informed pathways (e.g., Kaplan, 2000). There are also clear links to behavioral genetics here, since most quantitative behavioral genetics analysis is done using structural equation models. There has been a explosion of innovative statistical methods used in genetics in recent years (e.g., Thomas 2004). Given the complexities of these different types of modeling, it is essential that more substantial research teams working on understanding human behavior have integrated statisticians, as happens in much genetic and medical research. Moreover, the next stage should probably involve much greater attention to integrating the different approaches, since we badly need sensible approaches that enable the modeling of complex interplays among genes and environments (note the deliberate use of plurals here).

Others advocate moving away from statistical models (and the probabilistic nature of risks) altogether: in psychology this has become known as a "person-centered" approach, which involves distinguishing groups of individuals who are "alike." However, most cluster analytic approaches allocate individuals on a model-based probabilistic basis, making the claims of dealing with groups of real people hard to justify. Alternative approaches use recursive trees, which do genuinely divide the population into groups, but, again using only other statistical algorithms; the advantage of such recursive tree approaches in part lies in the greater emphasis on interactions (e.g., Hobcraft and Sigle-Rushton, 2005; Zhang and Singer, 1999) and an extension (to "recursive forests") has been used in identifying candidate genes (Zhang, Yu, and Singer, 2003).

A further problem that bedevils much modeling is that many simplifications may be virtually equally likely. Sometimes overly rigid theoretical straitjackets hide such model uncertainty, but more use could be made of Bayesian approaches to model uncertainty in exploring candidate genes or candidate environmental indicators and their interplays (see Sorensen and Gianola, 2006).

Conceptual Issues

Successful models require strong conceptual underpinnings or, put another way, statistical models are simply a tool to help inference and the crucial issues are involved in the formulation and underpinning of the models. In this section I touch on a range of issues that should be in mind when exploring the understanding of behavior over the life course, drawing particularly on some of the concerns of psychologists.

O'Connor (2003) provides a useful discussion of different conceptual models for early experience, although many of the issues are of broader life-course relevance. He distinguishes three different models of development: sensitive periods, experience-adaptive or developmental programming, and cumulative models. Each has different implications for how to study the process in question, and they are sometimes harder to separate analytically than conceptually. Sensitive periods or experience-expectant models mean that an input is necessary for development to proceed within a particular window of time. The clearest example is probably the development of the visual cortex, for which external stimulus is essential during a critical period. Experience-adaptive or developmental programming involves biological systems adapting to environmental stimuli in lasting ways: an example would be epigenetic effects, in which gene expression is changed in lasting ways as a result of some experience. The third route is through cumulative or chain models, in which lasting effects occur only if reinforced over time or early experiences set longer term pathways in train. Clearly distinguishing between these three is not always simple, and the first two can have longer term repercussions through the third.

Path dependence through the life course, whereby origins, endowments, and previous experiences shape current behavioral responses to current circumstances, is a crucial component in understanding human behaviors. The role of sequences or packages of experiences over time through developmental chains or cascades is an essential component. Intermediate elements in a causal chain are referred to as "mediators" in the psychological literature, whereas "moderators" alter the impact of another risk factor and involve contingent relationships (or interactions in statistical terminology). Gene-environment interactions are but one example of contingent relationships of this type. The consequences of a specific life experience for an outcome are extremely likely to vary among individuals: some will be more vulnerable to the impact of shocks than others, depending for example on their past experiences, their personality traits, or their cognitive style (see Rutter, 2002, 2004). Previous experience of adversity may steel the individual against a repetition or may sensitize them further (e.g., Rutter, 2004); related examples include the lasting legacies of childhood poverty or behavior and of early unemployment (among other factors, sometimes referred to as "scarring") for a wide range of adult outcomes (see Hobcraft, 2004). Rutter (2004) persuasively argues that many moderators only accentuate a pathway rather than cause an abrupt change, although there are sometimes claims that turning points exist, too. Sampson and Laub (1993) identify a good marriage or the discipline brought about by armed service as being critical turning points in the careers of criminals; however, a good marriage is hardly likely to be an exogenous shock and is at least as likely to be a marker in a process of some duration.

Just as there is a huge range of genetic or other biomarkers that might be considered in the context of understanding human behavior, there is also a wide range of environmental or contextual influences that can affect behavior or experience. In an extended discussion of contexts for understanding demographic behavior, I distinguished two classes: interplays of the person with other persons and with structures or institutions (Hobcraft, 2006). Examples of potentially important interplays with other persons include the partnership dyad, the mother-father-child triad, family networks (both kin and partner kin), peer groups, friendship or support networks, care or service providers, employers and workmates, and the local community. Examples of structural contexts include welfare regimes; labor markets; education and training systems; housing markets and neighborhoods; health systems; policy and benefit environments; norms and laws; economic, cultural, religious, and political institutions; and gender structures.

Faced with such a wide range of possible influencing factors, there is again a need for judicious simplification. One approach that can be advocated is to concentrate initially on more proximate factors, whether linked by the outcome being considered (e.g., health systems for health outcomes), or by proximity in likely causal chains (e.g., parent-child interplays for child development). I have advocated such a proximal approach for linking values or attitudes to reproductive behavior (Hobcraft, 2002, 2003). Cacioppo (Chapter 17 in this volume) takes a similar position regarding the genetic linkages to prostate cancer, and a similar case can be made for greater specificity in the genetic pathways to cardiovascular disease and longevity (see also Chapter 1 in this volume and Christensen, Johnson, and Vaupel, 2006). Perhaps the search for "generalist genes" (e.g., Kovas and Plomin, 2006; Butcher, Kennedy, and Plomin, 2006, on intelligence) is doomed to come up with few consistent results for many quantitative trait loci, each with small average effects, and much more specificity about pathways, epistatic or epigenetic interplays, or gene-environment interplays is needed (e.g., Rutter, Moffitt, and Caspi, 2006; for a specific example on ADHD and links to reading disabilities, see Stevenson et al., 2005).

The study of contextual (and perhaps particularly interpersonal) influences on individual behavior needs to pay attention both to persons seeking contexts compatible with their own characteristics and persons evoking a response from the context (e.g., other person). See the related literature on active and reactive gene-environment correlations (e.g., Plomin, 1994; Rutter and Silberg, 2002; Rutter et al., 2006) and also recall the discussions on the difficulties in separating agency and constraint above.

CONCLUSION

There is no doubt that human behavior is complex. Improving understanding of these behaviors necessarily requires exploring the feedbacks and interplays among genes, brain, mind, the person, other persons, and structures. Emotions, personality, decision making, endowments, experiences and networks all play some part in the processes too. Such research can be carried out only by transdisciplinary teams (e.g., Institute of Medicine, 2006) that bring together a wide range of disciplinary skills: molecular biology, genetics, neuroscience, behavioral science, and social science in its relevant guises would constitute a fairly essential list that might be supplemented by epidemiology and health sciences as well as many intermediate skill groups, such as behavioral genetics. The reward structures also need to change, so as to enable young scholars to work in these interstices while also gaining tenure. Much training to raise cross-disciplinary understanding and communication is also needed.

I have made a strong case that many social scientists might benefit from thinking more about these interplays and then sifting knowledge and empirically exploring so as to achieve enough simplification to make the challenge practicable. Most studies will address only fragments of the multiple levels and interplays involved, but I would still urge greater attention to pathways within the individual and their interplays with the processes and progressions whereby the individual interplays with multiple contexts over the life course. A concentration on chains or sequences of events, greater awareness of contingent relationships (interactions in statistical language), and elaboration of partial midlevel frameworks or mechanisms is also required (see Hedström, 2005; Hobcraft, 2006). As a result, models will become more complex, but also more realistic and less simplistic.

REFERENCES

Aassve, A., Burgess, S., Propper, C., and Dickson, M. (2004). *Employment, family union, and childbearing decisions in Great Britain.* (CASE paper 84). London: London School of Economics, Centre for Analysis of Social Exclusion.

Bachner-Melman, R., Zohar, A.H., Bacon-Shnoor, N., Elizur, Y., Nemanov, L., Gritsenko, I., and Ebstein, R.P. (2005). Link between vasopressin receptor AVPR1A promoter region microsatellites and measures of social behaviour in humans. *Journal of Individual Differences, 26*(1), 2-10.

Bartels, A., and Zeki, S. (2000). The neural basis of romantic love. *NeuroReport, 11*(17), 3829-3834.

Bartels, A., and Zeki, S. (2004). The neural correlates of maternal and romantic love. *Neuro-Image, 21,* 1155 1166.

Brocas, I., and Carrillo, J.D. (Eds.) (2003). *The psychology of economic decisions. Volume I: Rationality and well-being.* Oxford, England: Oxford University Press.

Brocas, I., and Carillo, J.D. (Eds.) (2004). *The psychology of economic decisions. Volume 2: Reasons and choices.* Oxford, England: Oxford University Press.

Brodie, E.D., III. (2000). Why evolutionary genetics does not always add up. In J.B. Wolf, Brodie, E.D., III, and Wade, M.J. (Eds.) *Epistasis and the evolutionary process.* Oxford, England: Oxford University Press.

Burton, P.R., Tobin, M.D., and Hopper, J.L. (2005). Genetic epidemiology 1: Key concepts in genetic epidemiology. *Lancet, 366,* 941-951.

Butcher, L.M., Kennedy, J.K.J., and Plomin, R. (2006). Generalist genes and cognitive neuroscience. *Current Opinion in Neurobiology, 16,* 145-151.

Cacioppo, J.T., Berntson, G.G., Adolphs, R., et al. (Eds.) (2002). *Foundations in social neuroscience.* Cambridge, MA: MIT Press.

Cacioppo, J.T., Berntson, G.G., Sheridan, J.F., and McClintock, M.K. (2000). Multilevel integrative analyses of human behaviour: Social neuroscience and the complementing nature of social and biological approaches. *Psychological Bulletin, 126*(6), 829-843.

Caldwell, H.K., and Young, W.S., III. (2006). Oxytocin and vasopressin: Genetics and behavioral implications. In R. Lim (Ed.), *Handbook of neurochemistry and molecular neurobiology, third edition.* New York: Springer.

Camerer, C.F., Loewenstein, G., and Rabin, M. (Eds) (2004). *Advances in behavioral economics.* Princeton, NJ and New York: Princeton University Press and Russell Sage Foundation.

Caspi, A. (2004). Life-course development: The interplay of social-selection and social causation within and across generations. In P.L Chase-Lansdale, K.E. Kiernan, and R.J. Friedman (Eds.), *The potential for change across lives and generations: Multidisciplinary perspectives.* New York: Cambridge University Press.

Caspi, A., McClay, J., Moffitt, T.E., Mill, J., Martin, J., Craig, I.W., Taylor, A., and Poulton, R. (2002). Role of genotype in the cycle of violence in maltreated children. *Science, 297,* 851-854.

Caspi, A., Moffitt, T.E., Cannon, M., McClay, J., Murray, R., Harrington, H., Taylor, A., Arsenault, L., Williams, B., Braithwaite, A., Poulton, P., and Craig, I.W. (2005). Moderation of the effect of adolescent-onset cannabis use on adult psychosis by a functional polymorphism in the Catechol-O-Methyltransferase gene: Longitudinal evidence of a gene x environment interaction. *Biological Psychiatry, 57,* 1117-1127.

Caspi, A., Sugden, K., Moffitt, T.E., Taylor, A., Craig, I.W., Harrington, H., McClay, J., Mill, J., Martin, J., Braithwaite, A., and Poulton, R. (2003). Influence of life stress on depression: Moderation by a polymorphism in the 5-HTT gene. *Science, 301,* 386-389.

Christensen, K., Johnson, T.E., and Vaupel, J.W. (2006). The quest for genetic determinants of human longevity: Challenges and insights. *Nature Reviews Genetics, 7,* 436-448.

Comings, D.E., Gade-Andavolu, R., Gonzalez, N., Wu, S., Muhleman, D., Chen, C., Koh, P., Farwell, K., Blake, H., Dietz, G., MacMurray, J.P., Lesieur, H.R., Rugle, L.J., and Rosenthal, R.J. (2001). The additive effect of neurotransmitter genes in pathological gambling. *Clinical Genetics, 60,* 107-116.

Conrad, D.F., Jakobsson, M., Coop, G., Wen, X., Wall, J.D., Rosenberg, N.A., and Pritchard, J.K. (2006). A worldwide survey of haplotype variation and linkage disequilibrium in the human genome. *Nature Genetics, 38*(11), 1251-1260.

Cook, T.D., Herman, M.R., Phillips, M., and Settersten, R.A., Jr. (2002). Some ways in which neighborhoods, nuclear families, friendship groups, and schools jointly affect changes in early adolescent development. *Child Development, 73*(4), 1283-1309.

Cordell, H.J. (2002). Epistasis: What it means, what it doesn't mean, and statistical methods to detect it in humans. *Human Molecular Genetics, 11*(20), 2463-2468.

Cordell, H., and Clayton, D.G. (2005). Genetic epidemiology 3: Genetic association studies. *Lancet, 366,* 1121-1131.

Craig, I., and Plomin, R. (2006). Quantitative trait loci for IQ and other complex traits: Single-nucleotide polymorphism genotyping using pooled DNA and microarrays. *Genes, Brain, and Behavior, 5*(Suppl. 1), 32-37.

Curtis, W.J., and Cicchetti, D. (2003). Moving research on resilience into the 21st century: Theoretical and methodological considerations in examining the biological contributors to resilience. *Development and Psychopathology, 15,* 773-810.

Davey Smith, G., Ebrahim, S., Lewis, S., Hansell, A.L., Palmer, L.J., and Burton, P.R. (2005). Genetic epidemiology 7. Genetic epidemiology and public health: Hope, hype, and future prospects. *Lancet, 366,* 1484-1498.

Davidson, R.J. (2004). Well-being and affective style: Neural substrates and biobehavioural correlates. *Philosophical Transactions of the Royal Society of London, Series B, 359,* 1395-1411.

Eaves, L., Silberg, J., and Erklani, A. (2003). Resolving multiple epigenetic pathways to adolescent depression. *Journal of Child Psychology and Psychiatry, 44*(7), 1006-1014.

Faraone, S.V., Perlis, R.H., Doyle, A.E., Smoller, J.W., Goralnick, J.J., Holmgren, M.A., and Sklar, P. (2005). Molecular genetics of attention-deficit/hyperactivity disorder. *Biological Psychiatry, 57,* 1313-1323.

Goldstein, H. (2003). *Multilevel statistical models (Kendall's Library of Statistics).* London, England: Hodder Arnold.

Granger, D.A., and Kivlighan, K.T. (2003). Integrating biological, behavioral, and social levels of analysis in early child development: Progress, problems, and prospects. *Child Development, 74*(4), 1058-1063.

Grigorenko, E.L. (2003). Epistasis and the genetics of complex traits. In R. Plomin, J.C. DeFries, I.W. Craig, and P. McGuffin (Eds.), *Behavioral genetics in the postgenomic era.* Washington, DC: American Psychological Association.

Guo, G., and Stearns, E. (2002). The social influences on the realization of genetic potential for intellectual development. *Social Forces, 80*(3), 881-910.

Guo, G., and Wang, J. (2002). The mixed or multilevel model for behaviour genetic analysis. *Behavior Genetics, 32*(1), 37-49.

Hajnal, J. (1955). The prospects for population forecasts. *Journal of the American Statistical Association, 50,* 309-322 (p.321 for quote).

Hammock, E.A.D., and Young, L.J. (2002). Variation in the vasopressin V1a receptor promoter and expression: Implications for inter- and intraspecific variation in social behaviour. *European Journal of Neuroscience, 16,* 399-402.

Hammock, E.A.D., and Young, L.J. (2005). Microsatellite instability generates diversity in brain and sociobehavioral traits. *Science, 308,* 1630-1634.

Harlaar, N., Butcher, L.M., Meaburn, E., Sham, P., Craig, I.W., and Plomin, R. (2005). A behavioral genomic analysis of DNA markers associated with general cognitive ability in 7 year-olds. *Journal of Child Psychology and Psychiatry, 46*(10), 1097-1107.

Hattersley, A.T., and McCarthy, M.I. (2005). Genetic epidemiology 5: What makes a good genetic association study? *Lancet, 366,* 1315-1323.

Heckman, J.J. (2000). Causal parameters and policy analysis in economics: A twentieth century retrospective. *Quarterly Journal of Economics,* 115, 45-97.

Hedstrom, P., and Swedburg, R. (Eds.) (1998). *Social mechanisms: An analytical approach to social theory.* New York: Cambridge University Press.

Hirschhorn, J.N., and Daly, M.J. (2005). Genome-wide association studies for common diseases and complex traits. *Nature Reviews Genetics, 6,* 95-108.

Hobcraft, J.N. (2002). Moving beyond elaborate description: Towards understanding choices about parenthood. In M. Macura and G. Beets (Eds.) *The dynamics of fertility and partnership in Europe: Insights and lessons from comparative research vol. I.* New York and Geneva, United Nations.

Hobcraft, J.N. (2003). Reflections on demographic, evolutionary, and genetic approaches to the study of human reproductive behavior. In National Research Council, *Offspring: Human fertility behavior in biodemographic perspective.* K.W. Wachter and R.A. Bulatao (Eds.). Washington, DC: The National Academies Press.

Hobcraft, J. (2004). Parental, childhood, and early adult legacies in the emergence of adult social exclusion: Evidence on what matters from a British cohort. In P.L. Chase-Lansdale, K. Kiernan, and R. Friedman (Eds.), *Human development across lives and generations: The potential for change.* New York: Cambridge University Press.

Hobcraft, J. (2006). The ABC of demographic behaviour: How the interplays of alleles, brains, and contexts over the life course should shape research aimed at understanding population processes. *Population Studies, 60*(2), 153-187.

Hobcraft, J., and Sigle-Rushton, W. (2005). *An exploration of childhood antecedents of female adult malaise in two British birth cohorts: Combining Bayesian model averaging and recursive partitioning.* (CASE paper #95). London: London School of Economics, Centre for Analysis of Social Exclusion.

Hopper, J.L., Bishop, D.T., and Easton, D.F. (2005). Genetic epidemiology 6: Population-based family studies in genetic epidemiology. *Lancet, 366,* 1397-1406.

Huppert, F.A., and Bayliss, N. (2004). Well-being: Towards an integration of psychology, neurobiology, and social science. *Philosophical Transactions of the Royal Society of London, Series B, 359,* 1447-1451.

Insel, T.R., and Fernald, R.D. (2004). How the brain processes social information: Searching for the social brain. *Annual Reviews of Neuroscience, 27,* 697-722.

Insel, T.R., and Young, L.J. (2001). The neurobiology of attachment. *Nature Reviews: Neuroscience, 2,* 129-136.

Institute of Medicine. (2006). *Genes, behavior, and the social environment: Moving beyond the nature/nurture debate.* Committee on Assessing Interactions Among Social, Behavioral, and Genetic Factors in Health. Board on Health Sciences Policy L.M. Hernandez and D.G. Blazer, Eds.. Washington, DC: The National Academies Press.

Jaenisch, R., and Bird, A. (2003). Epigenetic regulation of gene expression: How the genome integrates intrinsic and environmental signals. *Nature Genetics, 33,* 245-254.

Johnson, T.E., and Crow, J.F. (Eds.) (2005). *Research on environmental effects in genetic studies of aging*. Special Issue of *Journal of Gerontology: Social Sciences*, 60B.

Kahneman, D. (2003). Maps of bounded rationality: Psychology for behavioral economics. *American Economic Review, 93*(5), 1449-1475.

Kaplan, D. (2000). *Structural equation modeling: Foundations and extensions*. Thousand Oaks, CA: Sage

Kashi, Y., and King, D.G. (2006). Simple sequence repeats as advantageous mutators in evolution. *Trends in Genetics, 22*, 253-259.

Kendler, K.S. (2005). Psychiatric genetics: A methodologic critique. *American Journal of Psychiatry, 162*, 3-11.

Kosfeld, M., Heinrichs, M., Zak, P.J., Fischbacher, U., and Fehr, E. (2005). Oxytocin increases trust in humans. *Nature, 435*, 673-676.

Kovas, Y., and Plomin, R. (2006). Generalist genes: Implications for the cognitive sciences. *Trends in Cognitive Sciences, 10*(5), 198-203.

Leonardo, E.D., and Hen, R. (2006). Genetics of affective and anxiety disorders. *Annual Review of Psychology, 57*, 117-137.

Lieberson, S., and Lynn, F.B. (2002). Barking up the wrong branch: Scientific alternatives to the current model of sociological science. *Annual Review of Sociology, 28*, 1-19.

Lim, M.M., Hammock, E.A.D., and Young, L.J. (2004). The role of vasopressin in the genetic and neural regulation of monogamy. *Journal of Neuroendocrinology, 16*, 325-332.

Marchini, J., Donnelly, P., and Cardon, L.R. (2005). Genome-wide strategies for detecting multiple loci that influence complex diseases. *Nature Genetics, 37*(4), 413-417.

Mill, J., Xu, X., Ronald, A., Curran, S., Price, T., Knight, J., Craig, I., Sham, P., Plomin R., and Asherson, P. (2005). Quantitative trait locus analysis of candidate gene alleles associated with Attention Deficit Hyperactivity Disorder (ADHD) in five genes: DRD4, DAT1, DRD5, SNAP-25, and 5HT1B. *American Journal of Medical Genetics, Part B, (Neuropsychiatric Genetics), 133B*, 68-73.

Miller, W.B., and Rodgers, J.L. (2001). *The ontogeny of human bonding systems: Evolutionary origins, neural bases, and psychological manifestations*. Kluwer, Dordrecht.

Moffitt, R. (2003). Causal analysis in population research: An economist's perspective. *Population and Development Review, 29*(3), 448-458.

Moffitt, R. (2005). Remarks on the analysis of causal relationships in population research. *Demography, 42*(1), 91-108.

Moffitt, T.E. (2005). The new look of behavioral genetics in developmental psychopathology: Gene-environment interplay in antisocial behaviors. *Psychological Bulletin, 131*(4), 533-554.

Moffitt, T.E., Caspi, A., and Rutter, M. (2005). Strategy for investigating interactions between measured genes and measured environments. *Archives of General Psychiatry, 62*, 473-481.

Moffitt, T.E., Caspi, A., and Rutter, M. (2006). Measured gene-environment interactions in psychopathology: Concepts, research strategies, and implications for research, intervention, and public understanding of genetics. *Perspectives on Psychological Science, 1*(1), 5-27.

Mortimer, J.T., and Shanahan, M.J. (Eds.). (2003). *Handbook of the life course*. New York: Kluwer Academic/Plenum Publishers.

Munafò, M.R. (2006). Candidate gene studies in the 21st century: Meta-analysis, mediation, moderation. *Genes, Brain, and Behavior, 5*(Suppl. 1), 3-8.

Munafò, M.R., Clark, T.G., Moore, L.R., Payne, E., Walton, R., and Flint, J. (2003). Genetic polymorphisms and personality in healthy adults: A systematic review and meta-analysis. *Molecular Psychiatry, 8*, 471-484.

National Research Council (2001a). *Cells and surveys: Should biological measures be included in social science research?* Committee on Population, C.E. Finch, J.W. Vaupel, and K. Kinsella, Eds. Commission on Behavioral and Social Sciences and Education. Washington, DC: National Academy Press.

National Research Council. (2001b). *New horizons in health: An integrative approach.* Committee on Future Directions for Behavioral and Social Sciences Research at the National Institutes of Health, B.H. Singer and C.D. Ryff, Eds. Board on Behavioral, Cognitive, and Sensory Sciences. Washington, DC: National Academy Press.

National Research Council. (2003). *Offspring: Human fertility behavior in biodemographic perspective.* Committee on Population, K.A. Wachter and R.A. Bulatao, Eds. Division of Behavioral and Social Sciences and Education. Washington, DC: The National Academies Press.

National Research Council and Institute of Medicine. (2000). *From neurons to neighborhoods: The science of early childhood development.* Board on Children, Youth, and Families, J.P. Shonkoff and D.A. Phillips, Eds. Washington, DC: National Academy Press.

Numan, M., and Insel, T.R. (2003). *The neurobiology of parental behavior.* New York: Springer-Verlag.

O'Connor, T.G. (2003). Early experiences and psychological development: Conceptual questions, empirical illustrations, and implications for intervention. *Development and Psychopathology, 15,* 671-690.

Palmer, L.J., and Cardon, L.R. (2005). Genetic epidemiology 4. Shaking the tree: Mapping complex disease genes with linkage disequilibrium. *Lancet, 366,* 1223-1234.

Pastinen, T., Sladek, R., Gurd, S., Sammak, A., Ge, B., Lepage, P., Lavergne, K., Villeneuve, A., Gaudin, T., Brändström, H., Beck, A., Verner, A., Kingsley, J., Harmsen, E., Labuda, D., Morgan, K., Vohl, M.-C., Naumova, A.K., Sinnett, D., and Hudson, T.J. (2003). A survey of genetic and epigenetic variation affecting human gene expression. *Physiological Genomics, 16,* 184-193.

Plomin, R. (1994). *Genetics and experience: The interplay between nature and nurture.* Newbury Park, CA: Sage.

Plomin, R. (2001). Behavioral genetics: Psychological perspectives. In N.J. Smelser and P.B. Baltes (Eds.), *International encyclopedia of the social and behavioral sciences* (pp. 9058-9063). New York: Elsevier Science.

Plomin, R. (2003). General cognitive ability. In R. Plomin, J.C. DeFries, I.W. Craig, and P. McGuffin (Eds.), *Behavioral genetics in the postgenomic era.* Washington, DC: American Psychological Association.

Plomin, R. (2005). Finding genes in child psychology and psychiatry: When are we going to be there? *Journal of Child Psychology and Psychiatry, 46*(10), 1030-1038.

Plomin, R., DeFries, J.C., Craig, I.W., and McGuffin, P. (Eds.) (2003). *Behavioral genetics in the postgenomic era.* Washington, DC: American Psychological Association.

Plomin, R., Kennedy, J.K.J., and Craig, I.W. (2006). The quest for quantitative trait loci associated with intelligence. *Intelligence, 34*(6), 513-526.

Plomin, R., and Spinath, F.M. (2004). Intelligence: Genetics, genes, and genomics. *Journal of Personality and Social Psychology, 86*(1), 112-129.

Rowe, D.C., Jacobsen, K.C., and Van den Oord, E.J.C.G. (1999). Genetic and environmental influences on vocabulary IQ: Parental education level as a moderator. *Child Development, 70*(5), 1151-1162.

Rutter, M. (1994). Beyond longitudinal data: Causes, consequences, changes, and continuity. *Journal of Consulting and Clinical Psychology, 62*(5), 928-940.

Rutter, M. (2002). The interplay of nature, nurture, and developmental influences: The challenge ahead for mental health. *Archives of General Psychiatry, 59,* 997-1000.

Rutter, M. (2004). Intergenerational continuities and discontinuities in psychological problems. In P.L. Chase-Lansdale, K. Kiernan, and R. Friedman (Eds.), *Human development across lives and generations: The potential for change*. New York: Cambridge University Press.

Rutter, M. (2006). *Genes and behavior: Nature-nurture interplay explained*. Oxford, England: Blackwell.

Rutter, M., Moffitt, T.E., and Caspi, A. (2006). Gene-environment interplay and psychopathology: Multiple varieties, but real effects. *Journal of Child Psychology and Psychiatry, 47*(3/4), 226-261.

Rutter, M., Pickles, A., Murray, R., and Eaves, L. (2001). Testing hypotheses of specific environmental risk mechanisms for psychopathology. *Psychological Bulletin, 127,* 291-324.

Rutter, M., and Silberg, J. (2002). Gene-environment interplay in relation to emotional and behavioral disturbance. *Annual Review of Psychology, 53,* 463-490.

Sampson, R.J., and Laub, J.H. (1993). *Crime in the making: Pathways and turning points through life*. Cambridge, MA: Harvard University Press.

Seltzer, J.A., Bachrach, C.A., Bianchi, S.M., Bledsoe, C.H., Casper, L.M., Chase-Lansdale, P.L., DiPrete, V.J., Hotz, T.A., Morgan, S.P., Sanders, S.G., and Thomas, D. (2005). Explaining family change and variation: Challenges for family demographers. *Journal of Marriage and the Family, 67,* 908-925.

Shanahan, M.J., and Hofer, S.M. (2005). Social context in gene-environment interactions: Retrospect and prospect. *Journal of Gerontology: Social Sciences, 60B*(Special Issue 1), 65-76.

Shanahan, M.J., Hofer, S.M., and Shanahan, L. (2003). Biological models of behavior and the life course. In J. T. Mortimer and M. J. Shanahan (Eds.), *Handbook of the life course*. New York: Kluwer Academic/Plenum.

Singer, B.H. (2001). Longitudinal data. In N.J. Smelser and P.B. Baltes (Eds.), *International encyclopedia of the social and behavioral sciences* (pp. 9058-9063). Philadelphia, PA: Elsevier Science Ltd.

Sklar, P. (2005). Principles of haplotype mapping and potential applications to attention-deficit/hyperactivity disorder. *Biological Psychiatry, 57,* 1357-1366.

Sorensen, D., and Gianola, D. (2006). *Likelihood, bayesian, and MCMC methods in quantitative genetics*. New York: Springer.

Steele, F., Joshi, H., Kallis, C., and Goldstein, H. (2006). Changing compatibility of cohabitation and childbearing between young British women born in 1958 and 1970. *Population Studies, 60*(2), 137-152.

Steele, F., Kallis, C., Goldstein, H., and Joshi, H. (2005). Childbearing and transitions from marriage and cohabitation. *Demography, 42*(4), 647-673.

Stevenson, J., Langley, K., Pay, H., Payton, A., Worthington, J., Ollier, W., and Thapar, A. (2005). Attention deficit hyperactivity disorder with reading disabilities: Preliminary genetic findings on the involvement of the ADRA2A gene. *Journal of Child Psychology and Psychiatry, 46*(10), 1081-1088.

Teare, M.D., and Barrett, J.H. (2005). Genetic epidemiology 2: Genetic linkage studies. *Lancet, 366,* 1036-1044.

Templeton, A.R. (2000). Epistasis and complex traits. In J.B. Wolf, E.D. Brodie, III, and M.J. Wade (Eds.), *Epistasis and the evolutionary process*. New York: Oxford University Press.

Thapar, A. (2003). Attention deficit hyperactivity disorder: New genetic findings, new directions. In R. Plomin, J.C. DeFries, I.W. Craig, and P. McGuffin (Eds.), *Behavioral genetics in the postgenomic era*. Washington, DC: American Psychological Association.

Thomas, D.C. (2004). *Statistical methods in genetic epidemiology*. New York: Oxford University Press.

Turkheimer, E., Haley, A., Waldron, M., D'Onofrio, B., and Gottesman, I.I. (2003). Socioeconomic status modifies heritability of IQ in young children. *Psychological Science, 14*(6), 623-628.

Van den Oord, E.J.C.G., and Neale, B.M. (2004). Will haplotype maps be useful for finding genes? *Molecular Psychiatry, 9,* 227-236.

Van Gestel, S., and Van Broeckhoven, C. (2003). Genetics of personality: Are we making progress? *Molecular Psychiatry, 8,* 840-852.

Wang, W.Y.S., Barratt, B.J., Clayton, D.G., and Todd, J.A. (2005). Genome-wide association studies: Theoretical and practical concerns. *Nature Reviews: Genetics, 6,* 109-118.

Weaver, I.C.G., Cervoni, N., Champagne, F.A., D'Alessio, A.C., Sharma, S., Seckl, J., Dymov, R.S., Szyf, M., and Meaney, M.J. (2004). Epigenetic programming by maternal behaviour. *Nature Neuroscience, 7*(8), 847-854.

Wooldridge, J.M. (2002). *Econometric analysis of cross section and panel data.* Cambridge, MA: MIT Press.

Young, L.J. (2003). The neural basis of pair bonding in a monogamous species: A model for understanding the biological basis of human behaviour. In National Research Council, *Offspring: Human fertility behavior in biodemographic perspective.* Committee on Population, K.A. Wachter and R.A. Bulatao, Eds. Division of Behavioral and Social Sciences and Education. Washington, DC: The National Academies Press.

Young, L.J., Wang, Z., and Insel, T.R. (2002). Neuroendocrine bases of monogamy. In J.T. Cacioppo, G.G. Berntson, R. Adolphs, et al (Eds.), *Foundations in social neuroscience.* Cambridge, MA: MIT Press.

Zhang, H., and Singer, B. (1999). *Recursive partitioning in the health sciences.* New York: Springer.

Zhang, H., Yu, C.-Y., and Singer, B. (2003). Cell and tumour classification using gene expression data: Construction of forests. *Proceedings of the National Academy of Sciences, 100*(7), 4168-4172.

Appendix

Biographical Sketches of Contributors

Lise Bathum is a specialist in clinical biochemistry and chief physician in the Department of Clinical Biochemistry at Slagelse Hospital in Region Zealand, Denmark. Previously she was associate professor of epidemiology at the University of Southern Denmark and also biobank director of the Danish Twin Registry (2002-2006). She has conducted numerous studies in the field of pharmacogenetics. Her main research fields are twin studies and genetic research on aging-related phenotypes in order to estimate the impact of genes on aging. She has an M.D. and a Ph.D., both from the University of Southern Denmark.

Daniel J. Benjamin is assistant professor in the Department of Economics at Cornell University. His research focuses on psychological economics, incorporating ideas and methods from psychology into economic analysis. Currently his work includes an empirical analysis of the importance of politicians' charisma (as measured by laboratory subjects) in determining election outcomes and a theoretical analysis of how individuals' concern for fairness affects the efficiency of economic exchange. Ongoing work addresses how economic preferences are influenced by psychological/biological factors, such as cognitive ability, social identity (ethnicity, race, and gender), and specific genes. He has an M.Sc. in mathematical economics from the London School of Economics and Political Science and a Ph.D. in economics from Harvard University.

Gary G. Berntson is professor of psychology, psychiatry, and pediatrics at Ohio State University. He is coeditor of the Social Neuroscience Book Series, the *Handbook of Psychophysiology*, and the *Handbook of Neuroscience for the Behavioral Sciences*. He is an officer and board member of the Society for Psychophysiological Research and a fellow of the American Psychological Society, the American Association for the Advancement of Science, and the International Organization for Psychophysiology. His research interests include functional organization of brain mechanisms underlying behavioral and affective processes, multiple levels of organization and analysis in neurobehavioral systems, bottom-up and top-down processes in autonomic regulation, and the social neuroscience of health and disease. He has a Ph.D. in psychobiology and life sciences from the University of Minnesota and spent two years as a postdoctoral researcher at Rockefeller University.

John T. Cacioppo is the Tiffany and Margaret Blake distinguished service professor at the University of Chicago and director of the University of the Chicago Center for Cognitive and Social Neuroscience. He is also president of the Association for Psychological Science. His current research is in the area of social neuroscience, with an emphasis on the effects of social isolation and the mechanisms underlying effective versus ineffective social connection. Among his many awards, he received the Distinguished Scientific Contribution Award from the American Psychological Association, the Campbell Award from the Society for Personality and Social Psychology, and the Troland Award from the National Academy of Sciences. He is the former editor of *Psychophysiology* and a former associate editor of *Psychological Review, Perspectives on Psychological Science*, and *Psychophysiology*. He has a Ph.D. from Ohio State University.

Christopher F. Chabris is assistant professor in the Department of Psychology at Union College. Previously he was lecturer and research associate in the Department of Psychology at Harvard University. His research interests include individual differences in human cognition and their relationship to brain function and structure, molecular genetics of cognition and decision making, and behavioral economics and cognitive biases. His work has been published in such journals as *Nature, Psychological Science*, and *Neuropsychologia*, and has been covered by news media worldwide. He has a Ph.D. in psychology from Harvard University.

Ming-Cheng Chang is chair and professor in the Institute of Healthcare Administration at Asia University and a scientific adviser to the Bureau of Health Promotion, Ministry of Health, in Taiwan. His past academic working experiences include senior associate, School of Hygiene and

Public Health, Johns Hopkins University; research fellow, Institute of Economics, Academia Sinica, Taiwan; and director, Taiwan Provincial Institute of Family Planning. He has a Ph.D. in demography from the University of Pennsylvania.

Kaare Christensen is professor of epidemiology in the Institute of Public Health at the University of Southern Denmark and senior research scientist at the Terry Sanford Institute at Duke University. He is also director of the Danish Twin Registry and deputy director of the international research program the Oldest-Old Mortality. He has served on numerous working groups and advisory panels of the National Research Council and the National Institute on Aging. He has conducted a long series of studies among twins and the elderly in order to shed light on the contribution of genes and environment in aging and longevity. He has a long-standing interest in the relation between early life events and later life health outcomes and is engaged in interdisciplinary aging research combining methods from epidemiology, genetics, and demography. He has M.D., Ph.D., and D.MSc. degrees from the University of Southern Denmark.

Lene Christiansen is associate professor at epidemiology in the Institute of Public Health at the University of Southern Denmark and director of the biobank in the Faculty of Health Sciences and the Danish Twin Registry. She also heads the associated genetic laboratory. Her main focus of research is in the study of the genetics of aging and longevity. She has an M.Sc. and a Ph.D. from the University of Southern Denmark.

Shah Ebrahim is an epidemiologist with a clinical background in geriatric medicine at the London School of Hygiene and Tropical Medicine. His research spans interests in the use of genetic polymorphisms to test the effects of environmental exposures (termed Mendelian randomization), the causes of heart disease and stroke in women, the effects of migration on obesity and diabetes in India, and the determinants of locomotor disability in old age. He is coordinating editor of the Cochrane Heart Group and coeditor of the *International Journal of Epidemiology*. He has an M.Sc. and a D.M., both from the London School of Hygiene and Tropical Medicine.

Douglas C. Ewbank is research professor at the Population Studies Center at the University of Pennsylvania. His current research involves the application of demographic methods to the study of differences in mortality by genotype. Previously he worked on African demography, American historical demography, and Alzheimer disease. He has served many professional organizations, including on the board of directors of the Population Association of America, the Steering Committee for the Health and Retirement Committee, and the National Institutes of Health Study Section SNEM-3. He has M.A. and Ph.D. degrees from Princeton University.

Elizabeth Frankenberg is associate professor in the Department of Public Policy Studies at Duke University. Her research interests include health and mortality, family decision making, developing economies, and Southeast Asia. She has authored and coauthored many publications and articles on such topics as health care, health and mortality, labor economics, and fertility and reproduction. She has an M.P.A from Princeton University, and a Ph.D. in demography and sociology from the University of Pennsylvania.

Edward L. Glaeser is the Fred and Eleanor Glimp professor of economics on the Faculty of Arts and Sciences at Harvard University, where he has taught since 1992. He is also director of the Taubman Center for State and Local Government and director of the Rappaport Institute of Greater Boston. He teaches urban and social economics and microeconomic theory. He has published papers on cities, economic growth, and law and economics. In particular, his work has focused on the determinants of city growth and the role of cities as centers of idea transmission. He edits the *Quarterly Journal of Economics*. He has a Ph.D. from the University of Chicago (1992).

Dana A. Glei is a research demographer at the University of California, Berkeley, where she currently serves as project coordinator for the Human Mortality Database project. She also works as a research consultant on the Social Environment and Biomarkers of Aging Study in Taiwan. Over the past 13 years, she has published articles on a variety of topics related to health, marriage and the family, and poverty. She has a Ph.D. in sociology from Princeton University.

Noreen Goldman is the Hughes-Rogers professor of demography and public affairs at the Woodrow Wilson School and acting director of the Office of Population of Research at Princeton University. She conducts research in areas of demography and epidemiology, and her current research examines the role of social and economic factors on adult health and the physiological pathways through which these factors operate. She has designed several large-scale health surveys in Latin America and Taiwan. She has served as a member of the board of the Guttmacher

Institute, a fellow at the Center for Advanced Study in the Behavioral Sciences, the Institute of Medicine's Board on Global Health, the National Research Council's Committee on National Statistics, and the Population Research Subcommittee of the National Institute of Child Health and Human Development. She has also served in various capacities at the Population Association of America and the International Union for the Scientific Study of Population. She has a D.Sc. from Harvard University.

Harald H.H. Göring is an associate scientist in the Department of Genetics at the Southwest Foundation for Biomedical Research in San Antonio, Texas. He has a Ph.D. from Columbia University, where he worked on statistical methods for gene mapping. He continued his training as a postdoctoral scientist at the Southwest Foundation, where he expanded his research focus to quantitative traits. His research now focuses on the localization and characterization of genetic variation influencing human traits, with an emphasis on the development of statistical methods and study designs. To complement his methodological work, he is involved in applied gene mapping studies with collaborators worldwide, focusing on a wide variety of human characteristics, including rare Mendelian disorders, complex diseases, and quantitative traits.

Tara L. Gruenewald is an assistant professor in the Department of Medicine/Geriatrics at the University of California, Los Angeles. Her research interests focus on psychological and social factors that impact functioning and health outcomes across the life span, including the biological pathways through which psychosocial variables influence health. She has a Ph.D. in psychology and an M.P.H. in health services, both from the University of California, Los Angeles.

Vilmundur Gudnason is director of the Icelandic Heart Association Research Institute and an associate professor in cardiovascular genetics at the University of Iceland. He worked for nine years as a senior research fellow at the Centre for Genetics of Cardiovascular Disorders at the University College London School of Medicine. For the past seven years, he has worked on obtaining detailed and extensive phenotypes in large population-based samples to study genetic contributions to traits of common complex chronic diseases. The work takes advantage of the homogenous Icelandic population. He is the principal investigator for the AGES Reykjavik study, based on the 40-year-long Reykjavik study, and for the REFINE Reykjavik study of younger generations. He has a medical degree from the University of Iceland and a Ph.D. in genetics from University College London.

Jennifer R. Harris is a senior researcher in the Department of Genes of Environment, Division of Epidemiology, at the Norwegian Institute of Public Health (NIPH) in Oslo. She is the founder of the NIPH twin study, which merges her interests in life-span development with genetics. Currently she is involved in several twin research projects of physical health, mental health, and epigenetics. She leads the European Union FP6 coordination action promoting the Harmonization of Epidemiological Biobanks in Europe and also led the Norwegian participation in the GenomEUtwin project, in which she focused on studies of body mass index and was in charge of the GenomEUtwin Ethics Core. She is also a special expert at the U.S. National Institute on Aging, developing research directions integrating genetics/genomics with behavioral and social research. She has a Ph.D. from the University of California, Berkeley.

Tamara B. Harris is chief of the Geriatric Epidemiology Section, Epidemiology, Demography, and Biometry Program, in the National Institute on Aging of the National Institutes of Health. Prior to this she worked at the Office of Analysis and Epidemiology at the National Center for Health Statistics. Her research interests focus on how the role of the geriatric epidemiology section integrates molecular and genetic epidemiology with interdisciplinary studies of functional outcomes, disease endpoints, and mortality in older persons. This includes identification of novel risk factors and design of studies involving biomarkers, selected polymorphisms, and exploration of gene/environment interactions. The section has been particularly active in devising methods to integrate promising molecular or imaging techniques in ways that begin to explore the physiology underlying epidemiological associations, including adaptation of imaging protocols to epidemiological studies. She has an M.S. in epidemiology from the Harvard School of Public Health, an M.S. in human nutrition from Columbia University College of Physician's and Surgeons, and an M.D. from the Albert Einstein College of Medicine.

John Hobcraft is professor of demography and social policy at the University of York. He is also a visiting research fellow at Princeton University. He chairs the Consortium Board and the International Working Group for the United Nations Economic Commission for Europe's Generations and Gender Programme. He is an elected member of Academia Europaea and was an editor of *Population Studies* for 25 years. His research interests include intergenerational and life-course pathways to adult social exclusion; understanding human reproductive behavior; the role of gender in human behavior; population policies; and understanding genetic, evolutionary, mind, brain; and endocrinological pathways and their interplays with behavior. He has a B.Sc. in economics from the London School of Economics and Political Science.

Christine M. Kaefer is a scientific information analyst in the Office of Centers, Training, and Resources at the National Cancer Institute (NCI) of the National Institutes of Health. Previously she was a presidential management fellow at NCI, and in this capacity she worked with a variety of NCI divisions, offices, and centers, including the Nutritional Sciences Research Group in the Division of Cancer Prevention and the Behavioral Research Program in the Division of Cancer Control and Population Sciences. Her interests include health communication, especially as it relates to disease prevention and health promotion through healthy lifestyles, consumers' perceptions of health risks, health-related decision making, and behavior change. She is a registered dietitian and has an M.B.A. from Virginia Polytechnic and State University and a B.S. in nutritional sciences from Cornell University.

David I. Laibson is professor of economics in the Department of Economics at Harvard University. Previously he was Paul Sack associate professor of political economy there. He is on the editorial board of the *Journal of the European Economics Association* and is co-organizer of the Russell Sage Foundation Summer School in Behavioral Economics. He is also advisory editor for the *Journal of Risk and Uncertainty*, as well as the *Q.R. Journals of Macroeconomics*. His awards and honors include the Johns Hopkins University Ely Lectures, a National Science Foundation grant, and a National Institutes of Aging grant, and he was keynote speaker at the Austrian Economics Association annual meeting. He has an M.Sc. in econometrics and mathematical economics from the London School of Economics and a Ph.D. in economics from the Massachusetts Institute of Technology.

Lenore J. Launer is chief of the Neuroepidemiology Section in the Laboratory of Epidemiology, Demography, and Biometry at the National Institute on Aging of the National Institutes of Health. Her main research interests are the metabolic, inflammatory, vascular, and genetic factors that interact and lead to pathological brain aging and function. She is one of the principal investigators on the Age Gene Environment Susceptibility–Reykjavik Study and principal investigator of the Action to Control Cardiovascular Risk in Diabetes: Memory in Diabetes trial, which is investigating the effects on the brain of standard versus intensive treatment of cardiovascular risk factors. She also collaborates closely on the Honolulu Asia Aging Study. She has a Ph.D. in epidemiology and nutrition from Cornell University.

Stacy Tessler Lindau is an assistant professor of obstetrics and gynecology and medicine-geriatrics at the University of Chicago. Her work
combines biomedical and social scientific techniques to study health and
health behavior in the population setting. Her primary interest is deci-
phering the biological pathways linking sexual relationships to health,
particularly in the context of aging and illness. She is one of the principal
investigators of the National Social Life, Health and Aging Project, which
has collected detailed questionnaire data in respondents' homes in conjunction with a unique panel of minimally invasive biophysiological data.
She also directs the Chicago Core on Biomarkers in Population-Based
Research at the Center on Economics and Biodemography of Aging. She
has an M.D. from Brown University and an M.A. in public policy from
University of Chicago.

Michael Marmot has led a research group on health inequalities for the
past 30 years. He is principal investigator of the Whitehall Studies of
British civil servants, investigating explanations for the striking inverse
social gradient in morbidity and mortality. He leads the English Longitudinal Study of Ageing and is engaged in several international research
efforts on the social determinants of health. He chairs the Department
of Health Scientific Reference Group on tackling health inequalities. He
was a member of the Royal Commission on Environmental Pollution
for six years. In 2000 he was knighted for services to epidemiology and
understanding health inequalities. He is a vice president of the Academia
Europaea, a member of the RAND Health Advisory Board, a foreign associate member of the Institute of Medicine, and chair of the Commission on
Social Determinants of Health set up by the World Health Organization
in 2005. He won the Balzan Prize for Epidemiology in 2004 and gave the
Harveian Oration in 2006. He graduated in medicine from the University
of Sydney and has an M.P.H. and a Ph.D. from the University of California, Berkeley.

Gerald E. McClearn is Evan Pugh professor of health and human development and biobehavioral health in the Department of Biobehavioral
Health at the Pennsylvania State University and director for developmental and health genetics. His research interests focus on the application of
quantitative genetic models to analysis of phenotypes relevant to health
and development. He has a Ph.D. from the University of Wisconsin.

Thomas W. McDade is the Weinberg College board of visitors research
and teaching professor at Northwestern University. He is also director
of the Laboratory for Human Biology Research and associate director
of Cells to Society (C2S): the Center on Social Disparities and Health

at the Institute for Policy Research. He specializes in population-based research on human physiological function and health, with an emphasis on biomarkers of immune function and inflammation. He has a Ph.D. in anthropology from Emory University.

John Milner is chief of the Nutritional Science Research Group in the Division of Cancer Research at the National Cancer Institute of the National Institutes of Health. His current research focuses on substances in foods and cancer prevention and the molecular mechanism by which bioactive food constituents influence cancer risk and tumor behavior. He recently received the David Kritchevsky Career Achievement Award in Nutrition from the American Society for Nutrition. This award recognizes researchers who devote their careers to promoting interaction among and support for nutrition researchers in government, private, and academic sectors. He has a B.S. in animal sciences from Oklahoma State University and a Ph.D. in nutrition, with a minor in biochemistry and physiology, from Cornell University.

Shaun Purcell is a member of the Psychiatric and Neurodevelopmental Genetics Unit, part of the Center for Human Genetic Research at Massachusetts General Hospital. His work focuses on developing statistical and computational tools for the design of genetic studies, the detection of gene variants influencing complex human traits, and the dissection of these effects in the larger context of other genetic and environmental factors. In particular, he currently works on whole genome association studies of bipolar disorder and schizophrenia and the development of tools for whole genome studies. He is an associate member of the Broad Institute of Harvard and the Massachusetts Institute of Technology and its Psychiatric Disease Initiative. He has degrees from the University of Oxford and University of London and a Ph.D. from the Social, Genetic and Developmental Psychiatry Centre in the Institute of Psychiatry at King's College London.

Sharon Ross is a program director in the Nutritional Science Research Group in the Division of Cancer Prevention of the National Cancer Institute (NCI) of the National Institutes of Health. She is responsible for directing, coordinating, and managing a multidisciplinary research grant portfolio in diet, nutrition, and cancer prevention. Previously she worked at the Center for Food Safety and Applied Nutrition in the Food and Drug Administration and was a cancer prevention fellow in the Division of Cancer Prevention and Control. Her doctoral dissertation research was carried out in the Laboratory of Cellular Carcinogenesis and Tumor Promotion at NCI, where her research topic concerned the effects of retinoids in growth, differentiation, and cell adhesion. She has an M.S. in nutritional sciences from the University of Connecticut, a B.S. in nutrition and dietetics from the University of New Hampshire, a Ph.D. in nutritional sciences from the University of Maryland, College Park, and a M.P.H. from Johns Hopkins University School of Public Health, with an emphasis in epidemiology.

Teresa Seeman is professor of medicine and epidemiology in the Schools of Medicine and Public Health at the University of California, Los Angeles. Her research interests focus on the role of sociocultural factors in health and aging, with specific interest in understanding the biological pathways through which these factors influence health and aging. A major focus of her research relates to understanding how aspects of the social environment, particularly socioeconomic status and social relationships, affect biology and through that overall health and aging. She has extensive experience with developing and implementing community-based collection of biomarker data and has been a leading figure in research on cumulative biological risk indices. In collaboration with Bruce McEwen and Burton Singer, she has taken a lead in empirical research on the new concept of allostatic load as a cumulative index of biological aging. She has bachelor's, master's, and doctoral degrees from University of California, Berkeley.

George Davey Smith is professor of clinical epidemiology at the University of Bristol, visiting professor at the London School of Hygiene and Tropical Medicine, and honorary professor at the University of Glasgow. He is also director of the Avon Longitudinal Study of Parents and Their Children and of the Medical Research Council's Centre for Causal Analyses in Translational Epidemiology. His main research interests relate to how socially patterned exposures acting over the entire life course shape health of individuals and populations and also influence long-term trends in health. He is particularly concerned with integrating genetic epidemiology into such life-course studies to strengthen the ability to make causal inferences in observational studies. He has also worked on sexually transmitted disease/HIV infection prevention in India and Nicaragua and on methodological issues in epidemiology. He is coeditor with Shah Ebrahim of the *International Journal of Epidemiology*. He has an M.Sc. in epidemiology from the London School of Hygiene and Tropical Medicine.

Andrew Steptoe is British Heart Foundation professor of psychology in the Department of Epidemiology and Public Health at University College London. Previously he was professor of psychology at George's Hospital Medical School in the University of London (1988-2000). He is past president of the International Society of Behavioral Medicine and of the Society for Psychosomatic Research. He was founding editor of the *British Journal of Health Psychology* and an associate editor of *Psychophysiology*, the *Annals of Behavioral Medicine*, the *British Journal of Clinical Psychology*, and the *Journal of Psychosomatic Research*. He is author or editor of numerous books, including *Psychosocial Processes and Health* and *Depression and Physical Illness*. His main research interests are in psychosocial aspects of physical illness, health behavior, and psychobiology. He graduated from Cambridge in 1972 and completed his doctorate at Oxford University in 1975.

Ronald A. Thisted is professor in the Departments of Health Studies, Statistics, and Anesthesia and Critical Care at the University of Chicago. He has been a faculty member in statistics at Chicago since 1976, and since 1999 he has also chaired the Department of Health Studies, the home of biostatistics, epidemiology, and health services research at Chicago. His research involves statistical computation, the design and analysis of clinical trials and epidemiologic studies, and development of new computational and statistical methods. He is a past editor of the *Current Index to Statistics* and was associate editor of the *Journal of the American Statistical Association*. He is a fellow of the American Statistical Association and of the American Association for the Advancement of Science. He has M.S. and Ph.D. degrees from Stanford University.

Duncan Thomas is professor of economics at Duke University. His research centers on the dynamics underlying individual, household, and family behavior, particularly in low-income populations. This work explores how resources are allocated in families and the responses of individuals and their families to unanticipated shocks. His research on the association between health and socioeconomic status includes designing and fielding a randomized treatment-control nutrition intervention in Indonesia, the Work and Iron Status Evaluation, to measure the impact of health on economic prosperity. He has a Ph.D. in economics from Princeton University.

Elaine B. Trujillo is a nutritionist in the Nutritional Science Research Group of Division of Cancer Prevention at the National Cancer Institute of the National Institutes of Health. She is responsible for promoting the translation of information about bioactive food components as modifiers of cancer. Previously she was a senior clinical and research dietitian in the Metabolic Support Service of Brigham and Women's Hospital, Harvard Medical School. She received the 2007 Huddleson Award by the American Dietetic Association Foundation for her article, "Nutrigenomics, Proteomics, Metabolomics, and the Practice of Dietetics." She has a B.S. in nutritional science from the University of Delaware and an M.S. in nutritional science from Texas Woman's University, where she also completed a dietetic internship.

James W. Vaupel is the founding director of the Max Planck Institute for Demographic Research in Rostock, Germany, and director of Duke University's Population Research Institute. He oversees multinational research initiatives in Germany, Denmark, the United States, Italy, Russia, Mexico, Japan, and China. He has coauthored or coedited seven books and has written numerous research articles published in refereed journals. He is best known for his research on mortality, morbidity, population aging, and biodemography, as well as for research on population heterogeneity, population surfaces, and other aspects of mathematical and statistical demography. He is a member of the National Academy of Sciences and of the National Research Council's Committee on Population. He has bachelor's, master's, and doctoral degrees from Harvard University.

George P. Vogler is director of the Center for Developmental and Health Genetics at Pennsylvania State University and professor of biobehavioral health. His research interests include genetic epidemiology of complex traits, quantitative trait loci mapping, cardiovascular disease, methodological issues in genetic models and structural equation models, and behavioral moderation of expression of biological traits. His applied interests span human development, with interests in childhood development of cognitive abilities, substance use and abuse in adolescence, cardiovascular risk factors in adulthood, and maintenance of functional abilities in aging. He has a Ph.D. from the University of Colorado, Boulder.

Kenneth W. Wachter is chair of the Department of Demography at the University of California, Berkeley. Previously he taught at Harvard University and has published numerous articles and books on mathematical demography, statistical analysis, historical demography, nutrition, aging, and kinship models. He is a fellow of the American Academy of Arts and Sciences and received the Mindel Sheps award in 1988 from the Population Association of America. He serves on the board of directors of the Social Science Research Council and has served on the special advisory panel on 1990 census adjustment for the U.S. Department of Commerce and on the National Science Foundation Panel on Measurement Methods and Data Resources. He is a member of the National Academy of Sciences and the current chair of the National Research Council's Committee on Population. He has a master's degree in applied mathematics from Oxford University and a Ph.D. in statistics from Cambridge University.

Robert B. Wallace is Irene Ensminger Stecher professor of epidemiology and internal medicine at the University of Iowa Colleges of Public Health and Medicine and director of the university's Center on Aging. He is a member of the Institute of Medicine, past chair of its Board on Health Promotion and Disease Prevention, and current chair of its Board on Military and Veterans Health. He is the author or coauthor of numerous publications and book chapters and has been the editor of four books, including the current edition of Maxcy-Rosenau-Last's *Public Health and Preventive Medicine*. His research interests are in clinical and population epidemiology and focus on the causes and prevention of disabling conditions of older persons. He has had substantial experience in the conduct of both observational cohort studies of older persons and clinical trials, including preventive interventions related to fracture, cancer, coronary disease, and women's health. He is the site principal investigator for the Women's Health Initiative and a co-principal investigator of the Health and Retirement Study. He has been a collaborator in several international studies of the causes and prevention of chronic illness in older persons. He has B.S.M. and M.D. degrees from Northwestern University and an M.Sc. in epidemiology from the State University of New York at Buffalo.

Maxine Weinstein is distinguished professor of population and health in the Graduate School of Arts and Sciences of the Center for Population and Health at Georgetown University, where she has been since 1987. She is also principal investigator, along with Noreen Goldman, of the Taiwan project, a study that explores the reciprocal relations among stress, health, and the social environment among the elderly. Her work explores the behavioral and biological dimensions of reproduction and aging. She is also heading up the MIDUS II biology substudy at Georgetown. She has M.A. and Ph.D. degrees in sociology from Georgetown University.

David Weir is research affiliate in Population Studies Center of the University of Michigan and research professor in the Survey Research Center. Previously he was research associate in the Harris School of Public Policy at the University of Chicago. His current research interests include the measurement of health-related quality of life; the use of cost-effectiveness measures in health policy and medical decision making; the role of supplemental health insurance in the Medicare population; the effects of health, gender, and marital status on economic well-being in retirement; and, the effects of early life experience on longevity and health at older ages. He has a Ph.D. from Stanford University.

Kenneth M. Weiss is Evan Pugh professor of anthropology and genetics and professor of biology in the Department of Anthropology at Pennsylvania State University. His research interests focus on genetic epidemiology and the genetic basis, nature, and evolution of complex biological traits. These areas also raise issues about the nature of knowledge when effects are rather weak and causation probabilistic and must largely be approached through observational rather than experimental studies. His research has concerned the amount and origin of human genetic variation and the genetic basis of complex patterning during development of such structures as the skull and dentition. He has a B.A. from Oberlin College and a Ph.D. from the University of Michigan.

Mary Jane West-Eberhard is a senior scientist at the Smithsonian Tropical Research Institute, resident in Costa Rica. She studied zoology at the University of Michigan, was a postdoctoral fellow at Harvard University, and has lived and worked since 1969 in Latin America (Colombia and Costa Rica). Her research interests include fieldwork on the social behavior of tropical wasps, the evolution of social behavior, kin selection theory, sexual and social selection, speciation, and developmental plasticity and evolution. She is a past president of the Society for the Study of Evolution and a member of Phi Beta Kappa, the U.S. National Academy of Sciences, the American Academy of Arts and Sciences, the National Academy of Sciences of Costa Rica, and the Accademia Nationale dei Lincei of Rome. She has B.A., M.S., and Ph.D. degrees, all from the University of Michigan.